The Structure of the World

In *The Structure of the World*, Steven French articulates and defends the bold claim that *there are no objects*. At the most fundamental level, modern physics presents us with a world of *structures* and making sense of that view is the central aim of the increasingly widespread position known as *structural realism*. Drawing on contemporary work in metaphysics and philosophy of science, as well as the 'forgotten' history of structural realism itself, French attempts to further ground and develop this position. He argues that structural realism offers the best way of balancing our need to accommodate the results of modern science with our desire to arrive at an appropriately informed understanding of the world that science presents to us. Covering not only the realism-antirealism debate, the nature of representation, and the relationship between metaphysics and science, *The Structure of the World* defends a form of eliminativism about objects that sets laws and symmetry principles at the heart of ontology. In place of a world of microscopic objects banging into one another and governed by the laws of physics, it offers a world of laws and symmetries, on which determinate physical properties are dependent. In presenting this account, French also tackles the distinction between mathematical and physical structures, the nature of laws, and causality in the context of modern physics, and he concludes by exploring the extent to which structural realism can be extended into chemistry and biology.

The Structure of the World

Metaphysics and Representation

Steven French

OXFORD
UNIVERSITY PRESS

OXFORD
UNIVERSITY PRESS

Great Clarendon Street, Oxford, OX2 6DP,
United Kingdom

Oxford University Press is a department of the University of Oxford.
It furthers the University's objective of excellence in research, scholarship,
and education by publishing worldwide. Oxford is a registered trade mark of
Oxford University Press in the UK and in certain other countries

First published 2014
First published in paperback 2016

Published in the United States of America by Oxford University Press
198 Madison Avenue, New York, NY 10016, United States of America

British Library Cataloguing in Publication Data
Data available

Library of Congress Cataloging in Publication Data
Data available

ISBN 978–0–19–968484–7 (Hbk.)
ISBN 978–0–19–877666–6 (Pbk.)

Preface

To many people the idea that the world is populated by objects, that have properties, that in turn are related in ways that the laws of science describe, seems unassailable. It can be characterized as a 'bottom-up' metaphysics obtained from our interactions with 'everyday'/'mid-sized white goods'/macroscopic objects and it amounts to little more than a prejudice, or as many philosophers are fond of saying, an intuition. It is no dramatic revelation to point out that it fails and fails miserably when it is exported away from the 'everyday', into the domain of modern physics, or indeed, as I shall suggest in my final chapter, into that of biological phenomena. I prefer an alternative approach—characterized, appropriately, in contrasting terms as 'top-down'—which at least has the virtue of taking the relevant science seriously in the sense that it urges that we read our metaphysical commitments more or less directly off our best theories. This alternative approach underpins a cluster of positions that have achieved some prominence in recent years under the collective label of 'structural realism' and this book represents an attempt to set out and defend a form of structural realism that maintains that the fundamental ontology of the world is one of structures and that objects, as commonly conceived, are at best derivative, at worst eliminable.

This form, known as 'ontic structural realism' (OSR), has already been articulated and defended, most famously by Ladyman (1998; French and Ladyman 2003; Ladyman and Ross 2007) and this work can be seen as in many respects complementary to his. However, whereas Ladyman has excoriated current metaphysics for its failure to accommodate the conclusions of modern physics, I think it can be plundered for appropriate resources that we can then use to articulate our structuralist ontology. I've called this the Viking Approach to metaphysics, with my friendly neighbourhood metaphysicians cast in the role of hapless peasants, upon whom the philosophers of physics periodically descend for a spot of pillaging, but a less brutal image has been suggested by Kerry McKenzie in which metaphysics is regarded as a toolbox, from which we can take various implements—'dependence', 'supervenience', and so on—to use in order to fashion an appropriate notion of structure.

My book begins by outlining three core challenges that realism must face—the Pessimistic Meta-Induction, Underdetermination, and what I call 'Chakravartty's Challenge'—and in Chapter 1 I indicate how structural realism deals with the first of these, drawing on the work of Cei and Saunders to show how the discussion can be extended beyond the usual consideration of, for example, Fresnel's equations and the theory of light, to case studies that bear on the transition from classical to quantum mechanics. In his now-classic paper setting out what is often referred to as 'epistemic' structural realism, Worrall offered the hope that this stance could encompass quantum theory and in effect the 'ontic' form tries to make good on that promise. However, in order to do so, it must obviously tackle the metaphysically most profound consequences of that theory. As far as many commentators (such as Cassirer and Eddington) were concerned, the most significant impact these consequences had was on the notion of object and they saw quantum statistics in particular as implying the elimination of objects, at least in so far as this notion was intimately tied to that of the object as an individual. However, as Décio Krause and I have argued, first of all, one can articulate—both formally and metaphysically—an appropriate notion of non-individual object within the domain of quantum physics; and secondly, one can show that quantum mechanics is in fact compatible with an appropriate notion of individual object (and the extent of what can be considered appropriate has recently been expanded by Saunders and Muller in their work on 'weak discernibility'). This then marks a significant break between the earlier structuralists, such as Cassirer and Eddington, and their modern-day descendants, such as myself. The former took the negative implications of quantum physics for the notion of an individual object as directly motivating their structuralism. Today's ontic structural realist takes the fact that the physics supports two metaphysical packages—of non-individual objects and of individual objects—as presenting a major problem for realism and regards this 'metaphysical underdetermination' as the prime motivator for her position.

And so, in Chapter 2, I consider this motivation in more detail, examining and rejecting ways in which the underdetermination might be broken or avoided. In this manner, by dropping objects from its metaphysical pantheon, OSR is a metaphysically more minimal position than standard, or 'object-oriented' views. However, some sort of balance must be achieved, lest OSR collapses into some form of metaphysically *most* minimal position, such as structural empiricism (as advocated, in different forms, by Bueno and van Fraassen). This is where the third challenge comes into play: as Chakravartty has emphasized, it is not enough, if one is a realist, to simply wave one's hands at the relevant theoretical posits or equations and declaim 'that is what I'm a realist about'! One needs to provide some sort of 'clear picture' or understanding, and that, I maintain, must be metaphysically informed. It is here in Chapter 3 that I adopt the 'Viking Approach' to metaphysics and argue that achieving that crucial balance between keeping the metaphysics to a minimum and filling in a metaphysically informed clear picture behind one's realism provides a further motivation for OSR.

This concludes the 'motivational' part of the book. The next chapter represents a historical 'pause' in which I try to retrieve some of the 'lost' history of structuralism, in the context of Cassirer's and Eddington's responses to the metaphysical implications of quantum mechanics. In general I argue that both advocates and critics of structural realism have conducted their debate under the shadow of Russell, whose classic tome *The Analysis of Matter* still holds considerable sway. However, although he displayed considerable mastery of relativistic physics, Russell's grasp of the newly emergent quantum theory was much more tenuous and if one is to look for antecedents of a form of structural realism informed by quantum mechanics, one should shift one's historical focus forward a few years, to the commentaries and reflective ontological work of the likes of Cassirer and Eddington. Here one finds what is missing in Russell, specifically forms of structural realism that are informed by the powerful mathematical framework of group theory that had been developed and applied to the new quantum mechanics by Weyl and Wigner. As Cassirer and Eddington both realized, one of the features that distinguished modern physics—both relativistic and quantum—from its classic forebear was the increased significance of the role of symmetry and it is this that group theory gives mathematical expression to. In particular, the way in which quantum statistics was seen to undermine the notion of object and thus motivate forms of structuralism, follows from the incorporation within the theory of the so-called 'permutation symmetry' that underpins the metaphysical underdetermination articulated in Chapter 2. Thus the form of structural realism presented in this book is informed by the role of symmetry and invariance in just the manner that Cassirer and Eddington advocated and a significant portion of the rest of the work is taken up in trying to articulate an appropriate metaphysics from such an informed perspective.

In Chapter 5, then, I begin to set out my answer to the question 'so, what is structure?' One response, again, is to wave one's hands at the relevant equations and symmetries of the theory and insist 'That, that is the structure of the world'. But, first of all, that does not satisfy Chakravartty's Challenge and give us a clear picture of what the structure of the world is like. And secondly, in responding to the Pessimistic Meta-Induction, and articulating how the relevant theories are interrelated in general, philosophers of science have represented those theories structurally, using the resources of the so-called 'Semantic Approach' to theories, for example. Indeed, Ladyman, in his classic 1998 paper setting out OSR, appealed to this approach on the grounds that it effectively wears its structuralist commitments on its sleeve. However, this has led some to infer that advocates of OSR take the structure of the world to be set-theoretic or, more generally, mathematical. Here I try to clarify our commitments and answer the earlier question by drawing on a useful distinction made by Brading and Landry: the structure of the world is *presented* to us in the theoretical context under consideration by means of the relevant laws and symmetries, as informed group-theoretically. As philosophers of science, we then *represent* that structure by means of various meta-level resources, such as the Semantic Approach. This is not the only such resource available, and indeed the post-Russell history of structural realism, particularly in its 'epistemic' form, is marked by the use of the Ramsey sentence formulation. As a mode of representation this itself has an interesting history, running through the work of Carnap, Lewis, and others, but it is bedevilled and, for some, fatally undermined by the so-called Newman problem which famously caused Russell to retract his structuralist claims. Eddington, however, was dismissive of the problem and I take it to have been more than adequately responded to by Melia and Saatsi. In particular, their emphasis on the intensional character of laws in side-stepping the problem points the way to an appropriate understanding of structure that I try to articulate in the rest of the book.

This is not to say that I think the Semantic Approach is the only adequate meta-level mode of representation in this regard. On this I'm happy to adopt a pluralist attitude—personally I think this approach has a number of advantages over others in appropriately capturing the kinds of features that we philosophers of science are interested in, but I'm quite prepared to acknowledge that other modes (such as category theory) have their positive features too.

This still leaves the issue of how we are to understand the presentation of the structure of the world in terms of the laws and symmetries of the relevant theories, where these are group-theoretically informed. In Chapter 6 I tackle some initial obstacles with such an understanding, arising, in particular, from the role of the mathematics of group theory in informing this picture, and of the specific nature of certain symmetries that feature in current physics.

With these obstacles overcome, I adopt the Viking Approach in Chapter 7 to indicate how an eliminativist stance towards objects need not have the devastating implications that some take it to have. In particular, I argue that we can still utter truths about, and in general talk of, physical objects, while eliminating them from our fundamental ontology in favour of structure. Now, I take that structure to be *physical* structure—a claim that might seem clear and straightforward but of course distinguishing the physical from the non-physical, and in this context in particular, from the mathematical, is problematic, as I indicate in Chapter 8. A number of comparisons have been drawn between structural realism and structuralism in mathematics, mostly to the detriment of the former, and as with the case of Russell's shadow, I think these comparisons have proceeded from an inappropriate basis. Of course, one significant difference between the mathematical and physical realms concerns the putative role of causality and in the bulk of this chapter I consider how this might be accommodated within OSR. Ultimately I urge that we should focus on the relevant dependencies underpinning the causal claims and exploring the nature of these dependencies takes up the next two chapters, where I set out a view of structure as primitively modal.

In Chapter 9 I consider the two main rivals to this view, namely Humean structuralism—which takes the structure to be categorical—and dispositional structuralism, as represented by Chakravartty's semi-realism—which takes the structure to effectively flow from or be grounded in an understanding of the relevant properties as dispositionally constituted. Both views are problematic, I argue. Humean structuralism faces well-known problems when it comes to its view of laws, and even with recent upgrades to the classic 'best system' accounts, I can't see those problems as being easily resolvable. Dispositionalism also faces problems, particularly when it comes to understanding fundamental properties in the context of modern physics. However, I do think that its general approach can be appropriated—again in the spirit of the philosophical Viking!—and effectively reverse engineered to yield a modally informed kind of structuralism. Since this move is so crucial, let me spell it out.

Once one has moved beyond the Humean stance and accepted that there is modality 'in' the world, the issue is where to place it, as it were. Here the difference between the object-oriented and the structural realist comes into play: the former reads her ontology off theories at some remove, by taking the laws and symmetries that the theories present to be underpinned by property-possessing objects to which we should be ontologically committed. The latter reads her ontology off these theories directly, by taking the very same laws and symmetries as features of the structure of the world. Now, whereas the dispositionalist, adopting the former stance, takes the laws to arise from or be dependent in some way upon the properties of those objects, I suggest that we should invert that order, taking the properties to be dependent upon the laws and symmetries. With this inversion, the associated modality is shifted along the line of dependence from the properties to the laws and symmetries themselves. Thus, instead of expanding our fundamental ontology with dispositions, thereby inflating our metaphysical commitments, I stay with the structure that we read off our theories and invest that with the requisite modality.

That in effect represents the final element in my answer to the question 'what is structure?' It is the laws and symmetries of our theories of contemporary physics, appropriately metaphysically understood via notions of dependence and taken as appropriately modally informed. In Chapter 10 I try to explicate that sense of modal informing by spelling out the sense in which laws and symmetries encode the relevant possibilities via the relevant models. I then consider three issues, to do with representation, fundamentality, and counterfactuals.

With regard to the first I suggest that the vehicle of representation should be thought of as extending beyond the immediate model used to describe a system and to involve modal features. When it comes to fundamentality, in the spirit of the Viking Approach again, I draw on recent work in metaphysics to suggest that laws, as determinables, are acceptable as elements of our 'fundamental base'. And with regard to the relationship between laws and counterfactuals, I argue that standard accounts of this relationship, and of the supposed necessity of laws, rely on an object-oriented picture that the structuralist should reject. It is the primitive modality that gives laws their modal stability as compared to accidents and which explains those counterfactuals that are not rejected as inappropriate.

The last two chapters represent further developments of this picture, first within quantum field theory (QFT) and secondly beyond physics, in the chemical and biological contexts. In Chapter 11 I examine the issue of unitarily inequivalent representations in QFT that have been raised as a fundamental problem for OSR. Here the issue of arriving at an appropriate ontology of QFT comes to the fore and I try to extend the earlier suggestions of French and Ladyman (2003) by showing how the problem of unitarily inequivalent representations can be deflated in various ways, and in particular by adopting the view of modality outlined in the previous chapter.

Finally, the supposed lack of laws in biology has been taken as a fundamental block on the development of forms of structural realism in this domain, but in Chapter 12, I draw on the work of Mitchell and others to explore the extent to which some kind of structuralist ontology can be articulated here as well. Of course, the motivations are different, as it is not clear that the Pessimistic Meta-Induction represents the same threat as it does for physics-based realism, nor is there anything like the kind of metaphysical underdetermination regarding individuality that I outline in Chapter 3. Nevertheless, Dupré and O'Malley have identified a 'Problem of Biological Individuality' and together with the heterogeneity of what counts as an organism in biology, this can be taken as a powerful driver towards a biology-informed form of OSR. Given the reactions to the papers on which this chapter is based I should perhaps emphasize that my intention is not to attempt an imperialistic extension of OSR but simply to consider to what extent something like it can be sustained within biology. Certainly, I would argue, it offers an interesting alternative to Dupré's 'Promiscuous Realism' in this regard.

And that concludes the book. In writing it, and the papers and essays it is based on, I owe a massive debt to many people—too many to acknowledge in full here. But I cannot end this preface without saying something about those folk whose support and criticisms have played such a significant role in shaping this work. The whole process has been book-ended by my former students. At the beginning there was James Ladyman, with whom I had the kind of relationship supervisors can only dream of. Our rants and declamations, speculations, and bursts of inspiration, often expressed at high speed while driving along the A1, have informed so much of my work in the period since. At the end there is Kerry McKenzie, who has helped keep me on the physical and metaphysical straight and narrow (or at least, has tried!) and whose clarity and insight have given me something to aim for in this work. In between there have been Otávio Bueno, Angelo Cei, Juha Saatsi, and Dean Rickles. My conversations with Otávio have spanned just about every aspect of the philosophy of science, and much of philosophy besides, and his robust and constant anti-realism has challenged my realist intuitions at every turn. Similarly Juha, although a firm realist, soon moved beyond the structural form to develop his own account and his arguments about how realism should be understood and supported have had a profound influence. Angelo and Dean, although closer to me in structuralist inclinations, have led me to think harder about both the relevant case studies from the history of science and the foundations of space-time theory and quantum gravity, respectively.

Others have adopted a more critical role that has been just as valuable. Anjan Chakravartty taught at the University of Leeds for a little while and through his own form of structural realism and his advocacy of object-oriented dispositionalism showed me how one might metaphysically beef up one's realist stance. Like Anjan,

Stathis Psillos is a firm believer in objects, but also, as with Anjan, his constructively critical engagement with structural realism has had an enormous impact on the development of my ideas (as should be clear from the number of references!).

Closer to the structuralist camp are a group of folk who, over the years, have been hugely supportive and just wonderful interlocutors in the discussion. Katherine Brading, Elena Castellani, Elaine Landry, and Tom Ryckman have been involved since the early days with a series of workshops on various aspects of structuralism, its history and its relationship to physics and have been unfailingly considerate and helpful in their consideration of my defence of OSR. A good chunk of this book owes its existence to the short but delightful time I spent at Notre Dame as Katherine's guest, where she organized a wonderful conference on OSR with contributions from Katherine herself, Otávio, Elise Crull, Don Howard, Elaine Landry, Kerry, Antigone Nounou, Bryan Roberts, Pablo Ruiz de Olano, Tom, Susan Sterrett, Ioannis Votsis, and Johanna Wolff. Even if it's not always explicit, those discussions in the autumn sunshine had a huge impact on this project.

As did similar but earlier conversations at the Banff workshop organized by Elaine Landry and featuring contributions from, again, Anjan, Antigone, Elaine, Elena, Ioannis, James, Katherine, Tom, and John Worrall, against the awesome backdrop of the Rockies (and we'll just leave to one side the fact that the last day's 'stroll' up a mountain brought certain well-known structuralists closer to heart failure than they've ever been before or since).

Some of my ideas crystallized further during a conference in Wuhan, China, organized by Tian Cao, with myself, Simon Saunders, and John Worrall. For me at least one of the most impressive features of this meeting was the enthusiasm and interest of the postgraduate students, some of whom had travelled ridiculous distances just to be there and engage with us.

More recently, my efforts to take structuralism forward into biology have been massively helped by critical yet friendly (I hope) discussions with Jordan Bartol, Ellen Clarke, Jon Hodge, Phyllis Illari, Greg Radick, Alirio Rosales, Emma Tobin, and Marcel Weber, most particularly at a one-day workshop on objects in biology organized by Angelo, Phyllis, and myself here at Leeds.

These are just the more prominent occasions for extended discussions of structuralism in general and OSR in particular. Others have taken place in locations as diverse as Amsterdam, Athens, Cologne, Florence, Lima, Montreal, Oxford, Paris, Toronto, and Wuhan, to mention just a significant subset. And in addition to the folk mentioned already, I must acknowledge the always helpful comments and remarks, often critical, and deservedly so, from Michel Bitbol, Jeremy Butterfield, Adam Caulton, Alberto Cordero, Laura Crossilla, Mauro Dorato, Michael Esfeld, Laura Felline, Holger Lyre, Ioan Muntean, Laurie Paul, Simon Saunders, Michael Stolzner, and in particular Fred Muller who made useful comments on an earlier version of the manuscript. There are others I'm sure, but if I've missed any names off the list, please accept a blanket 'thanks' and a pint next time we meet.

To all these people I am hugely grateful, for their comments, criticisms, and support and just for being such wonderful colleagues. Much of the book was written during two years of research leave supported by a Major Research Scholarship from the Leverhulme Trust and their refusal to adhere to the UK government's 'impact agenda' and overall willingness to fund 'blue skies' research in the humanities is a testament to the kind of academic independence that other funding bodies should emulate but sadly do not. I would also like to thank Martin Vacek for his help with the references and bibliography, the readers of Oxford University Press for their extensive and helpful comments, Javier Kalhat for his excellent copy-editing, and Peter Momtchiloff, also of Oxford University Press, for his unflagging support and encouragement.

However, I reserve my final but no less heartfelt acknowledgement of gratitude, of course, to Dena, Morgan, and a certain small dog, for keeping me balanced and whole these past several years.

Some but by no means all of the material presented here has its origin in one or more of the following papers or chapters:

'The Resilience of Laws and the Ephemerality of Objects: Can A Form of Structuralism be Extended to Biology?', in D. Dieks et. al. (eds), *Probability, Laws and Structures*. Dordrecht: Springer, 2012, 187–200.

'Handling Humility: Towards A Metaphysically Informed Naturalism', in A. Cordero and J.I. Galparsoro (eds), *Reflections on Naturalism*. Amsterdam: Sense Publishers, 2013, 85–104.

'Semi-realism, Sociability and Structure', *Erkenntnis* 78 (2013): 1–18.

'The Presentation of Objects and the Representation of Structure', in E. Landry and D. Rickles (eds), *Structure, Object, and Causality: Proceedings of the Banff Workshop on Structural Realism*. University of Western Ontario Series in Philosophy of Science. Dordrecht: Springer, 2012, 3–28.

'Unitary Inequivalence as a Problem for Structural Realism', *Studies in History and Philosophy of Modern Physics* 43 (2012): 121–36.

'In Defence of Ontic Structural Realism', with James Ladyman, in A. Bokulich and P. Bokulich (eds), *Scientific Structuralism*. Boston Studies in the Philosophy of Science. Dordrecht: Springer, 2011, 25–42.

'Shifting to Structures in Physics and Biology: A Prophylactic for Promiscuous Realism', *Studies in History and Philosophy of Biological and Biomedical Sciences* 42 (2011): 164–73.

'Metaphysical Underdetermination: Why Worry?', *Synthese* 180 (2011): 205–21.

'The Interdependence of Structures, Objects and Dependence', *Synthese* 175 (2010): 89–109.

'On the Transposition of the Substantial into the Functional: Bringing Cassirer's philosophy of Quantum Mechanics into the 21st Century', with A. Cei, in M. Bitbol,

P. Kerszberg, and J. Petitot (eds), *Constituting Objectivity, Transcendental Perspectives on Modern Physics*. Western Ontario Series in Philosophy of Science. Dordrecht: Springer, 2009, 95–115.

'Symmetry, Invariance and Reference', in M. Frauchiger and W.K. Essler (eds), *Representation, Evidence, and Justification: Themes from Suppes*. Lauener Library of Analytical Philosophy, vol. 1. Frankfurt: Ontos Verlag, 2008, 127–56.

'Looking for Structure in all the Wrong Places: Ramsey Sentences, Multiple Realizability, and Structure', with Angelo Cei, *Studies in History and Philosophy of Science* 37 (2006): 633–55.

'Realism about Structure: The Semantic View and Non-linguistic Representations', with Juha Saatsi, *Philosophy of Science (Proceedings)* 78 (2006): 548–59.

'Structure as a Weapon of the Realist', *Proceedings of the Aristotelian Society* 106 (2006): 167–85.

'Scribbling on the Blank Sheet: Eddington's Structuralist Conception of Objects', *Studies in History and Philosophy of Modern Physics* 34 (2003): 227–59.

I am grateful to both my co-authors and the relevant publishers for permission to slice and dice this material, Frankenstein fashion.

Contents

1

Theory Change

From Fresnel's Equations to Group-Theoretic Structure

1.1 Introduction

Within the philosophy of science, the debate over scientific realism is one of the most vigorous and long lasting. In one camp are the scientific realists, of various hues; in the other are the critics, some of whom defend well-developed forms of anti-realism. How one characterizes scientific realism is itself a matter of contention, and thus so is what counts as a viable form of anti-realism, but generally speaking the scientific realist accepts that there is a mind-independent reality 'out there', that we can have knowledge of such a reality, and that science provides us with the best form of such knowledge. How, then, can this knowledge be extracted? Here's a fairly simple recipe: first, take our best current scientific theories. What do we mean by 'best'? There may be some debate about the relevant list of attributes here, but they will surely include being empirically successful, explanatorily powerful, simple (although characterizing that attribute is particularly problematic), and so on. Secondly, read off the relevant features of those theories. Which features? Those which are responsible for the empirical success, that feature in the relevant explanations, and so on. What is meant by 'read off'? One might take the theories as expressed in the 'natural language' of the scientists themselves—i.e. a mixture of mathematics and English (or Portuguese or whatever); or one might insist on casting the theory within a particular formal language, such as first-order or, more plausibly, second-order logic. Finally, take those features to stand in the appropriate relationship to aspects of the (mind-independent) world. What kind of relationship? One might take them to refer or to denote those features, or to correspond to them in a way that supports the correspondence theory of truth, or, more broadly perhaps, to represent them.

Of course, these questions can be answered in different ways, producing realisms of different flavours, but this is the basic recipe offered by scientific realism. Three challenges then have to be faced: the Pessimistic Meta-Induction (PMI); Underdetermination; and what I shall call 'Chakravartty's Challenge'. The first two are well known; the third less so but I shall suggest that unless it is answered, scientific realism risks lacking content. And I shall use all three challenges to motivate that flavour of realism known as 'structural realism'. It is now standard to see this as coming in two varieties, Epistemic Structural Realism (ESR) and Ontic Structural Realism (OSR), each expressed in slogan form as follows:

ESR: all that we *know* is structure
OSR: all that there *is*, is structure

The former allows for the existence of 'hidden' entities about whose nature we must, at best, remain agnostic but which lie beyond, or 'under', or in some way support, the relevant structure; whereas the latter dismisses any such entities and reconceptualizes the relevant objects in structural terms, where this reconceptualization can be regarded (weakly) as yielding a 'thin' notion of object, whose individuality is grounded in the relevant structure, or (strongly) as eliminating objects entirely. We shall return to these distinctions later on.

An immediate question is 'what is meant by structure here?' and it is the overall aim of this book to attempt to answer that question. Doing so will involve issues of presentation and representation, the content of realism, and the role of metaphysics and I shall be covering those in subsequent chapters. Before we get there, however, let me lay out the first of the three challenges just introduced, indicate how different forms of realism respond to them—or fail to—and articulate the distinction between ESR, OSR, and related views.

1.2 Challenge No. 1: The Pessimistic Meta-Induction (PMI)

Like many well-known arguments and claims in philosophy, how one should understand the PMI is itself a matter for debate but here is a useful reconstruction of it for my purposes:[1]

Premise 1: Entity *a*, posited in historical period p1, was subsequently agreed not to exist.
Premise 2: Entity *b*, posited in historical period p2, was subsequently agreed not to exist.
Premise 3: Entity *c*, posited in historical period p3, was subsequently agreed not to exist.
Premise n: Entity *i*, posited in historical period pn, was subsequently agreed not to exist.
(Inductive) Conclusion: The entities posited today will subsequently be shown not to exist.

The standard response to this induction is to argue, via detailed case studies, that those entities that were subsequently determined not to exist (the most well-known examples are phlogiston, caloric, and the ether) were in fact referred to by terms in the relevant theories that can be deemed 'idle', in the sense that they were not

[1] For an alternative presentation in the form of a reductio see Saatsi 2005.

responsible for the empirical success of those theories (see, for example, Psillos 1999). I shall take the response along these lines that has been articulated by Psillos as representative of 'standard' realism. He argues that,

a) the realist should only take as referring those terms which play an appropriate role in explaining the given theory's success and
b) the appropriate theory of reference in such cases is a form of causal-descriptive account, according to which reference is fixed via a 'core causal description' of those properties which underpin the putative entity's causal role with regard to the phenomena in question (1999: 295);
c) in addition, what this secures is reference to individual objects and their properties, and thus, Psillos insists, 'the world we live in (and science cares about) is made of individuals, properties and their relations' (2001: S23).

Psillos' articulation has the virtue of making explicit that which other accounts keep tacit—the commitment to a metaphysics of objects expressed in (c). For this reason I shall refer to this form of 'standard' realism as 'object-oriented' (OOR). It provides a useful contrast against which we can measure the virtues of structural realism that, broadly put, urges that we shift our ontological attention from the objects posited by theories, to the structures in which they feature (or, according to one form of this view, in terms of which they are constituted), which are retained (in a sense to be explicated) through the kinds of changes drawn upon by the PMI. In particular, I shall claim, OOR cannot respond adequately to the PMI nor accommodate the implications of modern physics as represented by the underdetermination challenge, nor can it respond appropriately to Chakravartty's Challenge.

Consider, as a specific example, the case of the optical and luminiferous ethers, which featured in successful theories of light and electromagnetism.[2] How is the realist to deal with the fact that current scientific theories no longer feature these terms? One option is to argue that they in fact refer to the same 'thing' as certain current terms, where 'sameness' here may be understood as fulfilling the same causal role. In other words, it is claimed, the luminiferous ether performed the same causal role as the electromagnetic field and hence was not actually abandoned after all (Hardin and Rosenberg 1982). However, this is a problematic move, not least because the theory of reference that underpins it is too liberal since just about any entity, now abandoned, can be said to have fulfilled the same causal role as some current entity.[3] Furthermore, by relying entirely on the causal role of the entities involved, this strategy effectively detaches the reference of the term to the relevant aspect of reality from its theoretical context and entails that 'we can establish what a theory refers to independently of any detailed analysis of what the theory asserts' (Laudan 1984: 161).

[2] The following is taken from da Costa and French 2003: 170–3.

[3] Thus, the 'natural place' of Aristotle may be said to fulfil the same causal role as the 'gravity' of Newton and the 'curved space-time' of Einstein; Laudan 1984.

An obvious alternative is to offer the kind of hybrid account of reference suggested by Psillos, which includes descriptive elements, drawn from the theoretical context, as well as causal roles (Psillos 1999: 293–300). The central idea here is that reference becomes fixed via a 'core causal description' of those properties which underpin the putative entity's causal role with regard to the phenomena in question (Psillos 1999: 295). The overall set of properties is significantly open to further developments, so that new properties get added around the core as science progresses. Of course, some of these latter properties may subsequently be deleted, as science progresses, but as long as there is significant overlap via the core set, continuity of reference through scientific change can be maintained and the PMI fails to get any grip.

In terms of such an account, one can then say that the term 'luminiferous ether' referred to the electromagnetic field (Psillos 1999: 296–9). In this case the 'core causal description' is provided by two sets of properties, one kinematical, which underpins the finite velocity of light, and one dynamical, which ensured the ether's role as a repository of potential and kinetic energy. Other—typically mechanical—properties to do with the nature of the ether as a medium were associated with particular models of the ether and the attitude of physicists towards these, of course, was epistemically much less robust. The core causal description was then taken up by the electromagnetic field, so that one can say that 'the denotations of the terms "ether" and "field" were (broadly speaking) entities which shared some fundamental properties by virtue of which they played the causal role they were ascribed' (Psillos 1999: 296). It is then a small step to conclude that the terms referred to the same entity. Finally, it is claimed that this avoids the previously noted problems associated with the PMI. First of all, not just any old entity can fulfil the same causal role as the current one since there needs to be a commonality of properties as represented by the core causal description. Secondly, it is only through a detailed reading off a theory that we can pick out the relevant properties in the first place; thus reference is not detached from the theoretical context.

However, the following concern arises: if the mechanical properties are shunted off to the models, as it were, in what sense can we still say that today's scientists, in talking about the electromagnetic field, are referring to the ether as an entity? The question is important because separating off the kinematical and dynamical properties from the mechanical ones in this way may obscure precisely that which was taken to be important in the transition from classical to relativistic physics. As well as the properties mentioned previously, and in virtue of its role as an absolute frame of reference, the ether also possessed certain 'positional' properties (Psillos 1999: 314 n. 9). If these are included in the core, then there can be no commonality of reference with the electromagnetic field. However, if they are not included in the core, then the perspective on theory change offered by this approach to reference may seem too conservative. The point is that whereas the ether was conceived of as a kind of substance, possessing certain mechanical qualities and acting as an absolute reference frame, the electromagnetic field was not (or at least not as a kind of substance in this sense). The metaphysical natures of the ether and the electromagnetic field, as entities, are very different and the claim might be pressed that, given this difference, there is no commonality of reference.

Now, an obvious response is to insist that in so far as these metaphysical natures do not feature in the relevant theories, the standard realist is under no obligation to accommodate them in her theory of reference or her position as a whole. In other words, she might insist that when she, as a realist, insists that the world is as our best theories say it is, that covers the relevant scientifically grounded properties only and not these metaphysical natures. But then the question is: what is it that is being referred to? It cannot be the ether/electromagnetic field qua entity, since this entity-hood is cashed out in terms of the metaphysical natures. Thus what is being referred to must be only the relevant cluster of properties which are retained through theory change. But now this response to PMI looks very different from what we initially took it to be. Instead of claiming that the ether was not abandoned—when scientists referred to it they were actually referring to the electromagnetic field—what is actually being claimed is that reference to the ether was secured via a certain cluster of properties which also feature in reference to the electromagnetic field. Now this response to the pessimistic meta-induction amounts to the claim that the ether as an entity was indeed abandoned, but that certain properties were preserved and retained in subsequent theories, where they feature in or are the subject of the relevant laws.[4] Thus, the theoretical elements that have been delineated can no longer be taken to be the relevant entities in a way that supports object-oriented realism.

This is not enough to push us towards structural realism of course, since that requires further steps that involve the articulation of the relevant properties in structuralist terms. A significant part of this book will be devoted to such an articulation. However, one might resist proceeding through these steps and insist that the properties themselves can form the ontological foundation for a viable form of realism.

1.3 Semi-Realism and Property-Oriented Realism

This is the core idea underlying Chakravartty's 'semi-realism', which rests on a crucial distinction between 'detection' properties and 'auxiliary' properties. The former are 'causally linked to the regular behaviours of our detectors' (2007: 47), and thus are those 'in whose existence one most reasonably believes on the basis of our causal contact with the world' (2007: 47); whereas the latter have an unknown ontological status, since detection-based grounds are insufficient to determine whether they are causal or not. It is in terms of the retention of clusters of detection

[4] It can't be claimed that the relevant cluster delineates the ether, on the basis of some form of bundle theory of objects, since, as already noted, certain properties that might legitimately be said to be part of the relevant bundle have been dropped.

properties that Chakravartty can respond to the PMI and indeed, he insists, one must retain such properties, or something like them, if one is to retain the ability to make decent predictions (2007: 50). Semi-realism thus captures the central features of those forms of realism that want to retain talk of entities, as well as of the kinds of structuralist positions we will be looking at here: it is in terms of the detection properties that we come to identify the putative entities, and it is these properties that provide the minimal interpretation of the mathematical equations favoured by the structural realist, as we shall shortly see.

There are two features of semi-realism that I find problematic and although I shall consider these in more detail later, I'll just mention them here. First of all the properties that semi-realism focuses on are causal properties and Chakravartty argues that such properties must be 'seated', as it were, in objects, metaphysically conceived. Thus, semi-realism is also object-oriented in a certain respect. Secondly, Chakravartty (rightly) provides a metaphysics for these properties, one that is articulated in terms of dispositions: according to the dispositional identity thesis (DIT), the identity of causal properties is given by the dispositions they confer. As I'll try to argue in Chapter 9, dispositional accounts are problematic in the context of modern physics and I shall suggest that when it incorporates an appropriate understanding of laws and symmetry principles in this context, semi-realism slips into the form of OSR that I favour.

Returning to the case study, consider the shift from Fresnel's ether-based theory of light to Maxwell's theory of electromagnetism. Here we go from conceiving of light in terms of wave propagation in an underlying ether to understanding it in terms of electromagnetic fields. The issue then is whether we can find sufficient continuity to be able to respond to the PMI. Worrall (1989) famously defended ESR by locating the continuity in the shift from Fresnel's ether-based theory of light to Maxwell's theory of electromagnetism in Fresnel's equations which express the relative intensities of reflected and refracted polarized light (we shall consider it in more detail later). These equations can be derived from Maxwell's and although it is this derivation that underpins this claim of continuity, the extent to which the derivation draws on the existence of certain properties and relations has been disputed. As far as Chakravartty is concerned, Fresnel's equations describe the relations that hold between certain dispositions in terms of which the relevant detection properties can be identified. This explains why Fresnel's theory was successful in making the right predictions about the behaviour of light: it was because they encoded the disposition of light to behave in certain ways under certain conditions.

However, Saatsi has argued that this fails to account for how Fresnel's false theoretical assumptions about the nature of light allowed him to latch onto these dispositions in the first place (2005). Furthermore, as he points out, in certain cases, Chakravartty's emphasis on the role of causal relations in distinguishing detection properties from auxiliary ones presents too narrow a construal of the relevant features that contribute to the explanatory success of the theory (2005). While it is certainly the case that the ether, qua entity, can be ruled out as not contributing to this success, merely focusing on the relevant properties, although a step in the right direction, is not sufficient since in the case of the Fresnel derivation, at least, it is certain 'higher-level' properties that we should be looking at. Thus Fresnel was able to predict the intensity of reflected and refracted polarized light on the basis of apparently false presuppositions because he had identified certain high-level boundary and continuity conditions for certain quantities that do the real work in the relevant derivation (for details see Saatsi 2005). These 'minimal explanatory properties' can then be realized in different systems, such as Fresnel's and Maxwell's, providing the required continuity. And it is towards these 'higher-level' properties that a realist stance should be adopted.

The crucial distinction now holds between these higher-level, multiply realizable properties that do all the explanatory 'heavy lifting' and the lower-level properties that represent one of the possible realizations in the context of the relevant theory (2005: 535). In particular,

> the explanatory ingredients are properties identified by their causal-nomological roles, and most (if not all) such properties are higher-order multiple realisable in the sense that these properties are instantiated by virtue of having some other lower-order property (or properties) meeting certain specifications, and the higher-order property does not uniquely fix the lower-order one(s). (2005: 533)

This 'property-oriented' stance is a core feature of Saatsi's own, 'eclectic' realism.[5] In so far as this represents a clear move away from object-oriented realism and further, in so far as the level of these multiply realizable properties lies close to the level of the laws and symmetry principles that are a central feature of the form of structural realism I favour, I regard this as a step in the right direction. My principal concern—which I will return to in Chapter 3—is that its explicit ontological neutrality and metaphysical minimalism raises concerns as to whether we obtain the clear understanding of how the world is that we associate with scientific realism.[6] In particular, an obvious concern has to do with the status of these properties as elements of our metaphysical pantheon. As things stand, they seem to be free-floating entities that have no metaphysical grounding. Both the object-oriented realist and the semi-realist will insist that they have to be associated with, at least indirectly (via inter-level instantiation perhaps), the relevant objects (which may then threaten Saatsi's whole

[5] For criticism see Busch 2008; and for a clarificatory response, Saatsi 2008.

[6] However, Saatsi has made it clear that the balance should tip towards the epistemological rather than metaphysical aspects of realism and that it is the former that he is primarily concerned with (the notion of 'Explanatory Approximate Truth' is central to his view). My view, which threads throughout this book, is that the realist cannot rest content with epistemology but must seek an understanding articulated in metaphysical terms. That articulation will then push the property-oriented realist towards one or other of OOR, ESR, or OSR.

project, since if he is not to fall into the clutches of the PMI, he will have to adopt one or other of the manoeuvres deployed by Psillos and Chakravartty respectively). The structural realist, on the other hand, will urge that they be understood as features of the relevant structure (however that is conceived!).

1.4 ESR and 'Hidden' Natures

Indeed, it has been argued (Busch 2008) that property-oriented realism, appropriately interpreted, is no different from epistemic structural realism (ESR). As already noted, this focuses on the relevant equations in the Fresnel–Maxwell example and since Fresnel's equations drop out as a special case of Maxwell's equations, the advocate of ESR insists both that this is where the level of continuity lies that allows us to respond to the PMI and that this continuity should be understood in terms of that of the relevant structures involved, with the ontological 'nature' of light vanishing from the picture (Worrall 1989):

> From the vantage point of Maxwell's theory, Fresnel was as wrong as he could be about what waves are (particles subject to elastic restoring forces and electromagnetic field strengths really do have nothing in common beyond the fact that they oscillate according to the same equations), but the retention of his equations (together of course with the fact that the terms of those equations continue to relate to the phenomena in the same way) shows that, from that vantage point, Fresnel's theory was none the less structurally correct: it is correct that optical effects depend on something or other that oscillates at right angles to the direction of transmission of the light, where the form of that dependence is given by the above and other equations within the theory. (Worrall 2007: 134)

Furthermore, it is claimed, Maxwell's equations are then retained in the 'photon' theory of light.[7] And so, optimistically, we can expect this form of continuity to continue.[8] ESR, and structural realism in general, is tied to a 'cumulativist' approach to science and the emphasis on the retention of structure can also be found articulated in such approaches. Thus Post, for example, famously offered a political analogy for these shifts in science: although the government (ontology) might come and go, the civil service (structure) remains broadly the same (Post 1971); or, as the structuralist says, 'it doesn't matter who you vote for, the structure always gets in' (Ladyman 1998).

Hence ESR can be summed up in the slogan,

> [t]here was continuity or accumulation in the shift, but the continuity is one of form or structure, not of content. (Worrall 1989: 117)

[7] It might be better to say they are 'quasi-retained' given the relationship between quantum and classical physics where theories of the latter are obtained from the former at some kind of limit; Post 1971; Pagonis 1996.

[8] For a useful discussion of what has been called the 'structural continuity argument', see Votsis 2011.

The position can be characterized as 'epistemic' because the central claim is that all that we know is this 'form or structure', whereas the ontological content of our theories is unknowable.

Two immediate questions then arise:

1) How are we to appropriately characterize this structure?
2) How are we to characterize (ontological) content?

Let me consider each in turn. With regard to the first, the Fresnel example, although accessible and much used, as we have seen, can be a little misleading because it has led to the impression that structural realism is wedded to a consideration of explicitly mathematized theories only and cannot offer much comfort to the realist when it comes to qualitatively expressed, or only partially mathematized, theories, such as we find in the biological sciences, for example. I do not agree, although a discussion of how structural realism might be extended to biology will have to wait until Chapter 12. Let me also sketch a distinction that will be further developed in Chapter 6, namely that between the presentation of structures at the level of the scientific theory itself and the representation of those structures at the 'meta-level' of the philosophy of science. With regard to the former, mathematical equations offer one way in which the relevant structures can be distinguished and identified but this is not the only way. One might, for example, identify certain families of relations as particularly significant within a given theoretical context and take these as a presentation of the relevant structure. When it comes to the representational aspect, philosophers of science have a range of tools and devices that they can deploy, depending, in part, on how they think theories themselves should be represented. Here I'm going to adopt a broadly pluralist stance and rather than advocate a particular such form of representation, suggest that there are various options, although some may be more suitable for certain purposes than others; again, I shall return to this in Chapter 5.

Thus, according to the so-called Received View of theories, the appropriate representation is in terms of a 'syntactic' logico-linguistic formulation. Within such a formulation, a syntactic form of structural realism was given by Maxwell (1970a) who argued that the 'cognitive content' of theoretical terms was exhausted by the structure, expressed—crucially—by the well-known Ramsey sentence of the theory. Represented thus, structural realism is widely perceived to fall foul of the so-called Newman problem—something I shall also consider in more detail in Chapter 5—a perception that is vigorously resisted by Worrall 2007 and in Zahar 2007.

Alternatively one might adopt the so-called 'semantic' or model-theoretic approach to theories, which represents them in terms of families of set-theoretic models. The extension of this approach to incorporate 'partial structures' allows it to capture, in a natural fashion, both the relationships that hold between theories, horizontally as it were, and those that hold vertically between a theory and the data models (da Costa and French 2003). With regard to inter-theory relationships partial structures can capture precisely the element of continuity through theory change that is emphasized by the structural realist (da Costa and French 2003: ch. 8). In particular, it offers the possibility of accommodating examples of such continuity that have been described as 'approximate' or partial. Thus Worrall refers to the shift from Newton to Einstein, from classical to relativistic mechanics, and suggests that 'there is approximate continuity of structure in this case' (Worrall 1989: 121).[9] He continues, '[m]uch clarificatory work needs to be done on this position, especially concerning the notion of one theory's structure approximating another' (Worral 1989: 121).[10] The partial structures approach can contribute to this clarification by indicating how such inter-theoretical relationships can be represented by partial isomorphisms holding between the model-theoretic structures representing the theories concerned (Bueno 1997; French and Ladyman 1999; da Costa and French 2003). For these reasons, in part, Ladyman advocated this approach in his now classic defence of the 'ontic' form of structural realism (1998). As I said, we will return to this issue in Chapter 5.

Of course, having identified the relevant structure and the way it is presented at the level of the theory and then adopted a particular mode of representation for one's purposes as a philosopher of science, there is still the issue of how to understand that structure in realist terms, namely as part of some conception of how the world is. Indicating how one might do that is, in large part, the purpose of this book. Again, the Fresnel example has perhaps misled some people in this regard as certain critics have suggested that the focus on mathematical equations implies that the structural realist takes the structure to be essentially mathematical and must therefore be some kind of Pythagorean in taking the world to be ultimately mathematical. This is certainly not the case. It is through the mathematical presentation of the relevant features of scientific theories that the structures we are interested in can be identified and thus, at that level, the mathematics is only playing a representational role, rather than a metaphysically constitutive one. The metaphysical nature of the

[9] Post refers to this case as an example of what he calls 'inconsistent correspondence', since classical mechanics agrees only approximately with the relativistic form, in the sense that the latter asymptotically converges to the former in the limit and the former asserts a proposition that only agrees with the latter in that limit (1971: 243). For further discussion see Pagonis 1996.

[10] Bueno has suggested that allowing for approximate correspondence may fatally weaken structural realism since it apparently grants that there may be structural *losses*, in which case a form of pessimistic meta-induction may be reinstated (private discussion). This is an important point. However, the problem is surely not analogous to the one that the realist faces with ontological discontinuity since the realist is claiming that we ought to believe what our best scientific theories say about the furniture of the world in the face of the fact that we have inductive grounds for believing this will be radically revised, whereas the structural realist is only claiming that theories represent the relations among, or structure of, the phenomena and in most scientific revolutions the empirical content of the old theory is recovered as a limiting case of the new theory. Another way of dealing with Bueno's point would be to insist that not all structures get carried over, as it were, but only those which are genuinely *explanatory*. We could then avail ourselves of Post's historically based claim that there simply are no 'Kuhn-losses', in the sense of successor theories losing all or part of the explanatory structures of their predecessors (Post 1971: 229).

structure of the world should not be identified with its mode of presentation. Likewise, just because we (as philosophers of science) choose to represent the relevant structures in set-theoretic terms does not mean that we take the structures themselves, as elements or aspects of how the world is, to be set-theoretic in a fundamentally constitutive sense.

Turning now to the second of our two questions, namely how the notion of non-structural content might be explicated, Worrall has famously drawn on a historical precedent for his epistemic form of structural realism (SR) in the work of Poincaré. The latter famously and lyrically expressed the view that theoretical terms 'are merely names of the images we substituted for the real objects which Nature will hide forever from our eyes. The true relations between these real objects are the only reality we can ever obtain' (1905: 162). Note the commitment to 'real objects' here. Unlike OOR, however, these are hidden from us, because—it is claimed—the only epistemic access we have is to the 'true relations'. In particular, scientific theories do not give us knowledge of the intrinsic natures of the unobservable 'real' objects. One can find similar sentiments expressed by Russell: 'although the relations of physical objects have all sorts of knowable properties... the physical objects themselves remain unknown in their intrinsic nature' (1912: 32–4; I shall return to both Poincaré's and Russell's views in Chapter 4).

According to this form of ESR then, there are such real objects but we cannot know them. More recently Worrall has moved to an alternative, 'agnostic' form, according to which there may or may not be such objects, but we cannot know either way, and if there are such objects we cannot know them (2012; see also Votsis 2012). I shall return to these two forms in the context of responding to Chakravartty's Challenge in Chapter 3, but note that the second form of ESR must involve what in the religious context would be called 'strong' agnosticism, which holds that it is *impossible* for us to know whether objects exist, rather than just that they are currently unknowable. One might then be tempted to deploy standard arguments against religious agnosticism to this case: one might argue, for example, that there are no good reasons to posit such hidden objects and good reasons not to posit them. The latter arise from the underdetermination argument that we shall consider in the next chapter; when it comes to the former, the agnostic may feel that we need objects to underpin the relations but I shall argue that such feelings are misplaced.

With regard to the 'hidden' aspect of these objects, critics have objected that this represents a return to a 'scholastic' philosophy that is out of step with the tenor of modern science. Thus, Psillos (1999: 155–7), in his defence of 'standard' realism, offers an alternative understanding of the 'nature of real objects'. He argues that this 'nature' should be understood solely in terms of the 'basic' properties of the objects together with the equations that describe their behaviour. Any talk of natures over and above this, he claims, is reminiscent of talk of medieval forms and substances, which were decisively overthrown by the scientific revolution. The understanding of 'nature' is hence essentially structural and there is no more to 'natures' over and above this structural description. Hence, he claims, the crucial distinction underpinning ESR collapses, fatally undermining the position as a whole.

This is an interesting line to take but there are a number of concerns that arise. First of all, articulating Poincaré's natures in terms of the set of basic properties of the relevant objects is not enough to yield structuralism and collapse the underlying distinction behind ESR. At the very least, these properties will need to be understood in structuralist terms (which is what I shall be arguing). Secondly, Worrall could appeal to an understanding of natures in terms of something other than forms and substances.[11] An obvious option is that by the 'nature' of these objects we mean their individuality (French and Ladyman 2003). Consider, for example, what many would take to be one of the more notable achievements of 19th- and 20th-century science, namely the rise of atomism. How was the content of atomism cashed out? Or, equivalently, how was the 'nature' of atoms understood? Briefly and bluntly put, atoms were understood as individuals where the metaphysical nature of this individuality was typically explicated in terms of substance or, more usually in the case of physicists at least, in terms of the particles' spatio-temporal location (see French and Krause 2006: ch. 2). Thus, one of the most prominent and ardent defenders of atomism, Boltzmann, incorporated such an understanding of the nature of atoms in terms of their individuality in Axiom I of his mechanics. The content of atomism was thus cashed out explicitly in terms of the metaphysical nature of atoms.[12] It is this 'nature' that Worrall could insist, following Poincaré, is hidden from our eyes, or more pertinently perhaps, which lies beyond our empirical and theoretical access.

This possibility is considered by Psillos in the three options for ESR that he sets out in his (2001):

> (A): We can know everything but the individuals that instantiate a definite structure; or, (B): We can know everything except the individuals and their first-order properties; or, (C): We can know everything except individuals, their first-order properties and their relations. (2001: S19)

Proceeding in reverse order, under option C structural realism would claim that only the higher-order properties of physical properties and relations would be knowable (Psillos 2001: S21).[13] All we can know in this case are the formal properties and relations of the structure. However, as Psillos notes, such a claim is trivial and unexciting (at least to the realist) since any set-theoretic representation of the world will yield such formal properties. Furthermore, in the scientific context we aim to describe more than just formal structure (Psillos 2001: S21) and Worrall himself would certainly not accept this as an appropriate understanding of ESR.

[11] However, Psillos is right in suggesting that he needs to appeal to some such understanding!

[12] Of course the 'ground' of the atoms' individuality could be some kind of Lockean substance, a form of haecceity, spatio-temporal location, or some relevant subset of properties (see French and Krause 2006).

[13] This is not the same as Saatsi's property realism.

Option B yields a 'relation description' according to which objects are described as standing in relations to other objects, but without further specifying the properties of those objects (Psillos 2001: S20). But, as Psillos notes, it seems implausible to insist that in the relevant physical situations we know only the relations between objects but not their first-order properties; as he argues, from the relations between electrons we can surely infer certain first-order properties such as their (rest) mass and charge. Thus if ESR were to adopt this option it would be committed to a principled cut between relations and first-order properties that in fact cannot be sustained.

Finally, turning to option A, this implies that the realist should accept that if there were two interpreted structures that were exactly alike in all respects except the relevant domain of individuals, then there would still be a fact of the matter as to which is the correct structure of the world. However, Psillos maintains, the only possible issue that remains is to name the individuals in the domain and this cannot be a substantive issue, because for each individual in either of the domains there is one in the other domain that performs the same causal role (since the individuals in each domain instantiate the same interpreted structure; Psillos 2001: S19–20). Whether A is a viable option then depends on issues to do with the metaphysics of individuality and in particular whether performing the same causal role implies that the relevant elements are in fact the same individual. In this context, as Psillos admits, ESR interpreted via option A would offer a metaphysically less costly alternative (in a principled sense) to standard realism but only if the latter is taken to accept the principle that two individuals can share all their properties (and hence causal roles) and yet still be different, something that Psillos regards as questionable.

But of course, in the classical context in which Boltzmann expressed the axioms of his mechanics, it had better be the case that performing the same role does not imply that we have the same individual, else the counting that underlies Maxwell–Boltzmann statistical mechanics will go awry. There are some subtle issues here (see, for example, French and Krause 2006: ch. 3; Huggett 1999a) and we shall return to them later, but basically, in classical statistical mechanics in order to get the right statistics (that then underpins our understanding of the Second Law of Thermodynamics and such), one must count permutations of otherwise indistinguishable particles (that is, particles that have the same 'intrinsic' properties, such as mass and charge and so on, and that also have the same state-dependent properties, so they play the same causal role). In effect the naming, or labelling of particles, is a substantive issue, since if one cannot do that, or if one cannot take the labels as meaningful in some sense, then one cannot apply the necessary permutations, or take them as meaningful and the wrong statistics will result. Far from being questionable, then, the aforementioned principle is critical in this context. In the quantum case, as is well known, the situation is different and there, to put it crudely, permutations do not count. This has been taken to imply that the relevant objects cannot be labelled and should not be regarded as individuals, but in fact one can maintain that quantum objects are individuals, albeit ones whose names or labels are effectively obscured by the relevant aggregate descriptions in terms of wave functions (French and Krause 2003: ch. 4). I shall return to the implications of this later, but clearly there is a substantive issue here.

Psillos' overall conclusion is that there are no in-principle restrictions on what we can know such that the distinction that Worrall seeks to establish with ESR can be maintained. However, as the previous discussion indicates, the advocate of ESR could insist on an understanding of hidden natures as distinct from structure in terms of the underlying individuality of the objects concerned. Nevertheless, I am sympathetic to Psillos' concern about the attempt by ESR to set some aspect of reality as beyond our epistemic ken, although for different reasons. As we shall shortly see, I shall argue that the situation regarding individuality (or lack thereof) in the quantum context pushes us to reject Worrall's hidden, or unknown, natures, conceived of in terms of objects for which we cannot say whether they are individuals or not, and understand structural realism in ontic terms.

Finally now, and setting Chakravartty's and Saatsi's concerns and those touched on here aside, one might wonder whether by forcing the collapse of the distinction underlying ESR, Psillos has also undermined the very basis of 'standard' realism: if there can be no in-principle distinction between relations and first-order properties, and if all the properties of objects are cashed out in structuralist terms, what is the content of standard realism itself? Ladyman has objected that standard realism without such 'natures' is nothing more than an 'ersatz' form of realism which draws on the plausibility of a structural description of theoretical objects whilst backing off from structural realism proper (Ladyman 1998). And, as we shall see, the 'proper' form of structural realism in this context is the 'ontic' form in which such objects are reconceptualized or eliminated altogether. The standard realist can't have it both ways: if she accepts the existence of objects, then she is going to have to face Poincaré-type arguments in the face of PMI that such objects do not feature in theory change and hence are hidden, or unknown; if she rejects such objects, then she has also given up standard realism and moved towards OSR.[14]

1.5 Another Case Study: the Zeeman Effect[15]

Returning to the challenge of responding to the PMI and the more general issue of accommodating theory change in general, there remains the concern whether the

[14] In Chapter 5 I shall briefly consider an alternative way of distinguishing between the structure and content of a theory in terms of multiple realizability in the context of the Ramsey sentence representation of structure.

[15] Crull presents the theory of the weak interactions as a further case study which, she argues, rules out Saatsi's property-based realism but can be accommodated by Worrall's ESR (Crull preprint). However, an important disanalogy exists between this and the Fresnel case: Fresnel's wave theory of light was not only empirically successful but generated a novel prediction in the form of the famous 'white spot', revealed through Arago's diffraction experiment. In the case of Fermi's theory of the weak interactions, when the relevant novel predictions were made, it wasn't Fermi's account per se that was responsible for them; rather it was the Standard Model in which elements of Fermi's account had been embedded. Given this lack of

kinds of approaches I have sketched earlier, and structural realism in particular, can accommodate other examples of such apparently radical ontological shifts. In other words, can these views be extended to other case studies in addition to the Fresnel–Maxwell example that has now become so well used as to seem hackneyed to some?

Here's one example: Cei (2005) has used the study of changing theories of the electron to argue that certain properties that played a crucial role in the explanation of certain phenomena were not represented by the appropriate equations. Following the prescription that we should be realist about those features of a theory that contribute to its empirical success, it is such properties that we should be embracing, rather than just the equations alone. Thus Cei takes this analysis to undermine ESR.

The phenomena concerned have to do with the Zeeman effect, discovered in 1896, whereby the lines of atomic spectra are split by a magnetic field. The theory that Lorentz put forward to explain the experimental results conceived of electrons as classically rigid bodies, interacting electromagnetically with the Maxwellian ether, and the mechanical properties of the electron turned out to be crucial, since the particle was treated as a harmonic oscillator. Now, as Cei notes, Lorentz's explanation was effectively a prediction: what Zeeman observed was a widening of the lines in the field, and Lorentz's account resolved this into a more complex pattern of splitting (2005: 1393). Furthermore, the relevant theoretical features then fed into further developments, leading to Larmor's famous precession formula, for example, which in turn is now derived within quantum mechanics and is important for understanding nuclear magnetic resonance. The conclusion Cei draws is that certain intrinsic properties play a crucial epistemic role in these developments and thus understanding ESR in terms of Psillos' option B (see previous section) is certainly not the way to go. More broadly, he argues, simply focusing on the relevant equations yields too restrictive a grasp of the underlying structures, and he takes this case study to motivate the move to OSR, which we shall consider in more detail in the next chapter.

1.6 Quantum Mechanics and Heuristic Plasticity

However, the developments that Cei maps out really only took place within what Kuhn called 'normal science' and cases of 'deep' revolutionary change might be expected to present a much more serious challenge to the structural realist, of whatever stripe.[16] We've already come across one such case previously, albeit briefly,

novel predictive success that can be attributed to the Fermi theory itself, surely the realist would be disinclined to regard that account (however it is delineated) as a successful theory requiring realist commitment in the first place.

[16] Another case that might also fall under 'normal science' is that of the development of field theories during the 20th century—including quantum field theory, General Relativity, and gauge field theory—as analysed in considerable and illuminating detail by Cao (Cao 1997; 2010; for a critical response see Saunders 2003a and b). Cao takes this study to support a form of structuralism according to which both

in Worrall's mention of the relationship between classical and relativistic physics. In that case he suggests that we have a kind of 'approximate' correspondence between the two in that we can recover the classical equations in the limit from those of Special Relativity as v/c tends to 0, where v is the velocity of the body under consideration and c is the speed of light (see, for example, Ballentine 1998: 388). However, in the quantum case things are less straightforward. First of all, there is the issue of which limit to take: as the principal quantum number n tends to infinity or as $\hbar$ tends to 0. The former underlies Bohr's correspondence principle. With regard to the latter, although both Bohr and Heisenberg emphasized the analogy between v/c tending to 0 and $\hbar$ doing the same, in the former case spatio-temporal trajectories are being recovered from spatio-temporal trajectories, with the difference being quantitative rather than conceptual; in the latter case one obtains well-defined trajectories as $\hbar$ tends to 0 only for certain kinds of states (Ballentine 1998: 389). Alternatively, one might try to recover the probability distributions for a classical ensemble from those of quantum mechanics via Ehrenfest's Theorem, but it turns out that satisfaction of this theorem is neither necessary nor sufficient to yield classical behaviour (Ballentine 1998: 391ff; cf. Post 1971: 233).[17]

In general, the theories look very different with regard to their theoretical content and the relevant mathematical representation. Nevertheless, as Saunders has indicated (1993), one should not exaggerate the extent of the divide and not only do there exist striking similarities between certain mathematical expressions on each side but these similarities and broader ones with regard to the structures on each side underpin the use of related techniques in each case. As Mehra notes, in certain respects, the difficulties were due 'not so much to a departure from classical mechanics, but rather to a breakdown of the kinematics underlying this mechanics' (Mehra 1987). Consider, for example, the well-known role of Fourier analysis in the history of quantum physics. Attempting to calculate the frequencies of atomic spectra using an oscillatory model, Heisenberg retained the classical equation for the electron, but dropped the kinematical interpretation of the quantity $x(t)$ as position. Instead he applied the standard Fourier transformation which decomposes the motion of the oscillator into a series, except he replaced the Fourier expansions for the spatial coordinates with what were recognized to be matrices, a move he justified by appeal to Bohr's correspondence principle. The rest, as they say, is history.[18]

One can also point to the 'bridge' provided by the Poisson bracket, which plays a central role in the Hamiltonian formulation of classical mechanics. I'll be touching on this formulation again in Chapter 2 but, briefly, the Poisson bracket allows for a

objects and structures mutually constitute one another, with the ontological priority of the former over the latter established once causal power is considered (for an early comparison with OSR see French and Ladyman 2003).

[17] For a survey discussion of this issue, see Landsman 2007.

[18] For an excellent account of the role of Fourier analysis in the development of quantum mechanics, see, for example, Bokulich 2010; also 2008.

convenient phase-space representation of the Hamilton–Jacobi equations of motion of classical mechanics. What it does, essentially, is take two functions of the generalized coordinates and conjugate momenta of phase space, and time, and produces a third function from them.[19] Its importance lies in yielding the relevant constants of motion, where a constant of motion for a system is a function whose value is constant in time, and hence whose rate of change with time is zero. If we form the Poisson bracket of such a function with the Hamiltonian for the system (where the Hamiltonian represents the total energy of the system; again we shall consider this in more detail in Chapter 2), then the function is a constant of the motion if and only if this Poisson bracket is zero, for all points in the phase space. And constants of motion represent quantities that are conserved throughout the motion, with prominent examples being energy, and linear and angular momentum. Furthermore, such conserved quantities correspond to symmetries of the relevant Lagrangian—which, again, we shall discuss in the next chapter but which basically encodes the dynamics of the situation—and so conservation of energy corresponds to symmetry in time, that of linear momentum to symmetry in space, and that of angular momentum to rotational symmetry.[20] And according to OSR, of course, symmetries are a fundamental feature of the structure of the world, so this 'bridge' offered by the Poisson bracket is intimately tied in to the theme of this book.

Of course, the bridge itself is not straightforward. As is well known, the Poisson bracket is strictly inapplicable in the quantum context and must be replaced by the appropriate commutator.[21] However, formally there is a relevant connection via the deformation of the underlying Poisson algebra to yield 'Moyal' brackets[22] which are the isomorphs in phase space of the commutators of observables in Hilbert space.[23] Historically of course, it was the apparent similarity between the Poisson bracket and the commutator that lead Dirac to his 'bra' and 'ket'

[19] By taking the partial derivatives of the functions and constructing a sum of their products, where each term in the sum contains one derivative of each function and one of the derivatives is with respect to the generalized coordinate and the other is with respect to the conjugate momentum and the terms change sign depending on what the derivative is with respect to.

[20] The correspondence is established by Noether's theorems (see, for example, Brading and Castellani 2008).

[21] The commutator of F and G is [F, G] = FG-GF. When F is the momentum operator (p) and G position (x) we obtain $[p, x] = \hbar$, which is what lies behind the Uncertainty Principle, of course.

[22] Formally a 'deformation' involves the change in some object (such as the Poisson bracket) in some space (such as phase space) as one changes the values in some parameter (a technical introduction can be found in Plfaum 2005). To obtain Moyal brackets one deforms the Poisson brackets with respect to the 'reduced' Planck's constant $\hbar$.

[23] The Moyal bracket yields the relevant Lie algebra which effectively gives the structure of the corresponding Lie groups (to every Lie group there corresponds a Lie algebra, but to every such algebra there may correspond more than one group, such that these groups are locally isomorphic). The term was introduced by Weyl, who will feature in our consideration of the history of structural realism in Chapter 4. Lie groups capture continuous symmetries, in particular those of the differential equations that are used in the presentation of certain of the fundamental laws of physics.

formulation of quantum mechanics and the heuristic role of the correspondence principle here is well known.[24]

There is a lot more to say here but the important point is that Saunders understands these examples as illustrating a fundamental 'heuristic plasticity' of the mathematics by means of which the structural features of classical dynamics are isolated, entrenched, and thereby preserved in subsequent developments.[25] In particular, he writes, certain of these features provide 'over-arching abstract frameworks…within which one dynamical structure may be embedded in another' (Saunders 1993: 308).[26] And the most important of these features are the group-theoretic ones by which the fundamental symmetries of the world can be presented. As indicated briefly earlier, in the classical domain the central symmetries are those in time, space, and rotational symmetry, but in quantum physics a further fundamental symmetry comes into play—that which is expressed by 'Permutation Invariance' and which not only yields Pauli's Exclusion Principle but the fundamental division of natural kinds between bosons and fermions.

Saunders' focus here is specifically on theories of dynamics, however, and he emphasizes the point that claims regarding apparent radical ontological changes that feature so prominently in not only the PMI but also broadly Kuhnian accounts of scientific revolutions are undermined by the entrenchment and preservation of these structural features. In general, as he later noted (2003a), the invariance group of symplectic geometry in Hamiltonian mechanics (that is, the group of canonical transformations) played a central role in most formulations of dynamics from the Hamilton–Jacobi theory to the Poisson bracket algebra and further, to the rules of 'old' quantum theory and it continues to do so to this day in quantum mechanics (particularly as defined by canonical and geometric quantization processes). Here, he insists, we see a progressive deepening of concepts[27] and it is in these terms, then, that we should understand the retention of the relevant explanatory elements within the structural realist context.

Thus if structural realism is to accommodate not just the more straightforward kind of continuity between theories represented by the Fresnel case in order to respond to the PMI, it needs to broaden its grasp of structure to include not just the kinds of equations that Worrall has highlighted, but the group-theoretic features that Saunders and others have emphasized. Now, although both ESR and OSR agree

[24] Dirac himself represented the fundamental underling discovery as occurring in a 'flash of insight' while out walking but he also had an excellent understanding of the Hamiltonian formulation, particularly as it was applied by Sommerfeld to atomic systems.

[25] Cassirer coined the term 'indwelling sagacity' (Spürkraft) for this feature of certain formulae and expressions.

[26] For an excellent account of the classical-quantum relationship in terms of the formalism of Lie group theory see Jordan and Sudarshan 1961. As they say, from this point of view the main difference between the two 'mechanics' lies in the choice of the Lie bracket.

[27] Likewise, he argues, the Standard Model incorporates (gauge) symmetries which are 'natural extensions' of those of classical electromagnetism.

in their commitment to the claim that science is progressive and cumulative and that the growth in our structural knowledge of the world goes beyond knowledge of empirical regularities, if ESR is to broaden its grasp of structure—if, indeed, it is going to make good on Worrall's promise (1989: 123) and encompass quantum mechanics—then it is going to have to incorporate the kinds of structures indicated here, and of course their group-theoretic presentations. But if it is going to do that, then it needs to incorporate the earlier symmetries—and not just the 'classical' ones, like rotational symmetry, but also Permutation Invariance.[28] But if ESR is going to do *that*, then it will have to take on the metaphysical consequences of this symmetry and those, I argue, lead us to abandon the notion of object, hidden or otherwise.[29] In other words, if structural realism is to broaden its grasp and seize the kinds of structures that modern physics actually presents to us, then it is going to have to shift from ESR to OSR.

Of course, the advocate of ESR might worry that this shift brings a certain tension with it: ESR is founded on a distinction between structure and objects, with only the former grounding the kind of continuity through theory change that the realist seeks. If, as OSR insists, there are no objects, then one cannot appeal to this distinction, or ESR's argument for scientific realism, based as it is on what Psillos has called the 'divide et impera' strategy. I think the core issue here has to do with how we establish eliminativism about objects and I shall respond to it by appealing to a kind of iterative move: the distinction ESR is based on is really one between structures and putative 'objects', such as electrons, protons, and so forth. In those terms, one can still make the desired structuralist claim about where the relevant continuity lies. However, and this is the next step in the iteration, the metaphysical consequences of, for example, Permutation Invariance, mentioned earlier, lead us to conclude that the putative objects should either be regarded as what we shall call (in the next chapter) 'thin' objects, at best, or should not be regarded as objects at all. Thus in the first stage of the iteration we begin with a putative distinction, one side of which we then discard in the second, leaving only the structure (or, at best, as we'll see, structure with 'thin' objects). The tension, I would suggest, can then be dissipated and the advocate of OSR can have all the advantages of ESR when it comes to accommodating theory change and the history of science but with a minimalist or, indeed, eliminativist, metaphysics.

[28] The standard view for many years held that Permutation Invariance is a peculiarly quantum symmetry. However, Saunders has argued that classical mechanics is also permutation invariant (2006a). I think the support for this claim partly depends on how one delineates 'classical' mechanics (see French and Krause 2006: 144–6).

[29] In fairness, the advocate of ESR could argue that no ontological weight should be given to these symmetries on the grounds that they are merely 'by-products' of the relevant laws and not further features of the world, as it were. I shall consider the 'by-product' view in Chapter 11 but it is not clear that one could adopt such a view about Permutation Invariance and so the consequences for our notion of object would remain, and even if one did treat other kinds of symmetries in this way, I would suggest that incorporating the laws that they are by-products of will still push one towards OSR.

In the next chapter I shall outline the metaphysical consequences mentioned earlier, recall how they create a form of metaphysical underdetermination, and argue that the appropriate response to that underdetermination is to adopt OSR. I shall also set out what I have called 'Chakravartty's Challenge', which has to do with our understanding of theories and argue that this also motivates a shift to OSR.

2

Mixing in the Metaphysics 1

Underdetermination[1]

2.1 Introduction

We recall the three challenges that the realist must face: PMI, underdetermination, and what I have called 'Chakravartty's Challenge'. In the previous chapter I indicated how responding to the first takes the realist towards structural realism. In this chapter and the next we will shift our focus from theory change and what is retained, to metaphysical concerns and again I shall suggest that these concerns should push the realist in the structuralist direction. Let us begin with a discussion of underdetermination in general and how it might be tackled, before considering what I have called elsewhere, the 'underdetermination' of metaphysics by physics (see French and Krause 2006: 189–97).

2.2 Challenge Number 2: Underdetermination

The underdetermination of theory by evidence (UTE) occupies a central place in the realism–antirealism debate. There is an issue as to the appropriate formulation of the thesis but one clear expression of it goes as follows:

> Suppose that two theories T_1 and T_2 are *empirically equivalent*, in the sense that they make the same observational predictions. Then [according to the UTE thesis] no body of observational evidence will be able to decide conclusively between T_1 and T_2. (Papineau 1996: 7)

The consequences for standard realism are clear: if UTE is correct then the realist is unable to determine which of T_1 and T_2 is more worthy of belief, where 'belief' here is understood as 'belief that... is true' and truth is explicated in the correspondence sense. Thus UTE becomes a powerful tool in the hands of the anti-realist.[2] However, other forms of underdetermination can also be articulated; consider the following passage, for example:

[1] Much of this chapter is taken from French (2011a), which in turn was based on a paper presented at the Düsseldorf 'Theoretical Frameworks and Empirical Underdetermination Workshop' organized by Gerhard Schurz and Ioannis Votsis, to whom I am grateful for the opportunity to present my work.

[2] Constructive empiricism, as is well known, urges us to abandon the view that theories should be believed to be true in favour of what, according to the empiricist, is the epistemically more secure position of accepting them as empirically adequate only (van Fraassen 1980).

> The phenomena underdetermine the theory... The theory in turn underdetermines the interpretation. Each scientific theory, caught in the amber at one definite historical stage of development and formalization, admits many different tenable interpretations. What is the world depicted by science? That is exactly the question we answer with an interpretation and the answer is not unique. (van Fraassen 1991: 491)

Now, with regard to the fundamental issue of which of T_1 and T_2 the realist should be committed to, the structural realist will urge ontological commitment to the underlying structure that is common to both theories. Of course, the objection will run that there may not be any such common structure but then the matter has to be determined on a case-by-case basis and until the anti-realist can provide some plausible cases in the first place, the whole issue is moot. However, rather than respond to this form of underdetermination any further here (see, for example, da Costa and French 2003: ch. 8), I would like to consider the further variety that van Fraassen has identified, with regard to the interpretations of a theory. Of course what one means by an 'interpretation' and how one distinguishes such from a theory itself are matters for discussion but here I shall identify two senses in which one might identify different interpretations associated with a given theory: the first arises from the existence of different formulations of the 'same' theory; the second is concerned with different metaphysical 'understandings' of the same theory. Both appear to raise problems for both the object-oriented realist and the epistemic structural realist.

A number of examples of the first kind have been given in Jones' powerful critique of realism, which 'envisions mature science as populating the world with a clearly defined and described set of objects, properties, and processes, and progressing by steady refinement of the descriptions and consequent clarification of the referential taxonomy to a full-blown correspondence with the natural order' (1991: 186). He gives a series of examples in which one has different empirically equivalent interpretations, each of which offers different sets of objects, properties, and processes; hence, he concludes, this realist vision cannot be achieved. A particularly prominent example is that of the Hamiltonian and Lagrangian formulations of classical mechanics. Thus if we were to open a standard undergraduate textbook in classical mechanics, we would typically be presented with, not just objects and forces, but potentials, 'action', and so forth. This yields nothing less than different sets of world-furniture, on Jones' view, arising from different formulations of the theory and thus necessitating different ontological commitments for the realist.

Now the Hamiltonian equations are straightforwardly obtained from Newton's equations and, put simply, are as follows:

$$\dot{q} = \partial H / \partial p$$
$$\dot{p} = - \partial H / \partial q$$

where p represents the generalized momenta, q the generalized coordinates and H ($H(p, q, t)$), the Hamiltonian, represents the total energy of the system and effectively encodes the dynamical content.

The Lagrangian equations, on the other hand, are as follows:

$$d/dt(\partial L/\partial \dot{q}) = \partial L/\partial q$$

where L represents the difference between the kinetic and potential energies. These equations straightforwardly reduce to Newton's equations. Briefly comparing the two, we can say that the content of Newton's equations is encoded in the structures defined over certain spaces:

Hamiltonian: the relevant space is the space of initial data for the equations; that is, the space of possible instantaneous allowable states. The underlying structure is that of the relevant cotangent bundle.

Lagrangian: the relevant space is the space of solutions to the equations; that is the space of allowable possible worlds. The underlying structure is that of the tangent bundle. (see Belot 2006)

As is well known, applying the Legendre transformation to the Lagrangian yields the Hamiltonian and on this basis it is typically claimed that the two formulations are inter-translatable. So an obvious response from the structural realist to Jones' claim would be to insist that the underlying structure of these formulations is essentially the same and it is to this that we should be committed, as realists (or, in other words, it is *this* that we should regard as the furniture of the world).

However, a number of concerns have been raised about this move. First it has been argued that a better option for the realist is to break the underdetermination, although as we shall see, in most cases the 'breaking factor' raises deep concerns of its own, and where it might be justified, it still leaves room for a structuralist stance. Secondly, it has been argued that on most straightforward characterizations[3] of structure—such as, and in particular, the set-theoretic one favoured by those structuralists, such as myself, who adhere to the 'semantic' or 'model-theoretic' approach—different formulations such as those just presented give rise to *different* structures (Pooley 2006). Hence, in terms of the structuralist's own framework, the underdetermination would remain. A detailed elaboration of the presentation and representation of structure will be given in Chapter 5 but nevertheless I shall indicate how the structuralist might respond to this concern. Finally, and perhaps more problematically, it has been argued that establishing an interrelation between formulations is not enough (Pooley 2006). What is needed is a 'single, unifying framework' that can be interpreted as corresponding more faithfully to reality than the alternatives. In the absence of such a framework, the structural realist has no grounds for resolving the underdetermination by appealing to underlying structure on the basis of inter-translatability. Here I shall indicate how such a single framework might be constructed. Let us now consider these concerns in turn.

In the case of the underdetermination of theories by evidence, attempts to 'break' the underdetermination proceed by appealing to further factors. In many cases of

[3] I use this word deliberately, so as not to undermine the objection from the word go.

apparent underdetermination these will be largely empirical factors: in those cases in which we appear to have underdetermination of two (or more) theories in terms of the evidence currently known, the realist will urge a wait-and-see attitude, holding off on her commitments until further evidence has come in and broken the underdetermination. In some cases she may appeal to indirect evidence that accrues to a particular theory through, for example, the embedding of that particular theory in a broader theoretical context (da Costa and French 2003: ch. 8). However, such forms of underdetermination breaking are not available to the realist in the more problematic cases of underdetermination that embrace all possible evidence, and so she may appeal to extra-empirical factors such as simplicity or explanatory power. Of course, such moves will be less than compelling for the anti-realist who may well ask what simplicity or explanatory power have to do with truth! As we shall see, it is unclear whether the underdetermination between the Lagrangian and Hamiltonian formulations of classical mechanics can be broken by appealing to the relevant kinds of factors.

2.3 Breaking the Underdetermination$_1$: Appeal to Metaphysics

Musgrave responded to Jones' critique by appealing to appropriate metaphysical factors, where he insisted the latter do not amount to 'mere philosophical whim and prejudice' but are continuous with the relevant physical factors (Musgrave 1992). Thus, 'physics has to look to metaphysics to help decide (fallibly, of course) between experimentally undecidable alternatives' (Musgrave 1992: 696).

However, there is the obvious concern regarding the justification for the metaphysical principles that are invoked in this regard. We shall touch on this when we consider 'metaphysical underdetermination', but it is not entirely clear what principles could be invoked to decide between the Hamiltonian and Lagrangian formulations and one's view of such principles will obviously determine whether one thinks the underdetermination can be resolved in this way or not.

More significantly, perhaps, there is the concern that much of modern metaphysics appears to have distanced itself from any grounding in modern physics and hence one might worry—perhaps in normative fashion—that appealing to principles drawn from this 'physics-free metaphysics' in order to break the underdetermination between different formulations or interpretations of theories could lead to some potentially disastrous choices being made.[4] (Of course, *how* disastrous depends on how seriously one views these kinds of underdetermination when it comes to either our understanding of, or even the progress of, modern physics. At the very least, if one thinks the choice of formulation has heuristic significance—and I will return to

[4] cf. Morganti 2011.

this shortly—one might be entitled to worry!) Indeed, the anti-realist will run the same line as she did with simplicity, but perhaps even more strongly, asking how metaphysics can contribute to our understanding of how the world is. In a sense, we can respond to this concern, as we shall see, although not in the context of underdetermination breaking.

Even if metaphysics is seen to be continuous with physics, as Musgrave suggests, this doesn't really help break the underdetermination since an obvious circularity could arise: if the metaphysics one appeals to is continuous with (a particular formulation of) the physics, then it may end up simply determining that particular formulation. Appealing to metaphysics seems to leave us with a dilemma: either the metaphysics floats free of the physics and requires justification itself; or it is continuous with the physics but then it can't actually break the underdetermination on pain of circularity.

Let us consider an alternative appeal—to the heuristic fruitfulness of one formulation over the other.

2.4 Breaking the Underdetermination$_2$: Appeal to Heuristic Fruitfulness

The idea, then, is that we should prefer that formulation which is more heuristically fruitful (see da Costa and French 2003: ch. 6), in some sense, where that sense can be broadly characterized, strongly, as leading to, or, weakly, as indicating (again in some sense!) an empirically successful theory (cf. Pooley 2006). Now, one might immediately wonder whether it is even possible for a *formulation*, as opposed to a theory per se, to give rise to a new theory. Of course this raises again the issue of the distinction, if any, between theories and formulations, but the thought is twofold: first and generally, there is the question whether formulations and theories are the kinds of things between which there can hold the sorts of interrelations that come to be established following certain heuristic moves; secondly, and more particularly, there is the question whether the well-known kinds of moves that one can discern as leading from one theory to its successor also hold between a formulation and a future theory (where it is not yet clear whether 'successor' is the appropriate term here).

At this point one could simply retreat and defend a notion of heuristic fruitfulness in the still broad sense of leading to a better, deeper, or whatever, *understanding* of the given theory—that is, a new formulation—but that seems a less than conclusive way of breaking the underdetermination. Here what one wants is some set of criteria for what counts as underdetermination breaking, conclusive or not, in this case. The (realist) intuition (carried over from the standard form of theory-theory underdetermination) that establishing that one 'horn' of the underdetermination leads to an empirically successful theory, whereas the other does not, certainly counts in this regard. However, establishing that one formulation rather than the other yields a new formulation and, consequently, better *understanding*, appears not so decisive, since we don't have that crucial factor of empirical success in this case.

Refusing to retreat would mean insisting that, first of all, the relevant interrelations can hold between formulations and theories, however characterized, and secondly, that appropriate heuristic moves can be made leading from one to the other. The former may not be a problem if one insists either that there is no in-principle distinction between 'theories' and 'formulations', or that any such distinction is blurred. The latter requires further detailed investigation, effectively doing for 'formulations' what the likes of Post (1971) did for theories. But one can at least make a first pass and note that, for example, the Lagrangian formulation is typically regarded as the 'natural' way to extend Newtonian particle dynamics to fluids[5] and the extension to quantum field theories is well known, with, for example, the Lagrangian density being straightforwardly related to Feynman diagrams. Here it is the fact that the Lagrangian density is a locally defined, Lorentz scalar field that makes it so useful for relativistic theories. A quick scan of the relevant physics literature will show Lagrangians all over the place, in quantum chromodynamics, quantum black holes, etc. Nor is their ubiquity a mere matter of pragmatics: Wallace has argued that although much foundational analysis in quantum field theory (QFT) has focused on algebraic QFT, with its clear set of axioms, 'naïve' Lagrangian QFT is sufficiently well delineated as a theory that it too can serve as the jumping off point for foundational considerations (Wallace 2006), a claim that I shall consider in more detail in Chapter 11.

Of course, in the quantization of a classical field the Hamiltonian (obtained, as noted, from the Lagrangian via the Legendre transform) plays a crucial role. And the central importance of the Hamiltonian for quantum mechanics hardly needs emphasizing. What does deserve more careful attention are the moves that led to this central role, and here we recall Saunders' point about the heuristic plasticity of the relevant structures, with the Poisson bracket providing a kind of bridge that allowed for a convenient phase-space representation of the Hamilton–Jacobi equations of motion.

Clearly both formulations can claim some degree of heuristic fruitfulness. What one would then have to do for underdetermination to be broken via these sorts of considerations would be to evaluate and compare the 'heuristic plasticity' of the relevant entities in the two formulations, in an attempt to weigh the one against the other. But even before we embark upon such an enterprise, further doubts might creep in as to whether heuristic fruitfulness is really sufficient to break the underdetermination. Consider: suppose we were evaluating the promise of the Lagrangian and Hamiltonian formulations at some point prior to the development of quantum

[5] Interestingly, 'conservation of particle identity' is fundamental to this approach, where fluid 'particle' identifiers—such as position at time t, or relevant thermodynamic properties—are treated as independent variables, although a form of indistinguishability also holds since the dynamics remain unchanged through permutation of 'particles' of the same mass, momentum, and energy.

mechanics, in the late 1890s say. At that time, any determination of the fruitfulness of one approach over the other, or the plasticity of certain elements as compared with others, could only act as a kind of 'promissory note', since it could be that the plasticity leads to a dead end and the fruitfulness withers away to nothing. Of course, looking back, we can take a realist stance and say these developments were in some sense inevitable, because that's how the world is (so, for example, the structural realist might insist that the structure of the world corresponds to, and is hence best represented by, some form of Lie algebra), but at the time we have no such guarantee. Is such a promissory note, presented in modal terms as it has to be, sufficient to push us to select one formulation over the other? Surely not; at best, any such selection must itself be tentative.

But now consider this: suppose we were to evaluate these formulations from a perspective reached after the relevant developments have taken place. Looking back, of course the promise of one over the other may become clear but, equally, the relevant developments will also be clear, as will the new theory led to by these heuristic moves. In this situation, there will no longer be any underdetermination, because theoretical developments have effectively made the choice for us. Of course, in the case of the Lagrangian and Hamiltonian formulations, one can justifiably claim that each demonstrated a degree of fruitfulness, and the relevant elements an associated degree of plasticity, so in this case one can't even make a retrospective determination. But the point is that even if one could, even if it were clear which formulation turned out to be more fruitful than the other, such considerations are really no help in breaking the underdetermination at all: either they are mere promissory notes, or there is no underdetermination to break![6]

These sorts of considerations will crop up again when we consider the issue of surplus structure but let's move on to another underdetermination-breaking move.

2.5 Breaking the Underdetermination$_3$: Appeal to Less Structure

We recall the well-known attempts to break 'standard' theory-theory underdetermination through appeal to simplicity, explanatory power, and their dismissal by anti-realists on the grounds that it is unclear what these factors have to do with the truth that the realist supposedly aims for, and that they should be treated as 'pragmatic' only. Recently, a similar appeal has been made in structural terms through the claim that that formulation should be preferred which incorporates less structure in some sense. Of course, the anti-realist will hardly be convinced by such an appeal but leaving them and their concerns to one side, realists of various

[6] As it turns out, and as we shall shortly see, a powerful argument has been given to the effect that the Lagrangian formulation should be preferred, on the grounds that it more naturally captures the core features of classical mechanics.

stripes, and structural realists in particular, will be interested in this move. However, as we shall see, it comes at a cost, related to the considerations already canvassed.

In a recent work, North has argued that the Hamiltonian formulation should be preferred over the Lagrangian on the grounds that the former involves less structure than the latter (North 2009). Essentially she reminds us that whereas the underlying framework of the latter is configuration space with a (Riemannian) metric structure and associated distance measure, that of the former is phase space with a symplectic structure and associated volume element. The symplectic structure, she claims, is sufficient for the relevant physics, so the choice is less structure (Hamiltonian) over more (Lagrangian). The idea, then, is that since metric structure determines, or presupposes, a volume structure, but not vice versa, the former adds another level of structure to what's needed to express the Hamiltonian equations of motion. Furthermore, the metric structure appears to be essential for the Lagrangian formulation, given the way the generalized coordinates feed into the Lagrangian.

Now this difference in the structures has implications for the earlier claim that we can straightforwardly transform from one formulation to the other, as North notes. In essence it implies that such transformations are only possible within certain constraints; we shall return to this point shortly.

In general, then, North's approach meshes nicely with a broadly structuralist perspective:

> I think modern physics suggests that realism about scientific theories is just structural realism: realism about structure. Modern geometric formulations of the physics suggest that there is such a thing as *the* fundamental structure of the world, represented by the structure of its fundamental physics. There is an objective fact about what structure exists, there is a privileged carving of natures at its joints, along the lines of its fundamental physical structure. (North 2009: 81–2)

Furthermore she gives the following recipe for obtaining the structure of the world:

> Take the mathematical formulation of a given theory. Figure out what structure is required by that formulation. This will be given by the dynamical laws and their invariant quantities (and perhaps other geometric or topological constraints). Make sure there is no other formulation getting away with less structure. Infer that this is the fundamental structure of the theory. Go on to infer that this is the fundamental structure of the world, according to the theory. (North 2009: 78)[7]

[7] Elsewhere she writes, 'Infer the least *structure* to the world needed for the mathematical formulation of its fundamental physics... The idea behind the general rule is simple. If the fundamental laws can't be formulated without implicitly referring to some structure, then that's a reason to think the structure represents real features in the world. For the laws presuppose the structure; they require it in order to be true. If the laws can be formulated without some structure, then that's a reason to think it is excess, superfluous structure; an artifact of the formalism, not something in the world... More generally, a match in structure between the dynamical laws and the world is evidence that we have inferred the correct structure to a world governed by those laws' (North 2013).

Of course, taking the mathematical formulation of a given theory should not be understood as meaning that the mathematics should be taken as uninterpreted, on pain of being accused of being a Pythagorean (see Chapter 8)! However, the admonition to make sure there is no other formulation 'getting away' with less structure is more problematic. How do we compare different 'amounts' of structure, such that we can say there is more or less in specific cases? Here the example of mathematical structure is turned to again:

> In building up a mathematical space, some objects will presuppose others, in that some of the mathematical objects cannot be defined without assuming others. Starting from a structureless set of points, we can add on different "levels" of structure. A bare set of points has less structure than a topological space, a set of points together with a topology (specifying the open subsets). A topological space has less structure than a metric space: in order to define a metric, the space must already have a topology. (Intuitively, a metric gives distances along curves by adding up the lengths of segments between nearby points; and without a topology, there is no sense of the "nearness," or neighborhoods, of points.) And so on. (North 2009: 65–6)

And in making these comparisons, symmetries generally mean less structure. Thus, adopting a realist stance towards those structures that involve appropriate symmetries may satisfy this methodological requirement of accepting those formulations that imply less structure.

Before I raise some concerns about North's prescription—particularly with regard to its application as an underdetermination breaker—let me just note that although she acknowledges Ladyman's 1998 paper on OSR, North insists that her account is different. Precisely wherein that difference lies remains a mystery. Perhaps we can see her approach as preliminary to adopting an appropriate realist stance: one begins by following her prescription (which of course is not particularly out of the ordinary, except for the insistence on 'less structure is better'), and having arrived at the relevant structure, one may then adopt a metaphysics of objects with 'hidden' or unknown natures, as in the case of ESR, or one of structure without objects, as in the case of OSR.

Of course, the anti-realist, such as the constructive empiricist, will be unmoved by this strategy, since if the relevant theories are not regarded as true, but only as empirically adequate, then the relevant commitment to structure is unmotivated. As Bueno puts it, 'The fact that the dynamics requires [the relevant structure] is not enough to justify ontological commitment to the latter on a view that does not take truth to be a norm for scientific discourse' (forthcoming: 2). In particular, in so far as 'the dynamics' is expressed via the dynamical laws of the theory, and in so far as North takes these to govern the fundamental level of reality (2010: 3), then the constructive empiricist, who does not adopt such a realist view of laws will be unwilling to follow her inferential move. I shall return to consider this idea of laws 'governing' the fundamental level of reality in Chapters 9 and 10 but let me just note that an empiricist of structuralist inclinations could follow North's strategy as she states it herself—that is, take the mathematical formulation of the given theory, figure out what structure is required by that formulation, make sure there is no other formulation getting away with less structure, infer that this is the fundamental structure of the theory...—but then stop before that final step of inferring that this is the structure of the world and simply insist that this is only the structure of how the world could be!

Returning to the case of the underdetermination between the Lagrangian and Hamiltonian formulations, North applies her prescription and concludes that the fundamental structure of the world is that which underpins the former, namely symplectic. As we have noted, the crucial step in this inferential procedure is the insistence on accepting that formulation that has less structure. We can reformulate this as follows: reject any formulation that can do the same job but with surplus, or in some way, superfluous, structure. Now this may seem straightforwardly plausible from a realist point of view, not least because one could obviously underpin such a move through considerations based on simplicity. However, taken as applying across the board, as it were, it is in tension with the previous suggestion regarding heuristic fruitfulness, since it may well be this very surplus structure that confers such fruitfulness. This was a point made by Redhead, some years ago, when he noted that a number of significant developments in theoretical physics were achieved through the appropriate interpretation of mathematical structures that are related to those in terms of which empirically grounded theories are couched (Redhead 1975).[8]

There are numerous examples of the fruitful role of such surplus structure. Consider Dirac's equation for the relativistic behaviour of an electron, from which spin emerges through the union of quantum mechanics and relativity theory. As is well known, the equation has positive and negative energy solutions. The latter were initially regarded as unphysical and hence as surplus mathematical structure, but subsequently came to be interpreted first in terms of protons and then as positrons.[9]

[8] Jumping ahead to the discussion in Chapter 5 one can characterize this notion of surplus structure as follows: we take the empirical sub-structures (representing the phenomena) to be embedded in theoretical structures, and the latter are understood to be related via partial homomorphisms to the relevant mathematical structures, which are related in turn to further structures which are then open to physical interpretation and hence being related to an extension of the theory, or a new theory entirely (Bueno, French, and Ladyman 2003). Redhead himself presented this idea in terms of a 'function space' characterization which can be understood, more or less straightforwardly, in terms of the standard set-theoretic formulation of the semantic approach to be presented in Chapter 5.

[9] Pashby (2012) suggests that this shift in interpretation yields a structural discontinuity that the structural realist may have difficulty in accommodating. This discontinuity has to do with the move from 'first quantised' quantum mechanics to quantum field theory: first of all, we lose conservation of particle number; secondly, we move from a kind of 'absence' (holes) to a 'presence' (anti-matter); and thirdly, we shift from unitarily equivalent representations to inequivalent ones. With regard to the first, this undermines the particle conception in QFT and if anything, gives further motivation to adopt a structuralist ontology; I'll come back to this in Chapter 12. As for the second, 'absence' and 'presence' have object-oriented connotations that a structural realist would surely reject. I shall return to the issue of inequivalent representations in Chapter 12 but as Pashby himself notes (2012: 469), these can be accommodated within my modally informed structuralism.

In effect, the positing of anti-matter derived from the ontological interpretation of mathematical surplus structure. Redhead also examined the significance of gauge symmetries within field theory from this perspective: understanding gauge transformations as acting non-trivially only on the surplus structure, he suggested that non-gauge-invariant properties can enter the theory via this structure, leading to further developments via the introduction of yet more surplus structure such as ghost fields, etc. (I shall come back to this in Chapter 6). One can also understand more recent cases presented to illustrate the apparent explanatory power of mathematics—such as the renormalization group, for example—as actually demonstrating the fertility of such surplus structure (Redhead 2001; Bueno and French 2012).

Numerous other examples can be given but what is important is the positive role played by such structure in these cases: eliminating it in the manner suggested by North's prescription would have been disastrous![10] In general, rejecting formulations that involve surplus structure may mean rejecting precisely that which could prove heuristically fruitful. This introduces an element of restraint when it comes to North's structuralist programme. Indeed, one might say that appealing to the formulation that has less structure not only carries with it all the standard problems that appeals to simplicity face, but in addition risks constraining heuristic fruitfulness.[11] As it turns out, there is a better way of breaking the underdetermination which supports the alternative claim—diametrically opposed to North's—that it is the Lagrangian formulation that should be preferred on the grounds that it is this that more naturally captures the core features of classical mechanics.

2.6 Breaking the Underdetermination$_4$: Appeal to the More 'Natural' Formulation

This is the claim made by Curiel in a rich and thought-provoking paper that argues first, that the geometric structures underpinning each formulation are not isomorphic and secondly, that classical systems evince the Lagrangian structure and not the Hamiltonian (Curiel 2014). He argues that given a plausible characterization of 'classical system' that does not beg any relevant questions in this context, the relevant state space naturally possesses the structure of (is isomorphic to) the tangent bundle of configuration space,[12] which meshes 'naturally' with the Lagrangian formulation.

[10] Her conclusion also has other associated costs, as she notes that within the Hamiltonian approach, momentum must be regarded as a fundamental property.

[11] Recall that we are talking about *formulations* here. Of course, in the case of underdetermined *theories* one might be reluctant to discard even quite extensive surplus structure without running the appropriate empirical tests first (this is just an expression of the usual dominance of empirical success over simplicity, however characterized).

[12] The tangent bundle associates with every point in the space the vector space of all vectors tangent to the space at that point (see Curiel 2014: 285).

Omitting the technical details (which are subtle and profound), the core of his argument rests on certain 'brute' empirical facts to the effect that when a classical system interacts with another, only certain ('configurative') quantities (associated with velocities) are directly 'pushed around'. The equations governing quantities that cannot be 'pushed around' can be thought of as kinematical constraints rather than equations of motion (Curiel 2014: 282) but according to Curiel not only does the Hamiltonian formulation of a system not allow one to express such constraints, it allows solutions to the equations of motion that violate them (Curiel 2014: 298 and 304–311). This is because the symplectic structure of that formulation induces a Lie algebra over the vector space and this is not isomorphic to the affine space in terms of which the family of kinematically possible evolutions of the classical system can be represented. Another way of putting this is to say that the (configurative) quantities just mentioned play no privileged role in this formulation (Curiel 2014: 303).[13] As a result, then, the relevant kinematic constraints must be put in by hand and hence there is a sense in which the relevant structure of a classical system does not possess the resources to construct its Hamiltonian formulation (Curiel 2014: 304).[14]

One might, for example, try to resist Curiel's conclusion and maintain that the two formulations are physically equivalent by virtue of yielding the same solutions to the equations of motion. However, this would be to adhere to a notion of physical equivalence that is more appropriate for an empiricist stance than a realist one (cf. Curiel 2014: 314) and in so far as it amounts to a claim of commonality of structure at the empirical level only, would not be suitable for structural realism. Alternatively, one might try to argue that having to apply constraints 'by hand', as it were, should not preclude a formulation from being regarded as 'natural' in whatever sense. As we'll note in later chapters, the constraint that particles of this world are either fermions or bosons, corresponding to anti-symmetric and symmetric representations of the permutation group, is imposed upon the theory of quantum mechanics from 'outside', as it were, as a background or initial condition. A defender of the Hamiltonian approach might urge something similar for the necessary constraints in classical mechanics. However, it is surely a point in favour of the alternative formulation if these constraints arise within it and indeed, as we shall note in Chapter 11, the constraint of permutation symmetry in quantum mechanics arises 'naturally' within a certain algebraic formulation of quantum field theory.

Alternatively, one might try to put pressure on the claim that the basis of the difference between the formulations, and of the advantages the Lagrangian bears over

[13] This goes beyond North's point, in note 10, that momentum must be taken as fundamental in the Hamiltonian approach.

[14] The familiar transformation between the Lagrangian and Hamiltonian formulations effectively 'wipes out' these kinematical constraints (Curiel 2014: 310).

the Hamiltonian, lie in 'brute empirical facts' of the sort alluded to previously. One might, for example, reject the distinction between those quantities that can be 'pushed around' and those that cannot as begging the question against the Hamiltonian formulation (see note 10 again). However, this hardly seems a fruitful line to take: after all, if these formulations are to be understood as formulations *of* classical mechanics, some basic characterization of what it is they are supposed to be formulations of needs to be given. One could deny this and insist that they amount to different theories, each equally entitled to the title 'classical mechanics' and from the perspective of the Hamiltonian theory, those brute empirical facts are in fact nothing of the sort, being theoretically informed representations of empirical phenomena, interpreted in terms of fundamental quantities that the advocate of the Hamiltonian theory would reject.[15] Of course, now the nature of the supposed underdetermination would be changed: instead of two formulations of the 'same' theory, where that is characterized along the lines that Curiel proposes, we return to the standard sense of underdetermination of two theories underdetermined by the same empirical phenomena. In that case, appealing to considerations of 'naturalness' would have the same force in the realist context as appealing to simplicity—that is, precious little, until and unless its truth-tracking status can be established.

That seems a hard row to hoe, but not an impossible one. If one took that line, the structural realist would then insist that what we should be realist about is the structure common to the two theories. I shall come back to how we might delineate that shortly. Alternatively, one might agree with Curiel and conclude that the underdetermination of formulations can be broken in the way he indicates. In that case, the structural realist would have to conclude that the structure of the world is not some structure common to the Lagrangian and Hamiltonian formulations, but simply that of the former.

Let me now move on to the final form of underdetermination that has been called 'metaphysical', to which the kind of forceful underdetermination breaking move that Curiel effectively suggests does not apply and which, partly as a result, pushes the realist towards OSR (or so it has been claimed).

2.7 Metaphysical Underdetermination

Examples of this form of underdetermination may seem too easy to find. Consider a chair, for example, and further, consider it as an object: is this to be cashed out, metaphysically, in terms of a bundle of properties only, or in terms of these properties plus something further in which they inhere, such as some substantial substratum? Relatedly, if the chair qua object is taken to be an individual, how is that

[15] And as Curiel notes (2014: 302–303) the Hamiltonian formulation imposes its own kinematical constraints among what it takes to be the 'natural' quantities.

individuality to be grounded? In terms of some qualitative difference, with the infamous Principle of Identity of Indiscernibles acting as an effective guarantor that there will always be some such difference? Or in terms of some non-qualitative 'thisness' or haecceity? And we can take these questions 'up a level' to the properties of the chair: are these to be understood as instantiated universals or as particulars, in the form of tropes for example? Should these forms of underdetermination trouble the scientific realist? It would seem that one can plausibly answer 'no', and that that is a good thing, not least because if we had to wait for the metaphysicians to settle their disputes over which of these views is better before adopting a realist stance, we'd have given up and ceded victory in the debate to the anti-realist long ago.

However, following Ladyman (1998; see also French and Ladyman 2003) I want to insist that there is a form of metaphysical underdetermination that should trouble the 'standard' realist, at least, and later I shall indicate why it is troublesome. This is the form that arises from quantum theory, or, more specifically from quantum statistics. The relevant details have been given elsewhere (see French and Krause 2006; van Fraassen 1989) but in brief they are as follows. Quantum statistics differs from classical statistics in the counting of arrangements of particles over states. With just two particles and two states we get the following arrangements (a useful analogy is to think of balls distributed between boxes, as in Illustration 1):

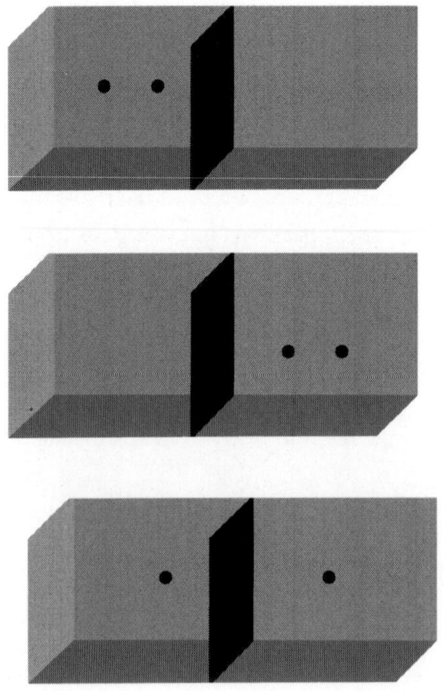

Illustration 1. The possible arrangements of two balls in two boxes

Now, consider the third arrangement corresponding to one particle in each state. Classical 'Maxwell–Boltzmann' statistics gives this arrangement a weight of 2, corresponding to the two ways it can be obtained from a permutation of the particles; that is, one arrangement corresponds to one particle in one state and the other in the other; and another, distinct arrangement corresponds to that obtained by the particles switching states. However, quantum statistics—whether of the Bose–Einstein or Fermi–Dirac kind—gives that third arrangement a weight of 1, since, to put it a little crudely, particle permutations are not counted here; that is, the arrangements corresponding to 'one particle in one state and the other in the other' and 'the same, but with the particles switched' are not counted as distinct arrangements. This is a consequence of the application of a form of symmetry known as Permutation Invariance (PI) that plays a fundamental role in quantum mechanics and that will crop up again in subsequent chapters.

Briefly, permutation symmetry is a discrete symmetry supported by the permutation group Perm(X) of bijective maps (the permutation operators, $\hat{P}$) of a set X onto itself.[16] When X is of finite dimension Perm(X) is known as the symmetric group S_n (where the n refers to the dimension of the group). For instance, X might be the set consisting of the labels of the two sides of a coin: heads 'H' and tails 'T'. Or perhaps the 'names' of n particles making up some quantum mechanical system, an He_4 atom for example. If we take the coin as our example, then X = {H,T} and Perm(X) is an order-two group, S_2, consisting of two elements (computed as having 2! elements via the dimension, n = 2, of the group): (1) the identity map, idX, which maps H to H and T to T; and (2) the 'flip' map (or 'exchange' operator), $\hat{P}$HT, which maps H to T and T to H.

Now, to say that some object (i.e. a set or the total state vector of a system of particles) is 'permutation invariant' means that it is invariant under the action of Perm(X): it remains unchanged (in some relevant sense) when it is operated upon by the elements (i.e. the permutation operators) of Perm(X), including (for $n \geq 2$) the elements that 'exchange' the components of the object (in this case the labels of the sides of the coin or the labels of the particles in a quantum system).

The coin clearly is not permutation invariant (i.e. does not satisfy PI), since we must distinguish 'heads' from 'tails'; that is, there is an observable difference between these two states of a coin. However, when we consider systems containing several indistinguishable particles, each with several possible states (particles such as electrons, neutrons, and photons), we find that they are indeed permutation invariant, and, as a result, different weights must be assigned to the relevant arrangements (such as 'one-particle-in-each-box').

[16] The fact that the set Perm(X) has the structure of a group simply means that: (1) we can combine any two elements ($\hat{P}1,\hat{P}2 \equiv$ Perm(X)) in the set to produce another element ($\hat{P}3 = \hat{P}1 \cdot \hat{P}2$) that is also contained within that set ($\hat{P}3 \equiv$ Perm(X)); and (2) each element $\hat{P} \equiv$ Perm(X) also has an inverse $\hat{P}_{-1} \equiv$ Perm(X).

The difference in the assignment of weights to these arrangements of particles and states can be explained in two ways. The first, which was for many years the 'received' view of the matter (again see French and Krause 2006), took the justification for the higher weight in the classical case to be the fact that the particles are regarded as individuals in that case, so that a permutation is significant; hence the loss of that significance and the reduction of the weight assigned to the corresponding arrangement in the quantum case is taken to imply that in that case the particles are not regarded as individuals. Philosophical reflection on the 'new' quantum mechanics was entwined with the development of the physics itself, and this view of quantum particles as 'non-individuals' was expressed by the quantum revolutionaries themselves in the papers that first presented the relevant technical details (see French and Krause 2006: 94–115). However, how one should understand this notion in both a formal and metaphysical sense was left unclear. Indeed, the mathematician Manin saw the issue of obtaining a formal framework suitable for accommodating such non-individuals as one of the fundamental problems of contemporary mathematics, writing, in the context of reflections upon Cantorian set theory,

> the [n]ew quantum physics has shown us models of entities with quite different behaviour. Even 'sets' of photons in a looking-glass box, or of electrons in a nickel piece are much less Cantorian than the 'set' of grains of sand. (1976: 36)[17]

Steps towards the resolution of this issue have now been taken with the formulation of forms of quasi-set theory and associated logical systems capable of accommodating this non-individuality (French and Krause 2006: chs 7 and 8). I emphasize these developments here because without them, this metaphysical position—of quantum particles as non-individuals—might not be treated as a viable 'horn' of an underdetermination argument at all.

The alternative horn is generated from a different explanation of the counting of arrangements. This is ultimately grounded in reflection on the role of Permutation Invariance (PI), the action of which can be understood as effectively dividing up the relevant Hilbert space into non-combining sub-spaces corresponding to irreducible representations of the permutation group.[18] The two most well known of these are the symmetric, corresponding to bosons, and the anti-symmetric, corresponding to fermions; other kinds of symmetry are also possible but do not seem to be applicable to any currently known kinds of particle.[19] On this view, the change in weight

[17] Just to emphasize the significance of this statement—it was expressed in the context of the 1974 meeting of the American Mathematical Society which was held to evaluate the status of Hilbert's famous list of 23 problems of mathematics, drawn up at the turn of the 20th century. As a result, the 1974 meeting drew up a new list of 'problems of present day mathematics' of which Manin's was the very first (see French and Krause 2006: ch. 6).

[18] Note that here of course I am using 'representation' in the formal group-theoretic sense.

[19] These possibilities include parastatistics. In the mid 1960s it was suggested that quarks might be paraparticles but they were subsequently re-described as fermions with an extra degree of freedom which

associated with the counting of permutations arises not because the particles are non-individuals, but because they have been assigned, or find themselves, in one sector—the symmetric, say—rather than the other. Thus they can still be regarded as individuals but subject to certain constraints on their behaviour as characterized by these restrictions to certain sub-spaces of Hilbert space, given by the action of PI (French 1989; van Fraassen 1989; French and Krause 2006). How this individuality is cashed out is a further issue, but various standard metaphysical options are available, including, for example, appealing to some form of primitive thisness or Lockean subtratum (French and Krause 2006).

We thus have two distinct metaphysical packages that are consistent with the physics: particles-as-non-individuals (described via quasi-set theory) and particles-as-individuals (subject to certain state accessibility constraints).

In his own book dealing with such issues, in which he offers an anti-realist understanding of quantum mechanics, van Fraassen presents this form of underdetermination as a challenge to what I have called object-oriented realism (van Fraassen 1989: 480–2). Here the underdetermination is taken to derive from the unnecessary metaphysical commitments of the realist. The fundamental flaw inherent in the latter is the combination of a form of 'minimal' naturalism that states that we should believe our best current theories, and hence take the world to be as these theories say it is, with a 'classical' metaphysics of individual objects. The existence of this kind of underdetermination is then taken to imply that physics cannot, in fact, tell us what the world is like when it comes to the most fundamental aspect of the nature of its objects—it simply cannot tell us whether they are individuals or not. But then a realism that insists on an object-oriented ontology but can't tell us whether those objects are individuals or not might legitimately be viewed as metaphysically deficient. The (ontic) structural realist offers a way of responding to the anti-realist's challenge by urging us to retract our metaphysical commitments, away from objects to the underlying common structures.[20]

Not everyone is convinced, of course. Chakravartty points to the 'everyday' metaphysical underdetermination previously alluded to and argues that if the realist

came to be known as 'colour' (French 1995; the beginnings of a history of paraparticle theory can be found in French 1985).

[20] To be clear, then, it is this underdetermination that I take to motivate (in part) OSR. Some, such as Morganti (2004), have taken the problematic status of the Principle of Identity of Indiscernibles (PII) within quantum mechanics as the driving force and have then argued that we can ground an appropriate notion of individual object by alternative means, via some form of 'hybrid' concept (Morganti 2004) or primitive individuality, for example; for critical comment see French 2010a. However, I am quite happy to accept alternative accounts of particle individuality, based on Quinean PII with 'weak discernibility', or substantival 'individual constituents' (Morganti 2004) or whatever (quite happy but not always massively so because I think some of these accounts are clearly deficient in various ways). The point is that these all simply serve to further articulate and strengthen the 'particles-as-individuals' horn, thereby reinforcing the metaphysical underdetermination as a whole and it is *that*, rather than the deficiencies of particular metaphysical approaches to object individuality, that I take to push us towards OSR.

is not expected to be concerned whether 'everyday' objects should be described as substances-plus-properties or bundles of properties, or whether the properties themselves should be described as instantiated universals or tropes, so she should not be at all concerned whether quantum particles should be described as individuals or non-individuals (Chakravartty 2003b).[21] In response, the structural realist can emphasize the differences between these two situations. In the case of everyday objects the issue is not whether they are *objects* or not, but rather, having already established that, how their objecthood should be conceived. Here the matter of access looms large: we have sensory-mediated access to 'everyday' entities in terms of which we can separate out those that count as distinguishable objects, by means of the relevant properties, or location in space-time and so on. Once we've established distinguishability, at least in principle, we can then go on to speculate as to the 'ground' of individuality, whether via properties within the scope of an appropriate form of Identity of Indiscernibles, or in terms of some form of 'primitive thisness', or whatever (cf. Gracia 1988).

When it comes to quantum particles, we lose that form of access and the danger of simply reading off the metaphysics from the physics is that our understanding of the latter may be infected, as it were, with the metaphysics of the everyday. Indeed, the very foundations of the mathematics we use to frame our theories is already so infected (we may recall Manin's view of set theory, already noted), requiring the genius of Weyl and his understanding of both those foundations and group theory to effectively 'twist' that everyday metaphysics to accommodate the new physics (French and Krause 2006: 261–3). Here we cannot establish distinguishability to begin with, and the choice the realist faces is not the apparently innocuous one of deciding between different metaphysical accounts of the individuality of objects, but that of deciding whether they should even be regarded as individual objects to begin with. This difference, I would suggest, is crucial, at least for the realist. I shall return to Chakrvartty's concern later, and shall present another perspective on the claim that the metaphysical underdetermination involving individuality is more serious for the realist than the 'everyday' form, but let me first consider two possible ways this more serious underdetermination might be broken.

2.8 Breaking the Underdetermination$_5$: 'Weak' Discernibility

One option is to try to break the underdetermination by appealing to certain principles, such as Quine's famous dictum, 'no entity without identity' and insist that since particles-as-non-individuals have no identity, they cannot actually be

[21] See also Morganti (2011), who insists that the advocate of OSR fails to offer appropriate methodological considerations of what is to count as 'proper' metaphysics in this situation. One response would be to urge that what counts as metaphysics here should be left as broadly delineated as possible, not least in order not to beg any questions against one horn of the underdetermination or the other.

entities in the first place. But of course, such principles may be rejected, as Quine's has been, by Barcan Marcus, for example, who responded with her alternative, 'no identity without entity'.[22] At the heart of this disagreement lies a fundamental issue to do with the status of identity (is it a relation that can only be said to hold once we have the relata (or relatum in this case), or is it constitutive of the entity?) and one's stance on that will effectively determine whether one thinks this metaphysical underdetermination can be resolved in this way or not. Furthermore, the development of quasi-set theory and 'Schödinger logics' goes some way to allaying the concerns of those who might wonder how we can formally accommodate the notion of particles whose identity is not well defined.

Alternatively, we might 'break' the underdetermination by considering how the particles-as-individuals package might be further supported. Typically, those who wish to restrain their metaphysical commitments when it comes to individuality have appealed to some form of the Principle of Identity of Indiscernibles (PII) in order to ground this individuality on some property of the objects concerned. Well-known concerns in the quantum realm have been taken to block this approach (again see French and Krause 2006 for a detailed account of this discussion), leaving—it would seem—Lockean substance, haecceity, or some form of primitive thisness as the only options if we are to regard quantum particles as individuals. As a way of breaking this metaphysical underdetermination these are seen as particularly costly, in ontological terms, and as leaving the realist wide open to anti-metaphysical criticism.

However, the approach based on PII has recently been revived with the claim that a relevant sense of individuality can be grounded in a notion of 'weak' discernibility applicable to quantum particles (Saunders 2006b). The central idea is to admit relations within the scope of PII and then to note that fermions in, for example, a singlet state can be weakly discerned via irreflexive relations such as 'has opposite spin to'. This weak discernibility can then ground a 'thin' form of objecthood that could then be invoked by the object-oriented realist.[23]

This result has also been extended to bosons (Muller and Saunders 2008; Muller and Seevinck 2009), although some of the formal details are contentious. More generally, however, it has been argued that what weak discernibility grounds is merely numerical distinctness, rather than the robust sense of discernibility that PII was originally concerned with (Bigaj and Ladyman 2010). If PII is understood as the claim that distinct objects must differ in some way, then, it is argued, weakly discernible objects do not differ in this sense and hence this PII-based approach remains blocked.

[22] The difference here has to do with the differences for things and objects; for Barcan Marcus object-reference is taken to be a wider notion than thing-reference, where the latter involves well-defined identity conditions, as well as other restrictions, such as spatio-temporal location.

[23] Although, as we shall see, it more plausibly forms part of the metaphysics of a non-eliminativist version of OSR.

Now, it is not clear whether such moves are sufficient to 'break' the metaphysical underdetermination, in favour of the object-oriented realist, particularly given the alternative to be presented shortly. Indeed, the approach of Saunders et al. can be seen as simply reinforcing it by offering a more plausible metaphysical alternative to haecceities and the like and thus supporting the particles-as-individuals horn. And of course, the 'force' behind any such break would be metaphysical, again, and this is not unproblematic. Admitting relations into the scope of PII and allowing them to yield a form of individuality in the sense discussed has long been seen as controversial on the grounds that a circularity threatens: in order to appeal to such relations, one has had to already individuate the particles which are so related and the numerical diversity of the particles has been presupposed by the relation which hence cannot account for it (see French and Krause 2006; Hawley 2006b and 2009). One response to this worry would be to question the underlying assumption that relata have the relevant ontological priority over relations and adopt a structuralist stance according to which either that priority is reversed or there is understood to be no priority of one over the other (for further discussion see French and Krause 2006; French and Ladyman 2011). The circularity is then avoided by situating this approach within a structuralist framework, with a concomitant 'contextual' notion of individuality[24] (Ladyman 2007; French and Krause 2006: 172). In this way the possibility of restoring a form of object-oriented realism is effectively neutered since the relevant objects (fermions discerned via irreflexive relations) are indeed 'thin' in so far as they are discerned and individuated only in structural terms.

In effect, then, these developments offer an alternative stance that the structuralist can take with regard to metaphysical underdetermination: rather than pulling back her ontological commitments in the face of the underdetermination, she can 'break' the latter via an appeal to weak discernibility and thin objecthood and still appropriately restrict her commitments. The difference from a non-object-oriented approach feeds into discussions over the various forms of structural realism currently on the table and in particular relates to the (possibly wafer thin) distinction between 'eliminativist' forms which attempt to remove the notion of object entirely from the metaphysical pantheon and those that accept an appropriately 'thin' characterization in the sense discussed here.

Finally, it has also been suggested that developments in physics may lead to new kinds of structure that offer the possibility of a role 'for individual particles or other entities' (Slowik 2012: 50) and this may undermine the underdetermination motivation for OSR. Furthermore, OSR would then be faced with the same sort of pessimistic meta-induction as the object-oriented realist, and nothing would have been gained.

[24] This is said to be 'contextual' in the sense of holding within a given structural context (see Stachel 2005; Ladyman 2007/2009; French and Ladyman 2011).

This is a worry that has been expressed before (Bueno 2000) and of course it has to be acknowledged that OSR, like all forms of realism that posit specific forms of ontology, is defeasible: it may well be that entirely new forms of mathematics will be proposed in terms of which physical theories are presented and that these will bear no relation to current forms. Note that last clause. It is only with that included that this concern can be taken as akin to that which lies behind the pessimistic meta-induction when it comes to object-oriented ontologies. But so far, no plausible examples of such radical structural change have been given. Furthermore, given the empirical success that is associated with the kinds of structures that OSR takes seriously, it is difficult to imagine any such example that would not recover these structures (perhaps in some limit), just as the equations of classical mechanics are 'recovered' from Special Relativity as v/c tends to 0.

Relatedly, any new structure must be able to yield—if only in some limit—PI in order to accommodate quantum statistics and the distinction between bosons and fermions. But then it is hard to see how it could incorporate a role for *individual* particles only. As before, this possibility remains as little more than a promissory note.

2.9 Breaking the Underdetermination$_6$: Non-Individuality and QFT

Indeed, the most well-known way of breaking the metaphysical underdetermination is to urge adoption of the particles-as-non-individuals package on the grounds that it meshes better with quantum field theory (QFT), where particle labels are simply not assigned right from the start (Redhead and Teller 1991 and 1992). In effect this is another appeal to the heuristic fruitfulness of one 'horn' of the underdetermination over the other. It is also a retrospective move, in so far as, having QFT to hand, we know now that there is such meshing, so it is not a mere promissory note. Still, the concern has been raised: why should appeal to a successor theory count in breaking the underdetermination associated with an earlier theory? Underlying this is the kind of modal issue alluded to previously and captured in the question: if we were faced with this underdetermination in the quantum context only, without the benefit of having QFT to hand, what weight would we give to such a promissory appeal?

In pursuing this approach, advocates have attacked the other 'horn'. We recall that according to the particles-as-individuals package, the weight given to the counting of arrangements is reduced because certain sub-spaces are regarded as inaccessible to particles of a certain kind. Thus for bosons the anti-symmetric sub-space associated with fermions is out of bounds, and vice versa. This has been criticized on the grounds that these inaccessible states represent unwanted 'surplus structure' and hence, again on what amount to grounds of simplicity, this package should be rejected in favour of the particles-as-non-individuals one (Redhead and Teller 1991 and 1992).

Again, however, this is a problematic line to take for the same reasons as before (once again see French and Krause 2006: 189–97): this surplus structure may prove to be heuristically fruitful in various ways,[25] and indeed it has in the case of 'non-standard' particle statistics such as those associated with paraparticles (see note 19) and anyons.[26] Drawing a line between such 'useful' surplus structure and the clearly redundant is notoriously difficult and adopting 'reject surplus structure' as a general methodological rule is crude at best, foolhardy at worst.

2.10 Don't Break It: Embrace It

Given the arguments that neither horn is to be preferred over the other, so that the underdetermination cannot be broken, one possibility is to simply accept it. Thus, one might argue that, 'an array of possible metaphysical interpretations enriches our understanding of quantum mechanics' (Howard 2011). To avoid this collapsing into the kind of scepticism that lies behind the constructive empiricist stance towards metaphysical underdetermination, one might then insist that the appropriate epistemic attitude in this situation is neither belief nor mere acceptance but a kind of Peircean 'pursuitworthiness' (Howard 2011).[27] So the idea seems to be that metaphysical underdetermination is to be welcomed since it presents a range of options that are worthy of pursuit, and by chasing them down, as it were, we obtain greater understanding. But presumably, by chasing them down, we will decide on one option rather than the other.[28] So taking these metaphysical packages to be pursuit-worthy would seem to be a preliminary attitude at best.

Alternatively, one might extend the agnosticism associated with ESR (as noted in the previous chapter) and insist that as a result we should keep both metaphysical options open, as it were. Thus, Slowik offers a 'liberal' form of ESR that includes both relations and relata as possible elements in its underlying ontology, but which takes the 'precise ontological details' to be epistemically inaccessible, where by such details Slowik means whether the ontology includes only relata, only relations, or both (Slowik 2011).[29]

Now, I shall return to discuss (and dismiss) the motivations for this liberal form of ESR in later chapters, but let me suggest here that it not only falls foul of the general methodological precept of 'avoid the positing of epistemically inaccessible ontology

[25] This was precisely the virtue that Redhead originally saw in surplus structure (Redhead 1975).

[26] Anyons are two-dimensional particles that obey non-standard statistics. They have proved useful in explaining the fractional Hall effect, although they are typically regarded as merely mathematical constructs (Camino, Zhou, and Goldman 2005).

[27] For discussion of the possibility of adopting a pragmatist stance towards the philosophy of science, and a Peircean one in particular, see da Costa and French 2003.

[28] Peirce of course agreed with the standard realist that in the long run, our beliefs would settle down and the array of possibilities would narrow down to just one.

[29] Thus this is related to but clearly different from Esfeld and Lam's 'moderate' form of OSR, also originally articulated in the space-time context (Esfeld and Lam 2008, 2010).

(where you can)' that I shall discuss in more detail in the next chapter, but when it comes to underdetermination, it sits too close to the kind of scepticism beloved by anti-realists for realist comfort. So we recall that the constructive empiricist, for example, remains untroubled by such underdetermination, given her generally sceptical stance regarding how the world is (van Fraassen 1989). 'Liberal' ESR takes this stance to the next level, as it were, by insisting that not only can we not know whether we have individual or non-individual objects, but we cannot know whether we have objects and relations, or just objects or just relations! At this point the structural realist will insist in return that this is just too much agnosticism to accept and that instead we should pull in our metaphysical horns as it were and reduce the degree of epistemic inaccessibility we have to accept.

2.11 Don't Break It: Seek the Commonalities

Thus she argues that we should not simply accept the underdetermination, nor try to break it by adopting one horn over the other, but undermine it by dropping the object-oriented stance to begin with and hanging our realist commitments on the relevant underlying structure.[30] In the case discussed here, that can be characterized as group-theoretical (French 1999). So the idea is that instead of conceiving our ontology in terms of objects, and then having to face the dilemma of whether to regard them as individuals or not, we focus on the relevant group-theoretical structures underpinning quantum statistics and reconceptualize (or eliminate) our putative objects in terms of these structures.[31] Elaborating the details of this conception will take up much of the book but in order to help clarify what I have in mind let me sketch a distinction that I shall come back to.

Consider how the realist 'reads off' her ontological commitments from a given successful theory. Putting things a little crudely, standard, object-oriented realists begin by identifying those features of the theory that are deemed to be responsible for its success (broadly following the 'divide et impere' strategy; Psillos 1999). These might be the relevant laws, expressed in mathematized form, like Fresnel's equation for example, plus symmetry principles, such as PI in the case of quantum statistics.

[30] For a corrective to this urging, see Brading and Skiles (2012). They argue that even if the underdetermination is conceded, further premises are required to obtain OSR but which the object-oriented realist can deny. One such is the assertion that object-oriented realism implies that there is a fact of the matter whether the objects are individuals or not. However, Brading and Skiles insist, a law-constitutive view of objects can be articulated, according to which what it is to be a physical object is to satisfy a certain system of physical laws, without, necessarily, satisfying what they call an 'individuality profile' (2012). I think this view takes us beyond what I have called object-oriented realism to an intermediate position between that and OSR. As they say, the law-constitutive view is neutral on structuralism but if one adopts a structuralist interpretation of laws, as I do, then their account offers a further route to OSR.

[31] Saatsi has argued (2009) that OSR simply presents us with a third horn and thus exacerbates the underdetermination. I disagree, since I maintain that OSR accommodates the common core of the competing 'particles-as-individuals' and 'particles-as-non-individuals' horns via its focus on group structure (cf. Saatsi 2009: 12).

These are then articulated, metaphysically, in terms of the associated relations holding between the relevant properties, such as charge, mass, etc., which are then understood as being possessed by or instantiated in the underlying objects. Thus the structures presented by the theory are used to infer the 'natures' of the objects that the realist believes in. Metaphysically proceeding in the opposite direction, it is these objects that are taken to be fundamental and thus as supporting the properties whose interrelations are described by the laws.

The structuralist also focuses on the relevant success-inducing structures presented by the theory. However, instead of taking these to be the metaphysical outcome of properties and their interrelations (and just what is meant by 'the metaphysical outcome' here is fleshed out in some detail by the dispositionalist, for example, as we'll see in Chapter 9) she takes these structures themselves to be fundamental (and again, I shall discuss how they might be so taken in Chapter 10), with properties as ontologically dependent on the structures and objects reconceptualized in these terms, or, perhaps more robustly, dropped from the ontology altogether as metaphysically unnecessary. I also want to emphasize that the structures I am suggesting should be taken as fundamental elements in our ontology are those that are *presented* at the level of scientific practice. Here I shall draw on a distinction between this presentation of the structure and its *representation*, at the level of the philosophy of science (Brading and Landry 2006). Whereas the former yields group-theoretic structure, for example, I shall argue that the latter is most appropriately effected in set-theoretic terms via the semantic or model-theoretic approach, a line I shall defend in Chapter 5.

However, this is not, of course, to suggest that the relevant structure, as our fundamental ontology, is to be regarded as set-theoretic, nor does it by itself imply that different formulations, as in the Lagrangian and Hamiltonian cases, give rise to different structures, in the sense of different elements of our fundamental metaphysics. This allows us to immediately respond to the first of Pooley's concerns discussed in section 2.2, namely that if the structure we are interested in is straightforwardly characterized set-theoretically, say, then different formulations will give rise to different 'structures', understood in those terms. That concern arises from a conflation of the characterization of structure at the level of its presentation within scientific practice, with its 'meta-level' representation in the philosophy of science. We may choose to *represent* the relevant structure set-theoretically, or via category theory, or however, but such meta-level representation does not *characterize*—in the sense of ontologically constituting—the structure. Of course, there remain the issues of how we can be sure there is such a common underlying structure in the Lagrangian and Hamiltonian cases, and, relatedly, of how we access it and characterize it. But the point is that having concluded there is such a common structure, and noted its presentation in mathematical and physical terms (e.g. via group theory), our different meta-level set-theoretic *representations* of the associated different formulations should not be accorded inappropriate ontological import. There are not different structures in this case, just *different representations* of the underlying structure.

Let us now consider Pooley's second concern and the issue of what the relevant structural commonalities might be in the case of the underdetermination between the Hamiltonian and Lagrangian formulations, for example. What we need to do is show how a 'single, unifying framework' is revealed by moving to some underlying structure. Belot, for example, has noted that,

> It is a fact of primary importance that for well behaved theories the space of initial data and the space of solutions share a common geometric structure—these spaces are isomorphic as symplectic manifolds. (Belot 2006: 17)

That is, the Lagrangian solutions can be mapped to the Hamiltonian initial data and in effect the actions of the groups implementing time translation (Lagrangian) and time evolution (Hamiltonian) can be considered as intertwined (Belot 2006: 17). Belot further suggests that a symplectic structure is the *sine qua non* of quantization, so again we might use this to advance a claim of there being appropriate structural commonalities between classical and quantum physics.

North, of course, wants to claim that it is the Hamiltonian structure that we should be realists about, primarily on grounds of simplicity. As we have seen, this is problematic. Not surprisingly, then, she rejects the above kind of commonality claim. Thus she agrees that 'if and when' both statespace structures are vector fibre bundles, they will be isomorphic *as vector spaces*. Nevertheless she insists that the two formulations differ in *relevant* structure, not least because the Hamiltonian statespace need not be a vector bundle, whereas the Lagrangian statespace must. Hence, she maintains, the Hamiltonian formulation is still to be preferred. Curiel, on the other hand, takes that very point to weigh in favour of the *Lagrangian* formulation, arguing that the most natural way to describe an abstract classical system is by a manifold and two families of vector fields with appropriate structure, corresponding to the Lagrangian formulation, rather than by fields with the structure of a Lie algebra based on a symplectic structure, as in the Hamiltonian case, where these fields are not isomorphic to one another.[32]

At this point, one could point out that it's the vector space structure that we need for our physics and that the structure we should be realists about in this context is something like the following: we begin with a symplectic manifold; the Hamiltonian is defined as a real-valued differentiable function on that manifold; one can then associate a Hamiltonian vector field with this function, where the integral curves of this field give the solutions of the Hamilton–Jacobi equations. Put briefly, what we have is a symplectic space of initial data, equipped with a Hamiltonian that generates

[32] More importantly, he maintains, neither are the relevant kinematical constraints in the sense that they do not encode isomorphic relations. However, the hoe-er of the line that these formulations actually amount to underdetermined theories will insist that one would not expect them to.

the relevant dynamics. On the Lagrangian formulation what we have is a symplectic space of solutions, on which one can define a function that assigns to each solution the total instantaneous energy. This function can be related to the Hamiltonian under the symplectic isomorphism by which solutions are mapped to initial data and which intertwines the action of the group implementing time translation in the Lagrangian formulation with the action of the group implementing time evolution in the Hamiltonian formulation (for details see Belot 2006: 38–9 in particular).

Of course, this would be anathema to Curiel who would deny that we should begin with a symplectic manifold in the first place. But it offers one way of exploring the idea of identifying the structure common to these two formulations or theories (depending on one's stance). And of course there is more to say (there always is) but this gives some indication of the way to proceed. And we can see how this way of resolving the underdetermination in this case bears comparison with that of dealing with a similar underdetermination between matrix mechanics and wave mechanics in quantum physics. In matrix mechanics, the classical Fourier series was replaced with what was identified as a matrix of coefficients, whose magnitudes represented the intensity of atomic spectra. In wave mechanics, on the other hand, the state of the system is described by a function whose time evolution is governed by a partial differential equation. As is well known, it was then shown that these two formulations could be understood as equivalent representations on an underlying Hilbert space, which is a complete vector space with an inner-product structure.[33]

There are two points I wish to emphasize. The first is that in so far as Hilbert space supports the relevant representations[34] of the groups that the ontic structural realist sets such store by, focusing on the 'common' structure will mean paying attention to the nature of these representations. I shall return to this point in subsequent chapters.

The second is that this is what is *presented* by the theory of quantum mechanics. It can then be *represented* in set-theoretic terms via the semantic approach, which allows us to capture the relevant interrelations between the various formulations and the underlying common structure (Muller 1997). If we were to pursue the analysis of the commonalities between the Lagrangian and Hamiltonian formulations we would have to do something similar on the representational side, but all I want to do here is convey the general strategy and move on to the further motivation for OSR.

[33] An outline of von Neumann's strategy can be found in Kronz 2004. Standardly, Schrödinger is taken to have made a first attempt at demonstrating the equivalence, obtaining partial results and von Neumann is regarded as having completed the job, introducing what is now known as Hilbert space (see Muller 1997). However, Perovic (2008) has recently argued that Schrödinger should be understood as having achieved an ontological and domain-specific equivalence in the context of the Bohr atom. These histories typically omit to note the important contribution made by Weyl (a point made by Ladyman 1998: 420–1), who wrote: '[T]he essence of the new Heisenberg-Schrödinger-Dirac quantum mechanics is to be found in the fact that there is associated with each physical system a set of quantities, constituting a non-commutative algebra in the technical mathematical sense, the elements of which are the physical quantities themselves' (Weyl 1931: viii; as noted in Ladyman 1998: 421).

[34] Again in the technical group-theoretic sense.

2.12 Concluding Remarks

So far, then, the realist has faced two challenges that, I claim, push her in a structuralist direction: in the face of the PMI, she should be a structural realist, whether of the ESR or OSR variety. In the face of underdetermination—particularly of the metaphysics of individuality—she should adopt OSR.[35] In the next chapter I shall present a third challenge and suggest that it adds further impetus to this conclusion, because OSR removes certain sources of metaphysical humility that would otherwise leave realism open to the anti-realist charge that the understanding presented of the world is seriously incomplete.

[35] Of course, there is the further worry, long expressed by Bueno for example, that there may be 'structural' cases of underdetermination that cannot be resolved by effecting the shift from objects to structures. Lyre (2011) suggests that a reconstrual of General Relativity in gauge-theoretic terms offers a 'live' example of such structural underdetermination, but concludes that given the small number of cases of underdetermination in general, this, and the other more well-known examples, can simply be dismissed as artefacts of our incomplete scientific knowledge. Indeed, one might perhaps speculate that by searching for the common structural core behind these gauge-theoretic cases, one might further progress the aim of achieving a viable form of quantum gravity, for example (see Rickles, French, and Saatsi 2006).

3

Mixing in the Metaphysics 2
Humility

3.1 Introduction

Having indicated how the realist should respond to the PMI, and how metaphysical underdetermination pushes her towards OSR, I shall now tackle what I have called 'Chakravartty's Challenge', and shall argue that an appropriate response will provide further impetus for this push.

Let us recall the realist recipe, given at the beginning of Chapter 1, for obtaining an understanding of how the world is: we choose our best theories; we read off the relevant features of those theories; and then we assert that an appropriate relationship holds between those features and the world. Is that enough? Some would say not.[1] Chakravartty, for example, writes,

> One cannot fully appreciate what it might mean to be a realist until one has a clear picture of what one is being invited to be a realist about. (Chakravartty 2007: 26)

But how do we obtain this clear picture? A simple answer would be, through physics which gives us a certain picture of the world as including particles, for example. But is this clear enough? Consider the further, but apparently obvious, question, are these particles individual objects, like chairs, tables, or people are? In answering this question we need to supply, I maintain, or at least allude to, an appropriate metaphysics of individuality[2] and this exemplifies the general claim that in order to obtain Chakravartty's clear picture and hence obtain an appropriate realist understanding we need to provide an appropriate metaphysics. Those who reject any such need are either closet empiricists or 'ersatz' realists (Ladyman 1998).

However, the example of the metaphysical underdetermination given in the previous chapter, which has at its core the question whether the particles of physics

[1] And the difference between those who say it is and those who say it is not corresponds to the distinction between 'shallow' and 'deep' realism (Magnus 2012). In a sense this chapter is a defence of the latter.

[2] Brading and Skiles (2012) disagree, as we noted in the previous chapter, but then I think their law-constitutive view of 'objects' (note the quote marks!) leads us straight to OSR once we adopt the kind of structuralist understanding of laws I outline in Chapter 10.

are individuals or not, illustrates a fundamental problem with this appeal to metaphysics, namely that the question of what metaphysics to adopt cannot be answered on the basis of the physics alone. I shall take this as an example of what is typically portrayed as a stance of humility that must be adopted and shall suggest that it imposes a critical constraint on the attempt to achieve a realist understanding of the world. Indeed it is this—how we might obtain such an understanding given the constraint imposed by this humility—that I shall call Chakravartty's Challenge. And I shall argue that to meet this challenge, we should adopt OSR.

My discussion in this chapter clearly bears on the thorny issue of the relationship(s) between science, metaphysics, and philosophy of science in general and I shall briefly sketch where things stand with regard to those relationships, before presenting the case for humility and considering various ways we might reduce it and thereby relax the constraint.[3]

3.2 The Viking Approach to Metaphysics

The history of the relationship between science, metaphysics, and philosophy of science is not a happy one, at least not when one considers the past 100 years or so. Carnap famously wrote that

> Most of the controversies in traditional metaphysics appeared to me sterile and useless. When I compared this kind of argumentation with investigations and discussions in empirical science or [logic], I was often struck by the vagueness of the concepts used and by the inconclusive nature of the arguments. (Carnap 1963: 44–5)[4]

And the current situation appears to some to present little in the way of improvement. In a recent collection in which metaphysicians apply the tools of their trade to their own field,[5] Price argues that

> What's haunting the halls of all those college towns—capturing the minds of new generations of the best and brightest—is actually the ghost of a long discredited discipline. Metaphysics is actually as dead as Carnap left it, but—blinded, in part, by [certain] misinterpretations of Quine—contemporary philosophy has lost the ability to see it for what it is, to distinguish it from live and substantial intellectual pursuits. (Price 2009: 323)

In this context many have felt that contemporary metaphysics has precious little to offer the realist, given its apparent lack of contact with modern science. In the opening chapter of their extended defence of structural realism, Ladyman and Ross

[3] For an alternative conception that insists that metaphysics has no such role to play, see Landry 2012.

[4] As Howard (preprint) has noted, Carnap went on to say that metaphysics could be seen as an expression of one's attitude to life and compared it to music, insisting, however, that '[m]etaphysicians are musicians without musical ability'. Interestingly, given the thesis defended in this book, he also wrote: '[p]erhaps music is the purest means of expression of the basic attitude because it is entirely free from any reference to objects' (Carnap 1963: 80).

[5] A development that Callender takes to be '[n]ever a good sign for a field' (2011: 35).

present an excoriating condemnation, insisting that 'Mainstream analytic metaphysics has . . . become almost entirely apriori' (Ladyman, Ross, et al. 2007: 24). Even that which pays lip-service to naturalism is 'really philosophy of A-level chemistry' (Ladyman, Ross, et al. 2007: 24).

However, one might well feel that there is reason to draw back from claims to the effect that a priori metaphysics is without purpose or that it should be 'discontinued'; whatever exactly the problem with contemporary metaphysics is taken to be, the appropriate reaction to it by philosophers of science has to be considered carefully. Thus, one could argue that even divorced from modern science as Ladyman and Ross feel it is, metaphysics might still offer an array of tools, moves, and manoeuvres of which the realist could avail herself. Such an attitude forms the heart of what I call the 'Viking Approach' to metaphysics: the products of analytic metaphysics can be regarded as available for plundering! Of course, some metaphysicians might baulk at being cast in the role of hapless peasants, happily tilling their fields of compositionality and ontological dependence, before being pillaged by ruthless realist marauders. Nevertheless, they might agree that it is only by moving to an appropriate level of generality, with a concomitant loss of contact with scientific concerns, that they can develop such broadly applicable tools and manoeuvres.[6]

Having said all that, when it comes to metaphysicians' claims about how the world is, based as they often seem to be on a view of that world as made up of little bits of matter banging around, or, in the context I am interested in, on Aristotelian concepts of substance, objects, and properties, one might feel some sympathy with Ladyman and Ross. Certainly, too many metaphysical positions are grounded in 'intuition' or reflection on 'everyday' objects and their properties and attempts to import these into the context of modern physics often prove disastrous. This is not to say that we should render metaphysics entirely dependent on science, for the reasons already given; indeed, it would be as problematic as doing the same for logic, say. In both cases we would lose the opportunity to explore new lines of enquiry unencumbered by already established worldviews, and generate the array of tools mentioned previously. This meshes with Callender's even-handed or, as he puts it, symmetric approach to science and metaphysics (Callender 2011) in which not only is the laying bare of the metaphysical assumptions of our best theories an important part of understanding the world, but metaphysical speculation itself (appropriately anchored in systematic theorizing) can be heuristically useful.[7] Like Chakravartty he takes metaphysics to help provide a crucial element of understanding when it comes to our theories and writes,

[6] Indeed, the relationship between philosophy of science and metaphysics might be usefully compared to that between physics and pure mathematics (see French and McKenzie 2012).

[7] Popper, of course, famously maintained that along with those metaphysical ideas that have impeded the progress of science, there are those that have aided it. Indeed, he maintained (1959: 16), scientific discovery would be impossible without the kinds of speculative ideas that one might call 'metaphysical'.

In slogan form, my claim is that metaphysics is best when informed by good science and science is best when informed by good metaphysics. (Callender 2011: 48)

But now the issue is how to understand that informing.

3.3 The Informing of Metaphysics by Physics

Two broad stances can be identified that one might adopt with regard to the possibility of metaphysics being informed by our best science, and physics in particular: the 'optimistic', which takes science to be capable of bearing upon metaphysical matters and helping drive progress in metaphysics; and the 'pessimistic', which insists that you only get as much metaphysics 'out' of a scientific theory as you put 'in', in the first place (Hawley 2006a). As an example of the first, consider the claim that Special Relativity shows presentism—crudely, the claim that the present has a distinctive ontological status—to be false (Sider 2001). Representing the second, take my claim about metaphysical underdetermination in the previous chapter (Hawley 2006a).

More specifically, these positions can be articulated as follows:

> (Optimism) There are actual cases in which the involvement of a metaphysical claim in an empirically successful scientific theory provides some reason to think that the claim is true.

The pessimist position can then be separated into two forms:

> (Radical Pessimism) The involvement of a metaphysical claim in an empirically successful scientific theory can never provide any reason to think that the claim is true; and
>
> (Moderate Pessimism) There is a kind of involvement in theory which, were a metaphysical claim to achieve this involvement, would provide some reason to think the claim is true; but there are no cases of metaphysical claims being involved in theory in this way. (Hawley 2006a)

One sense of involvement here is that which gives us reason to believe a claim about unobservable entities, from a realist perspective (2006a: 456). Thus, when a metaphysical claim is involved with scientific theories in this way, it can be taken to share responsibility for explaining the empirical success of the theory. However, according to radical pessimism, such involvement would not give us any reason to believe the claim, whereas the moderate pessimist accepts that it would but insists that metaphysical claims are never really involved with scientific theories in this way. Optimists, on the other hand, believe that such claims can be appropriately involved with theories and that this involvement gives us reason to believe the claims in question (Hawley 2006a: 456). As we'll see in a moment, this comparison with the involvement of unobservable entities is problematic.

How do these options line up within the realism debate? Well, the realist will accept that there are cases where the involvement of a claim about an unobservable entity in an empirically successful scientific theory provides reason to think that the claim is true. Of course, this is weaker than the standard characterization of realism as inferring the existence of entities, not least because it is compatible with structural realism (Hawley 2006a: 456). It is also more specific in focusing on the role of such claims in explaining the success of theories, something that Saatsi and others have emphasized.

Understood thus, scientific realism is incompatible with Radical Pessimism, because otherwise there would have to be some in-principle difference between claims about unobservables and metaphysical claims which could account for the former being confirmed via the relevant theory's success and the latter not, despite their both being 'integrated' into the theory. One option would be to insist that metaphysical claims are simply not truth-apt but this requires further motivation and the history of shifts from metaphysics to science without change in truth-apt status suggests any such motivation is going to be hard to produce. Alternatively, one might accept that metaphysical claims could be involved in this way, but in fact it just doesn't happen, or hasn't happened—in which case one would be a Moderate Pessimist, which is compatible with a realist stance. Now, both cases suppose a particular kind of relationship between metaphysics and science such that we can more or less cleanly distinguish metaphysical claims in theories from those involving unobservables and as should be apparent, I'm not sure that metaphysics and science stand in such a relationship and hence I have doubts whether such a clean distinction can be established. I'll come back to that shortly.

Anti-realists, on the other hand, might be comfortable with an attitude of Radical Pessimism, because they think that the involvement of a claim about the unobservable in generating predictive success is irrelevant to whether we should believe it; or they might prefer Moderate Pessimism, because they think that claims about the unobservable never do any work in generating novel success. Either way, the anti-realist cannot be an Optimist (Hawley 2006a).

With these taxonomic combinations out of the way, let us turn to the question: are metaphysical claims ever involved in scientific theories in this way? Or, to put it another way: can such claims stand in the kind of relationship to theories presupposed here such that the claims can be ruled in or out (putting it very generally) on the basis of the success of these theories?

First of all, it would appear that certain metaphysical claims can certainly be ruled out (Muller 2011). Consider for example Leibniz's Principle of Identity of Indiscernibles which states—again, broadly speaking—that (putative) entities which are indiscernible in some respect are in fact identical. There has been considerable discussion over many years whether the Principle should be understood as necessary or as contingent, with opinion shifting to the latter. Even as such it has been argued that it has been ruled out by quantum mechanics, on the most plausible understanding of what it is to be indiscernible in this context (French and Redhead 1988; for further discussion see French and Krause 2006: ch. 4). Such cases might be taken as providing grounds for a kind of 'falsificationist' Optimism: metaphysical claims can be ruled out by science and it is this possibility, I think, that motivates many of the negative attitudes towards metaphysics, since it may seem that in their prolific generation of metaphysical positions without regard to the impact of science, metaphysicians are unaware that many of these positions are metaphysical 'dead men walking'.

However, there are two things to note. First, the relationship in such cases is not best described as one of 'involvement'. It is not that the metaphysical claim is 'involved' in the theory in the way that a claim about unobservables is; rather the relationship is more akin to that between theory and disconfirming evidence as quantum mechanics is being used as 'evidence' to rule out this particular item of metaphysics. Secondly, just as apparently falsified theories may regain life as either the evidence or the theory itself are reinterpreted so the kind of rejection of metaphysics suggested here might be conditional on factors such as the formulation of the theory, its interpretation, the nature or formulation of the metaphysical posit, and so on (see Monton 2011). So, for example, one might reformulate quantum mechanics in such a way as to offer a different understanding of what counts as indiscernible, or put forward a different interpretation that also offers a different understanding.[8] Or one might reformulate the metaphysical posit concerned. Thus as we have already seen, Saunders has proposed a form of 'weak discernibility' in terms of which fermions, at least, can be understood as satisfying a form of Identity of Indiscernibles (Saunders 2003a: 289–307; Muller and Saunders 2008; Ladyman and Bigaj 2010). Of course, one could always insist that such reformulations generate different posits and so the original result stands, strictly speaking, but that's a hard line to hoe, not to mention a churlish one.[9]

But can metaphysical posits be ruled in? In other words, can at least some of these posits be involved in theories in the way indicated previously, such that they can share in the success of the theory? If not, then we will both have grounds for pessimism and face problems responding to Chakravartty's Challenge. We would then have to accept certain constraints on a realist understanding of the world. But if the answer is 'yes', then we must face the problem of metaphysics-induced humility.

Let's consider this problem in a little more detail. The claim is that there exists an extensive array of metaphysical 'facts' about which we can have no knowledge and towards which we must adopt an attitude of epistemic humility. Consider the example of intrinsic properties and the following argument: we can have knowledge of something only in so far as it affects us and so our knowledge is dependent on certain relations holding; these relations are not supervenient on or otherwise reducible to the intrinsic properties of things; hence we must remain ignorant of

[8] The Bohmian and modal interpretations both offer escape routes for the advocate of the Identity of Indiscernibles, for example; see French and Krause 2006: 160–6.

[9] Indeed, there are a number of different ways in which the advocate of the Principle might evade the above kind of 'falsification', although each has been deemed unsatisfactory (Hawley 2009).

and adopt a humble attitude towards these intrinsic properties.[10] Now there are various ways in which one could resist the force of such an argument (and indeed I shall suggest a number shortly) but consider the case of metaphysical underdetermination with regard to individuality given in Chapter 2, which can be construed as another example. Here, as I emphasized, the metaphysical 'packages' of objects-as-individuals and objects-as-non-individuals are both compatible with quantum mechanics and can be considered 'equally natural metaphysical doctrines' in this context (see Butterfield and Caulton 2012). Our only access to the relevant putative objects is via the theory but the theory underdetermines the metaphysics of individuality; hence, as things stand, it seems we must remain ignorant about which of these packages or doctrines holds and adopt an appropriately humble attitude.

What we have here is the conjunction of a justificatory claim and an ignorance claim (see Langton 2009): we must have some justification for positing the metaphysical 'facts' and yet we must be ignorant of them. In such cases the humility limits our realist understanding and unless eliminated or at least reduced, Chakravartty's Challenge cannot be fully met. Fortunately there are a number of fairly obvious ways in which the humility can be handled.

3.4 Handling Humility

The first is to accept our ignorance and acknowledge that we must be humble but insist that this is not in fact a problem. Thus it is certainly not a problem for the constructive empiricist who adopts a broadly sceptical position towards metaphysics. At best, the understanding provided just fleshes out the different ways the world could be. So, one way the world could be is that quantum particles are individual objects, and another way is that they are non-individual objects, but of course, we cannot tell which is correct on the basis of our physics (van Fraassen 1989).

It may also not be a problem for certain forms of realism. Thus one could accept our ignorance of these metaphysical features but still insist that the multiple metaphysical relativities they give rise to lead to greater understanding, as noted in Chapter 2. Here it seems that we achieve greater understanding at the 'meta-level', as it were, by surveying these various relativities, or ways the world could be, rather than by adopting a particular metaphysical package. So the idea seems to be that the array of metaphysical facts that generates humility is to be welcomed since it presents

[10] This is a crude condensation of the argument given in Langton (1998) which aims to show that Kant is not the kind of transcendental idealist we all thought he was but in fact he was a kind of realist who took our knowledge to be constrained by our limited access to, for example, intrinsic properties and hence things as they are in themselves. In a sense Langton portrays Kant as a kind of epistemic structural realist who adopts this attitude of epistemic humility towards the 'hidden' natures of things. Another argument for humility was given by Lewis (2009) based on the multiple realizability of properties; for the differences between the forms of humility in each case see Langton (2004). We shall return to consider Ramseyian humility in Chapter 5.

a range of pursuit-worthy options. However, as noted in the previous chapter, taking this line would seem to be a preliminary attitude at best.

Alternatively, one may retain the usual elements of the standard realist stance but insist that the forms of humility I have touched on here are innocuous. This is how we might understand Chakravartty's insistence that one should no more be worried about metaphysical underdetermination than scientific realists are regarding whether a chair is taken to be a substance plus properties, or a bundle of properties or whether those properties are regarded as instantiated universals or tropes or whatever (Chakravartty 2003b). In other words, we do not need to reduce our level of humility entirely to be a scientific realist: we can be a realist about chairs and other everyday objects, without feeling we have to resolve all metaphysical 'relativities' and likewise we can be a realist about quantum objects without having to resolve the metaphysical underdetermination.[11]

However, as I argued in the last chapter, there is a disanalogy in the case of chairs, or everyday objects more generally and quantum particles which blocks this easy acceptance of humility (see French and Ladyman 2003). Furthermore, there is a tension here with the requirement to supplement one's realism with some form of understanding. In the case of the chair, as realists we begin with a much clearer picture than we have of quantum particles and our relevant understanding is such that we can effectively 'live' with the level of humility associated with not knowing whether the chair is a bundle of properties or has a substantival metaphysical component. In the case of the particles, we do not have that level of understanding to begin with and the humility appears at a much more fundamental level. Indeed, it appears at the most fundamental level possible as far as the object-oriented realist is concerned, namely that of the objects towards which she is adopting her realist stance. But then the question is, how can such a stance be adopted towards something if one does not know whether it is an individual or not?

Let us move on to other ways of handling the humility. One might, for example, accept the existence of the relevant metaphysical facts but reject the claim that we must remain ignorant of which obtain and insist that we do have appropriate access to them.

Thus we might try to expand the relevant notion of 'cognitive access' in this regard and elaborate an account of knowledge that resolves our apparent ignorance of such facts, such as those regarding quiddities, for example (see Schaffer 2005). Just as a haecceity or 'primitive thisness' is taken to render an object the individual that it is (and thus provides one way of spelling out this notion in the quantum context; see French and Krause 2006: ch. 1), so a quiddity likewise underpins the identity of properties. The idea then is that the property of charge is the property that it is because of an underlying quiddity of chargeness, such that if this property were

[11] This corresponds to 'shallow' realism in Magnus' terms (2012).

instantiated in the absence of any other properties being instantiated, it would still be charge, just as if a given object existed in the absence of any other objects existing, it would still be an individual object by virtue of its haecceity.[12] The thought then is that scepticism about quiddities, say, should be regarded as just a form of scepticism about the external world in general and so whatever answer one offers to scepticism in general will thereby yield an answer to quiddistic scepticism.

So, one might adopt a broadly contextualist stance and loosen the standards for knowledge sufficiently that there is a sense in which one can say that we 'know' quiddities. One worry here, of course, is that too much loosening will let anything in, as it were, and if we're not careful, we'll lose any distinction between what can be known and what not. Another is that contextualism hardly seems the right way to go in this situation. It may be that within the metaphysical context, with the loose standards that are appropriate there, we can justifiably assert that we know quiddities, but the context we are concerned with covers both science and the philosophy of science, where it is, at least, unclear that such loose standards are appropriate. And if the standards are those that govern knowledge claims about entities such as electrons, or properties such as charge, then it would seem these are too tightly drawn to cover quiddities.

Taking a different tack, one might try to argue that we have 'direct perception' of such features of properties, in just the way that, it might be said, by putting one's finger in an electrical socket, one can directly perceive charge.[13] But this would be to ride roughshod over all sorts of distinctions in the philosophy of science between phenomena and theoretical entities and perhaps extend the notion of direct perception way too far. Consider: if we can directly perceive charge in this way, can we likewise directly perceive spin, or colour (the quark property, not the visual one)? If not, why not? And if there are barriers to perceiving spin, do these also apply to quiddities? But of course, even if one were to agree, madly, that one can directly perceive charge, as a property, it is quite another thing to insist that one can directly perceive metaphysical features of such properties, such as their quiddities. One might want to try the line that one directly perceives the quiddity of the property by virtue of directly perceiving the property itself—so one perceives the chargeness of charge when one perceives charge—but then I start to lose my grip on the distinction between the property and its quiddity. As a way of handling humility, this would collapse all kinds of distinctions and is a step way too far.

Alternatively, and more plausibly perhaps, one might suggest that we can have abductive knowledge of such metaphysical facts. Thus one might argue that quiddities offer the best explanation of the relevant 'phenomena' and hence can be known in just

[12] And so arguments for positing quiddities draw on a metaphysical manoeuvre that I shall return to and criticize in Chapter 9, namely that of imagining a 'sparse' possible world of, in this case, one instantiated property, or in the case of haecceities, one lonely object.

[13] Kids, don't try this at home!

the way that we can know theoretical entities and properties that are also offered in this way. Thus the way we treat metaphysical features would be put on a par with the way we treat theoretical ones. (Of course the empiricist would not be happy with such a move in either case but I suspect we've left her behind some while ago.)

Here we might usefully compare this move to similar ones that are made in the philosophy of mathematics, where it has been argued that mathematical entities offer the best explanation of certain phenomena (such as the periodic life cycles of cicadas or the structure of honeycombs) and hence should be regarded on a realist basis. However, care has to be taken in such cases, not least because such arguments leave it unclear whether the mathematics is truly playing an explanatory role and not just a representational or indexical one (see, for example, Saatsi 2007 and 2011). Likewise, we need to be clear on what explanatory role quiddities, for example, are supposed to be playing and what it is that they are supposed to be explaining. This takes us back to the argument that metaphysical terms can be treated like theoretical ones, but the former do not play the same role in theories as the latter. In particular, if we consider what is involved in generating predictions and yielding empirical success when it comes to scientific theories, then metaphysical terms cannot be considered as success-inducing in the same way as theoretical ones (e.g. Saatsi 2005). If one were to insist that terms like quiddities are not meant to play any role in explaining physical phenomena but do play such a role in the metaphysical context (assuming some appropriate notion of metaphysical 'phenomena' can be made out) then we are back to contextualism and the response that that's not the context we are concerned with here.

Relatedly, one might attempt to reject the claim of ignorance and break the metaphysical underdetermination by insisting that we should accept a metaphysical posit if it is essentially involved in a theory that generates novel predictions. But again, the involvement of metaphysical posits is not akin to that of theoretical ones and the underdetermination and consequent humility remain.

Here's a different tack: we might adopt a (broadly) Quinean approach (see Belot 2009) and posit the simplest total theory (involving the given metaphysical posit) that is consistent with the evidence, giving a nice parallel between 'the' scientific and metaphysical methods. And indeed, there is a flourishing field of 'meta-metaphysics', certain proponents of which advocate the view that theory choice in metaphysics should be modelled on the methodology of theory choice in science (see Chalmers et al. 2009). But of course, pinning down the latter itself is no easy matter! So, for example, it is more or less accepted that there is no argument that demonstrates that simplicity tracks the truth in the scientific case. And that, furthermore, the problem of characterizing what counts as a 'simple' theory is notoriously difficult (see, for example, Post 1960).[14] If that is the case for the mathematized theories of much of

[14] Having said that, interesting attempts have been made to capture this notion in certain formal contexts; see, for example, Dowe et al. 2007.

modern science, where one can at least take a crack at the problem by focusing on the number of variables, say, or the mathematical form of the theory, then how much more problematic is it going to be to determine what counts as a simple metaphysical theory? More profoundly, perhaps, in the scientific case the role of evidence in driving theory revision is crucial but there is nothing equivalent in the metaphysical case (or at least, not in the same relatively straightforward sense); in particular there is no evidence to wash out disagreements over simplicity. As Belot nicely puts it,

> If ontology follows a version of the scientific method, the relevant version is a degenerate case – and, I think, we should be suspicious of the credentials of its output. (Belot 2009)

But perhaps attempting to draw such a straightforward parallel between the methodologies of science and metaphysics is simply too quick. Perhaps a better and more sophisticated approach would be to adopt a framework within which the relationship between such metaphysical claims and the relevant scientific theories can be appropriately articulated. Indeed, this is what Ladyman and Ross do as part of their defence of OSR. In particular, they advocate the following 'Principle of Naturalistic Closure':

> [O]nly take seriously those metaphysical claims that are motivated by the service they would perform in showing how two or more hypotheses jointly explain more than the sum of what is explained by the two hypotheses taken separately. (Ladyman, Ross, et al. 2007: 37)

This is conjoined with what they call the Primacy of Physics Constraint:

> Special science hypotheses that conflict with fundamental physics, or such consensus as there is in fundamental physics, should be rejected for that reason alone. Fundamental physical hypotheses are not symmetrically hostage to the conclusions of the special sciences. (Ladyman, Ross, et al. 2007: 44)

Together, these yield positive and negative proscriptions regarding the role of metaphysics and its relationship to science. The positive is that metaphysics is now seen as 'the enterprise of critically elucidating consilience networks across the sciences' (Ladyman, Ross, et al. 2007: 28). And the negative is that we should reject any metaphysical hypothesis that conflicts with fundamental physics.

However, this framework has been criticized for being too liberal, and rejecting too little, and also for being too restrictive, and rejecting too much. It rejects too little because 'many contemporary scientific theories are *themselves* "neo-scholastic" in so far as they contain (naturalistically unjustified) metaphysical assumptions' (Dicken 2008: 291). Thus, in so far as current science incorporates metaphysical posits that do not satisfy the Principle of Naturalistic Closure, such posits should also be expunged, but doing so would remove many of the interpretive elements from the theories concerned. Underlying this criticism is the concern that there is an ambiguity in what is meant by 'fundamental physical hypotheses' in Ladyman and Ross's scheme: do we mean the hypothesis as formally given, or as interpreted? If the former, then we seem to be edging uncomfortably close to a positivistic understanding of theories; if the latter, then it is hard to see how one could include at least some metaphysics in such an interpretation. Here again we bump up against the considerations presented in section 3.3 in the context of deciding whether quantum physics rules out the Principle of Identity of Indiscernibles.

On the other hand, the Ladyman and Ross scheme rejects too much because it would rule out both the important heuristic role of metaphysics already noted (Dicken 2008) and the 'Viking Approach' suggested here (see also Hawley 2010). As we'll see in Chapter 7, a variety of metaphysical resources and tools can be laid out as available to help articulate the relationship that OSR posits as holding between putative objects and structures. The crucial point is that even where metaphysics has been developed in the absence of any relationship with current physics, or, as is more often the case, on the basis of only everyday examples or at best toy models, it may still prove useful.

Returning to the issue of handling humility, none of the approaches considered in this section seem to be adequate. Instead I suggest we deal with it by elimination. As Faraday asked,

> Why then assume the existence of that of which we are ignorant, which we cannot conceive, and for which there is no philosophical necessity?[15] (Faraday 1844: 291)

Now I shall construe existence here narrowly, in the sense that we should accept only such metaphysical posits as we minimally require to interpret our theories, along the lines suggested by Chakravartty. And I want to use this to push the claim that we do not minimally require *objects*, which generate unacceptable levels of humility via the metaphysical underdetermination regarding identity discussed previously.

In the specific context of a defence of OSR, the core idea is encapsulated in what I shall call 'Cassirer's Condition':

> Take the 'conditions of accessibility' to be 'conditions of the objects of experience'.

We shall return to consider Cassirer's neo-Kantian form of structuralism in Chapter 4 but by 'conditions of accessibility' I shall understand those conditions encoded in our best theories that give us *access* to the way the world is (on a realist construal). And by the 'conditions of the objects of experience' I shall understand those conditions that lay down *how* the world is, where, of course, we are taking 'objects' here in a broader sense than in the object-oriented stance.

If we adopt this condition, then

> there will no longer exist an empirical object that in principle can be designated as utterly inaccessible; and there may be classes of presumed objects which we will have to exclude from the domain of empirical existence because it is shown that with the empirical and theoretical

[15] By 'philosophical' here Faraday of course meant the term in its 'old school' sense that embraced the scientific.

means of knowledge at our disposal, they are not accessible or determinable. (Cassirer 1936/ 1956: 179)

This is how I view the objects posited by object-oriented realism and ESR: as not accessible, via our theories, nor determinable, in the sense of being able to specify well-defined identity conditions for them on the basis of those theories.[16]

The general attitude that underlies Cassirer's Condition crops up elsewhere. Thus Hawthorne, echoing Faraday, asks, '[w]hy posit from the armchair distinctions that are never needed by science?' (Hawthorne 2001: 369). And returning to the particular issue of positing quiddities, he writes,

If there were a quiddity that were, so to speak, the role filler, it would not be something that science had any direct cognitive access to, except via the reference fixers 'the quiddity that actually plays the charge role'. Why invoke what you don't need? (Hawthorne 2001: 368)

One finds this kind of humility-reducing manoeuvre being made in a variety of contexts that are amenable to structuralist approaches. More explicitly, Esfeld notes the gap that appears between metaphysics and epistemology if an attitude of humility is allowed (2012) and also urges the closing of this gap in the specific case of quiddities by denying their existence as underpinning the identity of properties.[17] The point, then, is that humility is handled by *eliminating* the 'inaccessible' posits whose existence opens this gap between metaphysics and epistemology.

Let me now sum up where we are with regard to the relationship between metaphysics and science.

3.5 Gaining Understanding while Reducing Humility

We recall 'Chakravartty's Challenge' and the demand to provide understanding of scientific theories by offering an appropriately metaphysically informed interpretation. Here we've looked at some of the obstacles faced by and dangers inherent in such an interpretation. Object-oriented realism, in particular, is hamstrung through being unable to ground the identity conditions of its objects in the relevant physics, and the metaphysical underdetermination regarding identity and individuality that we considered in the previous chapter introduces an unbridgeable gap between the relevant epistemology and metaphysics. That, in turn, brings with it a level of humility that, I would insist, is too much for any realist to swallow.

I'll come back to this shortly, but clearly what we need to do is to balance the gain in clarity and understanding that metaphysically informed interpretations can yield

[16] Some, such as Morganti (2004), see this as an unwarranted 'jump' from epistemology to metaphysics. As in similar cases of revisionary philosophy, this perhaps reveals a fundamental divide between attitudes over the relationship between epistemology and metaphysics. However, I take Cassirer's Condition as simply embodying the not unreasonable view that we should strive to bring our metaphysics in line with our epistemology, as far as we can.

[17] I shall discuss Esfeld's own form of dispositional structuralism in Chapter 9.

with an appropriate reduction in the level of humility that we have to accept as a consequence. We can do this if we follow something like the following process: we draw on metaphysics to respond to Chakravartty's Challenge and thus involve metaphysics in our interpretation of science, but only as much as necessary; we then reduce any associated humility in line with Cassirer's Condition regarding the reliance on what he called the 'conditions of accessibility', thereby minimizing the metaphysics, as much as possible. Indeed, Chakravartty himself expresses something along these lines, when he writes,

> we must turn to the equations with which we attempt to capture phenomenal regularities, and ask: what do these mathematical relations minimally demand? We must consider not what possible metaphysical pictures are consistent with these equations, but rather what kinds of property attributions are essential to their satisfaction—i.e. to consider not what is possible, but what is required. (Chakravartty 1998: 396)

For Chakravartty, what is required is a dispositional metaphysics of properties, albeit one that is reconfigured along structuralist lines. I'll discuss that in Chapter 9, but here I want to argue that OSR achieves just the right balance of gain in understanding with reduction in level of humility.[18]

3.6 Manifestations of Humility in the Realism Debate

So, let's begin with object-oriented realism, crudely summarized in the claim that reading off the relevant physics yields a picture of the world as composed of objects, that possess certain properties, enter into certain relations, etc. The question then is what sort of objects are these? More specifically, can we understand them in terms of our usual metaphysical frameworks regarding individuality and identity or not? Unfortunately, the metaphysical underdetermination outlined earlier in this chapter prevents us from giving a definitive answer to this question, at least on the basis of the physics itself. Likewise, how should we understand the relevant properties? In particular, is their identity given by quiddities or not? Again, we can't say, on the basis of the physics.

Here we have way too much humility! Indeed, it is surprising that the object-oriented realist has got away with such a high level of humility for so long but perhaps this is simply because the metaphysics behind her realism is typically not examined very closely, which in turn has to do with the continued failure to fully engage with the implications of quantum mechanics.

[18] Here my claim is very similar to that of Brading and Skiles, who, as we noted in the previous chapter, argue that when viewed as a proposal for distinguishing between those aspects of a formulation of a theory that are candidates for representing ontology, and those that should be regarded as mathematical artefacts, OSR is 'metaphysically more modest' than other forms of realism, and should be adopted on those grounds (2012).

What about Chakravartty's own view, semi-realism? Here we have a dispositional framework in which properties are understood in terms of causal powers, extended holistically to include relations in a way that meshes nicely with some of the central features of modern physics (although as I said, I shall criticize the dispositional basis of this account in Chapter 9). As far as properties are concerned, then, quiddities are excluded from the picture and hence the level of humility is correspondingly reduced. However, Chakravartty still retains objects as the 'seat' of these causal powers and thus still falls prey to the metaphysical underdetermination regarding individuality. Again, then, there is still too much humility in this respect, although, as I have also noted in the previous chapter, Chakravartty takes the underdetermination to be innocuous. Here I can press my earlier point that this should not be set on a par with and dismissed alongside the kind of underdetermination we find with regard to whether 'everyday' objects should be regarded as bundles of tropes or universals. In the latter case the level of humility, although high, can indeed be regarded as innocuous because these objects are not taken to be elements of our fundamental ontological base. In a sense it just doesn't matter that different metaphysical accounts can be given of them because there is a tacit understanding that they are dependent upon, or indeed eliminable in favour of, a more fundamental set of objects. These constitute, in some sense, the way the world is and here too much humility is an issue, as the gap between epistemology and metaphysics widens and we find ourselves buying into a picture—such as that of the object-oriented realist—where we have to accept elements that are simply not grounded in our best scientific theories.

What about Epistemic Structural Realism, with its claim that 'all we know (i.e. all we have epistemic access to) is structure'? Unlike most versions of object-oriented realism, here at least we're starting from the right epistemic point, with the structures presented to us by theories. But here again humility enters with the 'hidden natures'—indeed, we get an extra helping of humility by virtue of their hiddenness![19] At least the object-oriented realist's objects are intended to be out in the epistemic open, as it were, but here we have something utterly inaccessible that is posited solely to prop up the structures to which we do have access (thus assuming that they need such props).[20] 'Liberal' ESR (Slowik 2012) fares even worse, since here the veil is drawn over not just objects, but objects and relations, so even more humility is piled on the plate!

Sliding across the metaphysical spectrum, in his articulation of an 'eclectic' realism (Saatsi 2008; see also 2005), Saatsi questions whether even in the classical Fresnel–Maxwell case deployed by ESR it should be the *equations* that are the focus of

[19] Worrall's agnosticism can be seen as a further manifestation of humility.

[20] It has been suggested that I may be putting up a straw person here as the epistemic structural realist need not be committed to 'hidden natures'. However, if this means that she may take the natures not to be hidden, then I fail to see the difference between that and object-oriented realism; if, on the other hand, such hidden natures are eliminated, then we have OSR.

attention. Thus, Saatsi argues that the success-yielding features of theories can be identified in a metaphysically minimal manner with those theoretical properties, such as spin, charge, etc., which are actually involved in the relevant theoretical derivations, or more generally, which lie at the theoretical end point of the relationships between theory and phenomena. He demonstrates that the recovery of Fresnel's equations from Maxwell's theory can be articulated in terms of certain dispositional descriptions that are satisfied by those properties that feature in the solutions of Maxwell's equations. Here again we have a shift away from objects, but Saatsi is keen to steer clear of structuralism, arguing that the balance should tip towards the epistemological rather than metaphysical aspects of realism.[21]

Whether he can do so remains unclear. We recall the criticism regarding the tenability of ESR's distinction between nature and structure that we discussed in Chapter 1 (Psillos 1999): the object-oriented realist would insist that the relevant properties are those of an unobservable object, whose nature is ultimately playing the explanatory role Saatsi is concerned with. However, the structural realist would claim that if the 'nature' of these objects is cashed out in metaphysical terms, then the conclusion doesn't follow. If it is not, then 'nature' signifies nothing more than the relevant properties and the conclusion is empty. And in that case the structuralist can agree that Saatsi's principles tell us something about the relevant properties, where these are understood as aspects of structure. In the absence of such an understanding it is unclear how we are to regard them—in a sense, eclectic realism avoids humility but offers too little metaphysics and thus may fail Chakravartty's Challenge.

We come now, like Goldilocks, to Ontic Structural Realism and a balance between understanding and humility that, I would argue, is 'just right'. As in the case of ESR it proceeds from the appropriate epistemic base but avoids having to be humble about hiddenness.[22] And, of course, unlike object-oriented realism it overcomes the obstacle presented by metaphysical underdetermination by dropping the entities whose 'identity profiles' (to use Brading's phrase) remain detached from that epistemic base. Moreover, as in the case of semi-realism, it understands properties in terms of

[21] A notion of 'Explanatory Approximate Truth' is central to his view.

[22] Interestingly, Floridi provides an argument to the effect that meta-theoretical analysis also propels us from ESR to OSR (Floridi 2008). A crucial notion here is that of 'level of abstraction' (LoA; Floridi and Sanders 2004), where this involves commitment to certain types of putative objects in particular. According to ESR 'a theory is justified in adopting a LoA that commits it . . . ontologically to a realist interpretation of the structural properties of the system identified by the model that has been produced by the theory at the chosen LoA' (Floridi 2008: 231). This gives us first-order knowledge of the structural properties of the system and having committed ourselves to the relevant underlying structure, on the grounds that ESR sits upon, we are then entitled to perform a kind of transcendental inference to the effect that whatever the underlying objects are in themselves, they must be such as to allow the theory to appropriately model their structural properties. In other words, the commitments associated with OSR are what make ESR possible, in that the LoA adopted at this second-order level is one that involves commitment to an interpretation of the objects as themselves structural in nature (2008: 233). Of course on Floridi's view, since these differing commitments are associated with different levels of analysis, there is no incompatibility between ESR and OSR.

their nomic role and hence does away with quiddities (although, as we shall see in Chapter 10, it reverses the dependence relation that the dispositionalist takes to hold between properties and laws). Thus the principle sources of metaphysical humility are eliminated.[23]

Of course, further attitudes of humility may have to be adopted as the structuralist goes on to elaborate her ontology in terms of the metaphysics of structure. But I would hold that this humility would also be generated for the object-oriented realist, the semi-realist, and the epistemic structural realist in so far as they are obliged to engage in a similar exercise if they are going to take the structures presented by physics seriously (as they should). In particular, none of these views have, so far, properly engaged with some of the most ubiquitous, powerful, and important structures presented by modern science, namely those represented by the symmetries of contemporary physics. Here, as we shall see, considerable work remains to be done, but in so far as all forms of scientific realism are going to have to do such work, and outline an appropriate metaphysical understanding of these structures in order to meet Chakravartty's Challenge, I shall take any humility that has to be adopted as a result as applying across the board and not simply to OSR alone.

With these arguments in favour of OSR behind us, we can proceed to elaborate this structuralist ontology. The picture I shall set out can be sketched as follows: the ontology we should 'read off' our physics should be one of laws and symmetries, understood as features of the structure of the world. The laws characterize relations between properties, the identity of which is given by their nomic role. However, instead of taking these properties to be instantiated in metaphysically robust, or 'thick', objects, the advocate of OSR understands them to be dependent upon the relevant laws and symmetries. In so far as these encode the relevant range of physical possibilities, the structure of which they are features can be said to be modally informed.[24]

Colouring in this sketch will take up most of the rest of the book but before I embark on this, I shall pause to recall some of the forgotten history of structuralism and bring back into the light certain aspects that will inform my own elaboration of it.

[23] Votsis queries whether it is illegitimate not to achieve such a balance and advocates Worrallian agnosticism (2012). All I can say is that this renders the realist far too humble for my liking!

[24] Again there is a sense in which we have reduced the level of humility by understanding modality in this way rather than through dispositions, say. Of course, one could reduce it still further by adopting a Humean approach to modality but although a form of Humean structuralism has indeed been proposed, it faces well-known problems.

4

Scenes from the Lost History of Structuralism

4.1 Introduction

The history of structuralism deserves an entire book to itself. It is rich and complex and intersects with a wide range of developments in mathematics, philosophy, and science. Rather than attempt to sketch the whole of this history here,[1] I shall highlight those features I regard as particularly significant for the discussion to follow.[2] My aim is twofold: first, to bring into the light aspects of what I shall call the 'lost history' of structuralism and thereby reconstruct, at least partially, a historical context in which the developments I am interested in can be situated; and secondly, to emphasize certain accounts, methods, and manoeuvres in general that might be brought forward from this context and put to use in defence of current forms of structuralism.

In effect I shall apply a form of 'Viking Approach' to history itself but one that is moderated with the recognition that some mediation is needed between the past and the present in order for developments in the former to be used as resources to help shape the latter (see Domski 2013).[3] At the very least one needs to recognize that both the language and the aims of past views may be very different from that of our own and that if they are to be used as philosophical resources, they cannot simply be shoe-horned into the current debates. Bearing this in mind, I shall at least tip my hat to any significant context dependence where appropriate; or, to put it another way, acknowledge that certain features of these historical views can't be dragged forward, or at least not in their original form, precisely because they are too firmly tied to their context. This should become clearer with some examples, so let's move on to the historical episodes themselves.

[1] For attempts to cover certain aspects of this history see Gower 2000; Votsis 2004; Frigg and Votsis 2011; van Fraassen 1997. Howard also has important things to say about this history (talk given at the Workshop on Structural Realism, University of Notre Dame, November 2010).

[2] An important figure I shall have to leave out is Schlick, for example.

[3] Or as Howard has put it, we can see the relevant issues of today and the past as related via a kind of genealogy metaphor without making the figures of the past partners in our enterprise.

4.2 The Poincaré Manoeuvre

In his 1989 paper that revived the structuralist tendency in the context of the modern debate over scientific realism, Worrall sets out the historical antecedents of his epistemic structural realism quite explicitly: they lie with Poincaré's great work, *Science and Hypothesis*, 1905 (Dover 1952), and in particular the passage where he writes that theories

> teach us now, as they did [in the past], that there is such and such a relation between this thing and that; only the something which we then called *motion*, we now call *electric current*. But these are merely the names of the images we substituted for the real objects which Nature will hide for ever from our eyes. The true relations between these real objects are the only reality we can attain. (1952: 162)

Here we see the emphasis on both the relations, as the 'only (aspects of) reality' we have epistemic access to, and the underlying objects, which, although hidden, are nevertheless real.

However, there are two significant features of Poincaré's structuralism which tend to be glossed over in recent discussions: the first is the fundamental importance of group theory in representing these 'true relations' and expressing the kind of structure that is important for physics. Putting it crudely, certain significant aspects of this structure may be preserved under various symmetry transformations and these transformations form a group, in the mathematical sense (that is, subject to the axioms of group theory). The history of group theory and the way this history is entwined with developments in both mathematics and physics is nicely outlined in Bonolis (2004). Here Klein's 'Erlangen' programme occupies a significant place, as with the development of non-Euclidean geometries and the introduction of large numbers of dimensions (motivated at least in part by developments such as Maxwell's theory of electromagnetism), concerns arose about how to capture the central unity of geometry and classify its different forms. Klein's core insight was to apply the theory of infinite groups and reduce geometry to the study of invariances under the relevant group of transformations. What this insight yields, of course, is a structural conception of geometrical objects that shifts the focus from individual geometrical figures, grasped intuitively, to the relevant geometrical transformations and the associated laws. This conception, and the development of group theory in the work of Lie in particular, had a significant impact on Poincaré, who defended the group-theoretic approach to geometry in a paper for *The Monist* (1898).[4] Here he used it to explain the dimensionality of space (Crilly 1999: 12–14)[5] but more generally, it underpins the beginning of Poincaré's conventionalism, since if geometry is nothing but the study of groups, then the truth of Euclidean geometry is not incompatible

[4] I am grateful to Mary Domski for bringing this to my attention.

[5] Thus the 3-dimensionality of space is explained in terms of a representation of the Euclidean group of rigid motions acting on the conjugate space of rotation sub-groups.

with that of non-Euclidean geometries, because the existence of one group is not incompatible with that of any other (Torretti 2010). The latter is an example of a context-dependent feature that I certainly do not want to import into discussions about the viability of OSR. Of more interest to me is the way that Poincaré tackles the objection that in order to study such groups they need to be constructed but they cannot be constructed in the absence of material objects, and thus there is more to geometry than group theory. His response is to insist that 'the gross matter which is furnished us by our sensations was but a crutch for our infirmity', which serves only to focus our attention upon the idea of the group. In other words, the material objects whose movements and interrelationships appear to ground geometry are but a form of heuristic device via which we can arrive at group theory, but having reached our destination, we can then dispense with such heuristic 'crutches'. We shall come across this move again in the history of structuralism and it will be useful to express it in general terms as follows:

> Poincaré's Manoeuvre
> Although we might introduce the terminology, or perhaps better, symbology, of objects as part of our representation of the relevant structure, these should be regarded as mere devices that allow us to construct, articulate, or appropriately represent the relevant structure, and any representational priority they might have should not be taken to imply that they are ontologically foundational.

The second feature of Poincaré's structuralism I'd like to flag up concerns its Kantian flavour. We will encounter this again when I touch on Cassirer's views later but I will suggest we can largely leave it behind as far as fleshing out OSR is concerned, although as I acknowledge in the previous chapter, we can appropriate 'Cassirer's Condition' for realist purposes.[6] In Poincaré's case, although he cleaved to a Kantian view of mathematics as synthetic a priori, his conventionalism led him to reject the claim that Euclidean geometry was a priori imposed by intuition. Instead of the Kantian idea of an intuitive space (whose geometry is Euclidean), Poincaré adopted the more minimal a priori basis consisting of an intuitive idea of continuity (for an accessible discussion, see Folina 2010). Likewise, and relatedly, it is the general notion of 'group' that is given to us a priori, rather than some particular group itself. Thus in *Science and Hypothesis*, after noting again that the object of geometry is the study of a particular group, he writes that 'the general concept of a group pre-exists in our minds, at least potentially' (1905: 70)[7] and that the general concept of a group is 'imposed on us not as a form of our sensitiveness, but as a form of our understanding; only, from among all possible groups, we must choose one that will be the standard,

[6] Although some might want to import it into current debates and add to the burgeoning neo-Kantian movement in philosophy of science; see Massimi 2009 and Bitbol et al. 2009, for example.

[7] And later on he writes, 'In our mind the latent idea of a certain number of groups pre-existed; these are the groups with which Lie's theory is concerned' (1905: 87–8).

so to speak, to which we shall refer natural phenomena' (1905: 70).[8] Experiment does not dictate this choice in the sense of telling us what is the true geometry, but only which is the most convenient.[9] (According to the Helmholtz–Lie theorem, there are only three possibilities to choose from: Euclidean; Bolyai–Lobachevsky; and Riemannian.[10]) As I said, I shall not consider Poincaré's conventionalism here[11] since my concern is with the subjective element that enters structuralism on this view: for Poincaré, the notion of a group is something innate to us that *we* contribute to our knowledge of the world; a decision then has to be made as to which group, and hence which geometry, is most convenient for describing the world.[12]

Thus, as Domski has emphasized, Poincaré may not be the most suitable forefather to claim for Worrall's form of structural realism.[13] Nevertheless, one can render Poincaré's account, or at least a facsimile of it, compatible with a realist stance (see Folina 2010). In particular, in so far as the Kantian element has to do with the conceptual origin of the notion of a group, in the sense of where it should be situated, we can in effect hive it off and regard it as one of the descriptive resources that we can deploy in presenting structure at the level of theories.

I shall return to such 'disentangling' moves later but let me now briefly discuss the work that tends to dominate considerations of the history of structuralism, to the extent that other and, in some senses, more interesting forms of this tendency have been lost in its shadow.

4.3 The Analysis of Matter

In his more recent work (such as, for example, his contribution to Zahar's book on Poincaré, Zahar 2007), Worrall advertises an alternative historical antecedent for ESR, namely Russell's 'epistemological' structuralism. Indeed, many of today's structural realists, such as Redhead, would point to Russell as their philosophical forebear, possibly due to the influence of Maxwell (1962, 1970a, 1970b, 1972) who rediscovered and re-presented Russell's approach in the 1960s.

[8] Thus for Poincaré, the impact of the development of non-Euclidean geometry, underpinned by the group-theoretic conception, is to effectively shift the place of space from the sensibility to the understanding in Kantian terms.

[9] See also his discussion on pp. 87–8 where he returns to the consideration of the dimensionality of space.

[10] Where this last refers to Riemannian geometry of constant curvature; Riemann's theory of manifolds of variable curvature which underpinned General Relativity was not compatible with Poincaré's conception, as Poincaré himself recognized (see Friedman 1995).

[11] For a useful discussion that emphasizes the group-theoretic underpinnings and, as a consequence, the differences from the form of conventionalism adopted by the logical positivists, see Friedman 1995.

[12] For more on the role of the development of non-Euclidean geometry in motivating structuralism, see van Fraassen 1997.

[13] He also rejected truth as the aim of science; see Domski preprint. She offers the early Schlick as a more suitable candidate.

The usual point of reference here is his classic book, *The Analysis of Matter*, in which Russell attempted to construct an epistemology that was appropriate for the new physics of relativity and quantum theory. At its heart lies the 'Causal Theory of Perception', which, put simply, states that our experiences—represented by our 'percepts'[14]—are causally related to the relevant stimuli (Russell 1927). The assumption that differences in our percepts are brought about by differences in the stimuli,[15] together with spatio-temporal continuity, suffice 'to give a great deal of knowledge as to the *structure* of stimuli'[16] (1927: 227; his emphasis). However, the 'intrinsic characters' of the stimuli must remain unknown, so all that we know about the external world is its *structure*.

Thus, Russell invites us to consider a set of propositions about an electron, *E*. On the traditional view, which analyses such propositions in terms of an underlying substance within the framework of subject–predicate logic, we conclude that there is a certain substantive entity *E* that is mentioned in all statements about this electron. According to the new, structuralist analysis, however, what we obtain is

> a certain relation *R* which sometimes holds between events, and when it holds between *x* and *y*, *x* and *y* are said to be events in the biography of the same electron. (1927: 287)

In particular the formal properties of the relevant propositional functions will be the same, something we shall return to shortly. Notice how the 'new analysis' is presented in opposition to a substantivalist view of objects—this is a common theme of structuralism within this period and it bears on an important question that has been raised in current debates: what notion of object is the structuralist rejecting when she asserts the priority of structures over objects? For Russell, and, as we shall see, Cassirer and Eddington, it was a substantival notion, and, indeed, as Russell insisted, had to be replaced in favour of an events-based ontology:

> science is concerned with groups of 'events', rather than with 'things' that have changing 'states'. (1927: 286)

With the demise of substance in the context of modern physics, current forms of structuralism have tended to articulate their stance in opposition to a broader notion of object.

There are two further features of Russell's structuralism that bear a close resemblance to Worrall's epistemic structural realism. The first pertains to the hidden nature of that to which the symbols of our theories apparently refer. Indeed, how we understand these symbols is to a certain extent an arbitrary matter, by analogy with coordinates in General Relativity:

[14] These are the entities of which we have knowledge by direct acquaintance according to Russell.

[15] This is referred to by Psillos (2001) as the 'Helmholtz–Weyl Principle'.

[16] By stimuli, Russell meant events lying just beyond the reach of our sense organs, and which are connected via causal chains to physical objects.

We have, in fact, something more or less analogous to the arbitrariness of co-ordinates in the general theory of relativity. Provided our symbols have the same interpretation when they apply to percepts, their interpretation elsewhere is arbitrary, since, so long as the formulae remain the same, the *structure* asserted is the same whatever interpretation we give. Structure, and nothing else, is just what is asserted by formulae in which the meaning of the terms is unknown, but the purely logical symbols have definite meanings. (1927: 288–9)

The second feature has to do with the emphasis on the formulae that remain the same, expressed in the passage just quoted but even more explicitly here:

When we are dealing with inferred entities, as to which . . . we know nothing beyond structure, we may be said to know the equations, but not what they mean: so long as they lead to the same results as regards percepts, all interpretations are equally legitimate. (1927: 287)[17]

The similarity with Worrall's view here is obvious.

There are also further aspects of Russell's view that are worth noting in this context. The first is that he takes what Psillos calls the 'upward path' to structure (Psillos 2001), beginning with broadly empiricist premises and attempting to move upwards to a 'sustainable realist position' (2001: S13). However, the assumption which underpins the claim that differences in our percepts are brought about by differences in the stimuli is too weak to do the work required. Russell talks of a 'roughly one-one relation'[18] between percepts and stimuli and for that one needs the converse of that assumption, namely that differences in stimuli yield differences in our percepts.[19] However, Psillos argues, the realist should allow at least the possibility that the unobservable world might contain structure not manifested in the phenomena and hence the relevant relation should be an embedding rather than an isomorphism.[20] It is just such an embedding that the likes of van Fraassen propose (holding between the empirical sub-structures and theoretical structures) as a crucial feature of his structural empiricism, but this, of course, is not a realist position. However, the modern-day structural realist can resist this form of guilt by association. The defender of ESR in particular already admits that the domain of physical objects, together with the associated properties and relations, is not determined absolutely, but only up to isomorphism (Votsis 2005: 1367). Thus, similar to

[17] Votsis articulates a further principle underpinning the inference here, which he calls the 'Mirroring Relations Principle', to the effect that the relations of physics are not identical with those we perceive but rather 'mirror' them in the sense of having the same logico-mathematical properties as them.

[18] Psillos asks if it even makes sense to talk of a 'roughly' 1–1 relation in this context. Of course, one might try to capture such talk via the formalism of partial isomorphisms (da Costa and French 2003), although this will not help in the present context.

[19] However, as Votsis points out, Russell accepted that different stimuli may often lead to different percepts and suggests that this was why he refrained from saying that we can *know* the structure of the world and instead maintained only that we can 'infer a great deal' about it (Votsis 2005: 1365–6).

[20] As Votsis notes (2005: 1365), the initial assumption (the Helmholtz–Weyl principle) is not enough to yield even an embedding as things stand, since an embedding maps relations from one domain to another, and that requires that the domain of percepts, for example, already be appropriately structured.

structural empiricism, there is a kind of underdetermination that is constitutive of this position and articulating the relevant relationship between the theoretical and phenomenal 'levels' in terms of embedding would likewise seem to be appropriate.

Alternatively, one may, as Psillos notes, attempt to block this possibility of 'extra' structure at the unobservable level, by insisting, as Weyl did (see his 1963: 117) that there can be no diversity at the theoretical or unobservable level that is not manifested in diversity at the level of phenomena. On what grounds, however, could one insist on this? Weyl, of course, was no realist and took this insistence to be a central and constructivist feature of idealism that should be conceded. That is not an option for the structural realist, needless to say. Thus, Psillos concludes, Russell faces a dilemma: without the converse of the initial assumption he cannot establish the 1–1 relation (even roughly) between the structure of the phenomena and the structure of the unobservable world that would allow him to claim that we can know the latter on the basis of the former; with the converse, he runs the risk of conceding too much to the constructivist, or more broadly, idealist, view (Psillos 2001: S16).[21]

Now the modern-day structural realist might propose an apparently reasonable principle that is similar in spirit to Weyl's, namely that one should only adopt a realist stance towards (unobservable) structure that makes a difference at the level of phenomena, broadly construed.[22] Structure that makes no difference whatsoever can legitimately be dismissed as 'surplus', possibly arising from the mathematical descriptive framework being invoked (I shall return to this notion of surplus structure in subsequent chapters).[23] And such a principle can be related to, or at the very least meshes with, the more general claim that we should be realists only with regard to those elements of our theories that feature in the explanations of the relevant phenomena (Saatsi 2005). However, even granted the shift from percepts to a broader notion of phenomena, this will not be enough to clear a Russellian upward path to the structure we are interested in. What is needed is something more, and Votsis identifies this with what he calls the 'Mirror Principle', to the effect that relations between percepts have the same logico-mathematical properties as relations between their causes (Votsis 2005: 1362; see Russell 1927: 252). This, he argues, allows us to

[21] Psillos also considers and criticizes what he calls the 'downward path' adopted by modern-day structural realists, which instead of beginning with an empiricist basis, starts with fully fledged realism and then attempts to weaken it. I shall discuss his criticisms of this approach later on.

[22] Denying such a principle and allowing the existence of physical structures (that is, structures about which we should be realists) that make no difference runs the risk of attracting the same kind of opprobrium that motivated the pragmatist and verificationist theories of meaning (see, for example, Schlick 1932).

[23] Of course, the epistemic structural realist might accept 'extra' structure at the unobservable level, extending her view of objects as 'hidden' to certain structures themselves. This appears to be what Votsis has in mind in responding to Psillos' concern here (Votsis 2005: 1367). However, it is not clear why the proponent of ESR would need to take a realist stance towards such structure, and not just dismiss it as surplus as suggested here. After all, she does at least have grounds for positing objects, hidden as they are claimed to be, since, she maintains, they are needed to act as the relata for the relations in the structure, but it is not clear what the corresponding grounds would be for insisting on 'hidden' extra structure!

preserve any relations the set of causes may have (Votsis 2005: 1366). However, as Votsis acknowledges, Russell was unclear on the grounds for accepting such a principle and Votsis' own allusion to the general requirement of epistemological realism that there be some correspondence between language and reality, where this correspondence should be articulated in relational terms (Votsis 2005: 1366), is certainly not sufficient. One could insist that the Mirror Principle be taken as a minimum requirement, since if a given relation between the percepts does not have at least the same logico-mathematical properties as the putative corresponding relation then it can hardly be said to be the same, or indeed, the 'corresponding', relation. But the modern structural realist might demand more than this, since it correspondingly yields only a minimal notion of structure, namely that which captures the logico-mathematical properties of the relevant relations.

This brings us on to the second feature of Russell's position, which concerns the nature of this structure. By 'structure' Russell meant the class of relations that are isomorphic to a given relation (Russell 1927: 250), since our indirect epistemic access to the world, obscured as it is by the veil of percepts, means that we cannot uniquely identify the properties and relations that are possessed by and hold between physical objects (for discussion see Votsis 2005: 1362–3). Redhead identified this notion with that of abstract structure (Redhead 2001) in the sense of an 'isomorphism class' of structures that are isomorphic to a given structure <A, R>, where A is a set of elements and R a family of relations. In this sense, the domain of objects and associated relations are only specified up to an isomorphism. This notion can be contrasted with that of 'concrete structure', which picks out a specific domain of objects and the associated family of relations (we shall return to this distinction in subsequent chapters).

Unfortunately, Russell's 'abstract' structuralism was soon the subject of a powerful criticism from the mathematician Newman that has become the default basis for the rejection of structural realism in general (and I shall return to it in Chapter 5). At the core of the criticism is the following claim: if we know only the *structure* of the world, then we actually know very little indeed. The basis for this claim is straightforward: 'given any "aggregate" A, a system of relations between its members can be found having any assigned structure compatible with the cardinal number of A' (Newman 1928: 140). Hence, the statement 'there exists a system of relations, defined over A, which has the assigned structure' yields information only about the cardinality of A:

> the doctrine that *only* structure is known involves the doctrine that *nothing* can be known that is not logically deducible from the mere fact of existence, except ("theoretically") the number of constituting objects. (Newman 1928: 144; his emphasis)

Hence, for any given collection of objects, a variety of 'systems of relations' is possible, yielding the posited structure and hence a choice must be made. The problem then is how to justify such a choice. One might try to pick out a particular system as physically 'important', in some sense, but then as Newman himself points out, either the notion of 'importance' is taken as primitive, which seems absurd, or it must be grounded on what Russell calls the 'intrinsic characters' of the relata, but that introduces a non-structural element and undermines the very structuralism being defended (Newman 1928: 146–7; for a reprise of this point in the context of today's structural realism, see Psillos 1999: 63–5. I shall return to the issue of whether the introduction of such elements is always undermining at various points in this book). In a classic Homer Simpson 'Doh!' moment, Russell concedes the point and admits to Newman that he hadn't really intended to say what he did say, namely 'that nothing is known about the physical world except its structure' (Russell 1967, 1968, 1969: vol. 2, p. 176). This has recently generated a vigorous debate over the impact of the argument on current forms of structuralism (Demopoulos and Friedman 1985; see also Worrall's appendix in Zahar 2007 and also Worrall 2007 and 2012; Ketland 2004; Melia and Saatsi 2006; Votsis 2003); however, I shall leave further discussion until Chapter 5,[24] although I will return to Newman's point and Eddington's reaction to it shortly.

Let us now pause to consider what might be brought forward from the Russellian context (cf. Landry 2012). It seems that the defender of ESR would be willing to accept the existence of 'extra' structure at the theoretical level and, methodologically, something like the 'Mirror Principle' but the associated articulation of structure in abstract terms might be seen as problematic. Even if a response to the Newman objection can be given, this notion of structure seems a very thin peg on which to hang one's realist hat. Let's leave that concern for now, as I shall try to suggest that Russell's structuralism, and any form of ESR that draws heavily upon it, is inadequate on the grounds that it fails to take into account the implications of quantum mechanics.

Thus I want to shift the focus of the debate a little by suggesting a new way of looking at Russell's book, in terms of its status as a historical document, occupying a particular point in the entwined history of physics and philosophy. Consider the date of publication: 1927, the year Heisenberg formulated his famous Indeterminacy Principle and when he and Fermi and Dirac set the new quantum statistics into the formal framework we've inherited (namely that relating to the symmetry properties of wave-functions). It was written in 1926, the year Schrödinger published the last of his classic papers on wave mechanics and a year after Heisenberg, Born, and Jordan presented their alternative matrix mechanics. Russell's book thus sits on the cusp of the quantum revolution and given his stated intent to capture the essence of the new physics and construct a fitting epistemology for it, it can be regarded as a kind of literary lens through which the new quantum mechanics can just be seen emerging into the public sphere.

[24] Just to jump ahead, my own view is that Melia and Saatsi (2006) have successfully blunted the impact of this objection.

In particular, given the metaphysical implications of the theory, Russell records Heisenberg as having argued that electrons do not have 'the degree of immediate reality of objects of sense' but 'only the sort of reality which one naturally ascribes to light quanta' (Russell 1927: 45) and thus, given the new quantum physics, it is 'in principle impossible to identify again a particular corpuscle among a series of similar corpuscles' (Heisenberg, quoted in Russell 1927: 46). Having said that, the full implications of this new physics as expressed in what became the 'received' view that quantum particles should be regarded as non-individuals (French and Krause 2006: ch. 3) are nowhere apparent and even the aforementioned hints and glimpses have been overlooked by modern-day structuralists harking back to their Russellian past. In so doing, they have skipped over a whole history—a hidden history—of forms of structuralism which explicitly attempted to accommodate these implications. These forms have been effectively obscured by Russell's shadow.

If we fast-forward merely one year, for example, to 1928, we find Eddington explicitly incorporating the implications of quantum statistics into the group-theoretic (and 'subjective') structuralism he had developed in response to General Relativity. Before we turn to this variant of the structural tendency and consider what we might take from it, let me again briefly touch upon a further strand within our historical narrative.

4.4 Wigner, Weyl, and the Application of Group Theory to Quantum Statistics

The history of quantum statistics (see French and Krause 2006: ch. 3 for further details) can be traced back to Planck's original 1900 paper that began the whole revolution,[25] but its significance for us lies with the developments from 1925 to 1927. It was then that the accepted forms of these statistics[26]—namely Bose–Einstein and Fermi–Dirac, applying to photons and electrons, for example, respectively—were formally articulated in terms of the symmetry features of the relevant wave-functions (Bose–Einstein statistics arising from symmetric wave-functions for an assembly of particles, and Fermi–Dirac arising from anti-symmetric wave-functions).

As we noted in Chapter 2, the traditional, or 'Received', view of the difference between classical and quantum statistics is that whereas in the former case the

[25] Although Kuhn argued that Planck did not fully appreciate what he had wrought and that it was Einstein and Ehrenfest in 1905 and 1906 respectively, who understood that he had introduced something fundamentally different from the classical statistics of Boltzmann and in that sense, the beginning of quantum statistics can be identified with their work (Kuhn 1978).

[26] As I've already noted, other forms of statistics—known as parastatistics—are theoretically possible and were anticipated by Dirac but despite some interest in these following the suggestion in 1964 that quarks might be paraparticles, it is generally accepted that all quantum particles are either bosons or fermions (see French 1985 and French and Krause 2006: ch. 4 for further discussion of the history of parastatistics).

counting of permutations is metaphysically underpinned by the individuality of the particles, the fact that permutations must not be counted in quantum statistics indicates that quantum particles are, in some sense, *non-individuals* (for a formal and philosophical articulation of this notion, see French and Krause 2006). This Received View was expressed almost simultaneously with the birth of quantum statistics itself—by Born, for example, in his 1926 paper (Born 1926)—and it is an interesting historical exercise to trace its diffusion through the secondary literature, both physical and philosophical. As we noted, this non-individuality is not a necessary consequence of quantum theory since one could, in fact, maintain that quantum particles are individuals, albeit at a certain metaphysical cost (French 1989 and 1998; French and Krause 2006; van Fraassen 1989). Indeed, this is the basis of the metaphysical underdetermination that helps to motivate OSR, as we saw in Chapter 2.

Given the role of symmetry in formally articulating the new quantum statistics, it should come as no surprise that the history of the latter is intertwined with that of the development of group theory (the following is taken from French 1999; French 2000b; Bueno and French 1999; Bueno and French forthcoming). This group-theoretic strand can be decomposed into two programmes (Mackey 1993): the 'Weyl programme' was initiated by Weyl's 1927 paper which used group theory to provide a formal basis for the Heisenberg commutation relations and was generally concerned with the group-theoretic elucidation of the foundations of quantum mechanics in general. The 'Wigner programme', on the other hand, was more concerned with the solution of dynamical problems by focusing on the underlying invariances of the situation and thus applied group theory to the construction of quantum mechanical explanations of physical phenomena. Despite the name tags, both Weyl and Wigner contributed to each of these programmes,[27] with Wigner, for example, emphasizing the dual role played by group theory in physics: the establishment of laws—that is, fundamental symmetry principles—which constrain the laws of nature;[28] and the development of 'approximate' applications which allowed physicists to obtain results that were difficult or impossible to obtain by other means.

As Wigner subsequently emphasized, the initial stimulus for these developments was the work of Dirac and Heisenberg on quantum statistics (Wigner 1959: vi).[29] As I have just noted, this work emphasized the connection between such statistics and the symmetry characteristics of the relevant states of the particle assemblies, where such symmetry characteristics were associated with the non-individuality of the

[27] Although Wigner was emphatic that he never interacted with Weyl (1963).

[28] And we shall return to this role of symmetry principles later.

[29] For a useful discussion of the origins of Wigner's application of group theory to quantum mechanics, see, for example, Chayut 2001; further historical insights can be gleaned from his interview with Kuhn (Wigner 1963).

particles. As Weyl put it in his inimitable fashion, in the book that set out the 'Weyl programme':

the possibility that one of the identical twins Mike and Ike is in the quantum state *E*1 and the other in the quantum state *E*2 does not include two differentiable cases which are permuted on permuting Mike and Ike; it is impossible for either of these individuals to retain his identity so that one of them will always be able to say "I'm Mike" and the other "I'm Ike". Even in principle one cannot demand an alibi of an electron! (Weyl 1931: 241)[30]

This loss of an identifying 'alibi' was associated with a fundamental new symmetry property, namely invariance under permutation.[31]

Now, consider such an assembly of indistinguishable particles, such as electrons in an atom. The central problem in understanding the behaviour of such an assembly has to do with the effect of some (small) perturbation of the relevant Hamiltonian for the assembly on the known eigenvalues of that Hamiltonian. For 3 or fewer particles this problem could be solved by elementary means, but for greater than 3 Wigner noted that the theory of group representations as applied to the permutation group could be used to determine the splitting of the eigenvalues of the original Hamiltonian under the effect of the perturbation (Mackey 1993: 242–6). Multidimensional representations give rise to multiple eigenvalues of the appropriate Hamiltonian, which split under the effect of the perturbation.

This use of group theory hinges on the fundamental relationship between the irreducible representations of the group and the sub-spaces of the Hilbert space representing the states of the system. Under the action of the permutation group that Hilbert space decomposes into mutually orthogonal sub-spaces corresponding to the irreducible representations of this group. The symmetric and anti-symmetric representations are the most well known, corresponding to Bose–Einstein and Fermi–Dirac statistics respectively, but as already noted, others, corresponding to so-called 'parastatistics', are also possible, although not, it seems, exemplified in nature.

A further fundamental atomic symmetry is rotational symmetry (ignoring inter-electronic interactions). Again group representations can be appropriately utilized to label the relevant eigenstates and here Wigner appealed to results established by Schur and Weyl who had extended the theory of group representations from finite groups to compact Lie groups. Thus, in his three classic papers of 1925 and 1926, Weyl established the complete reducibility of linear representations of semi-simple Lie algebras. This allowed the irreducible representations of the three-dimensional

[30] As well as contributing to both the mathematics and the physics at this time, as Wigner also did, Weyl is also interesting because of the role that Husserl's phenomenology played in the development of these contributions. As Ryckman notes, there is a hidden history here that is generally unacknowledged but yet is crucial for understanding Weyl's account of General Relativity and his early articulation of the principle of gauge invariance (2003a and 2005; Bell and Korté 2011; see also Tonietti 1988).

[31] As Weyl himself emphasized in his non-technical presentation of 1929 (1968: 268) and also in his 1938 paper on symmetry (1968: 607–8).

pure rotation (or orthogonal) group to be deduced. Note, again, the relevant *dates* of publication here: not only was quantum physics under construction at this time, but so were the relevant features of group theory.[32]

In another 1927 paper, Wigner presented a systematic account of the application of group theory to the physics of the energy levels of an atom that covered both the permutation and rotation groups. In the following year, the newly proposed notion of spin was incorporated into the analysis using Weyl's 'double valued representations' of the rotation group (see Wigner 1927: 157–70, and Judd 1993: 19–21). These results were then presented in systematic fashion in Wigner's 1931 book *Group Theory and its Application to the Quantum Mechanics of Atomic Spectra*. It is to this work that elementary particle physicists returned in the 1950s when they 'rediscovered' Lie algebras and group-theoretical techniques in general.

Wigner's works are cited by Weyl in the latter's 1927 paper on group theory and quantum mechanics (Weyl 1927), and several years later, in 1939, Weyl refers to Wigner's 'leadership' in this context[33] (1968: 679). However, Weyl was careful to point out that his work takes a completely different direction from Wigner's. In the former's classic 1928 book, *The Theory of Groups and Quantum Mechanics*, one can find both the 'Wigner' and 'Weyl' programmes represented.[34] Thus, with regard to the latter, the central idea was to represent the 'kinematical structure' of a physical system via an irreducible Abelian group of unitary ray rotations in Hilbert space, with the real elements of the algebra of this group representing the physical quantities of the system (1931: 275). Heisenberg's formulation then follows 'automatically' from the requirement that the group be continuous and, in particular, the requirement of irreducibility gives the relevant pairs of canonical variables. Weyl also concludes that only one irreducible representation of a two-parameter continuous Abelian group exists, namely the one that leads to Schrödinger's equation. Thus, the fundamentals of quantum mechanics appear to simply drop out of the group-theoretic approach[35]

[32] Of course the physics was already articulated mathematically to a certain degree, although in non-group-theoretical terms. What this gave was a rather rough and ready collection of models, principles, and heuristic rules (including, for example, the 'Aufbauprinzip', Heisenberg's Uncertainty Principle, Pauli's Exclusion Principle, and so forth), which, as Weyl subsequently noted, could be brought under a unifying mathematical framework via group theory. An alternative framework was, of course, provided by von Neumann's introduction of Hilbert spaces. These contrasting developments are examined further in Bueno and French (forthcoming).

[33] Specifically with reference to the decomposition into irreducible invariant sub-spaces using Young's symmetry operators.

[34] It is probably fruitless to speculate which of these books, Weyl's or Wigner's, was the more influential. On the one hand, Eckart's important paper 'The Application of Group Theory to the Quantum Dynamics of Monoatomic Systems' (Eckart 1930) relies heavily on Weyl, and the latter's book is the only work cited by Dirac in the Introduction to his *The Principles of Quantum Mechanics*. (Thanks to James Ladyman for pointing this out.) On the other hand, many people found Weyl's work difficult to penetrate and the resurgence of group-theoretic considerations in the 1960s can be traced back to Wigner. Wigner himself offers a personal recollection of the rivalry between the two in Wigner 1963.

[35] For a discussion of the significance of Weyl's results and its connection with subsequent important work in group theory, see Mackey 1993: 249–51 and 274–5.

and, Weyl maintained, 'The theory of groups is the appropriate language for the expression of the general qualitative laws which obtain in the atomic world' (1968: 291).

There is of course more to say here, but the upshot is that the group-theoretic approach appeared to deliver an embarrassment of riches: on the Weyl side, it gave both the Heisenberg commutation relations and Schrödinger's equation; on Wigner's, it not only provided the classification of atomic line spectra, taking into account the exclusion principle and spin (Chapter 5),[36] but also a formal understanding of the nature of the homopolar molecular bond (1931: 341–2) and chemical valency in general (1931: 372–7).

The last in particular demonstrated the power of permutation symmetry. It was clear that the attraction between two hydrogen atoms could not be accounted for in terms of Coulomb forces. The solution to the problem lay with the non-classical exchange integral, introduced by Heisenberg. The understanding of this concept underwent a shift from the idea of a literal exchange of electrons to its conceptualization in terms of the application of the permutation group (Carson 1996a). On the basis of an understanding of the electrons as indistinguishable, Heitler and London noted that within the group-theoretic framework the electronic wave-function of the two-atom system could be written in either symmetric or anti-symmetric form. With the electron spins incorporated as anti-parallel, and the anti-symmetric form chosen (corresponding to the electrons occupying the relevant fermionic sub-space), one obtains a state of lower energy and hence attraction. Thus chemical valence and saturation could be understood and the 'problem of chemistry' solved, leading Heitler to declare, famously, now '[w]e can . . . eat Chemistry with a spoon' (Gavroglu 1995: 54).[37]

Of course, not everyone was so taken with this group-theoretic approach,[38] although its significance subsequently re-emerged in the context of post-war

[36] Referring to developments in spectroscopy, Weyl writes 'The theory of groups offers the appropriate mathematical tool for the description of the order thus won' (1931: 245). Wigner also did important work in the application of group theory to solid state physics.

[37] In his discussion of the physical basis of chemical valence Weyl presents the relationship between chemistry and quantum physics in terms of a hierarchy of structures (Gavroglu 1995: 266–75) and—interestingly, given what I say in Chapter 12—goes further by suggesting a structural 'mediation' between biology—in particular, genetic diagrams—and physics via 'the simplest combinatorial entity', namely the permutation group. Here the elements are the genes, of course, the different discrete states are the alleles, and union and partition then correspond to syngamy and meiosis respectively.

[38] Wigner notes that there was a 'certain enmity' at the time (1963). Interestingly, in the context of this book, Wigner also said that, 'most people thought, "Oh, that's a nuisance. Why should I learn group theory? It is not physical and has nothing to do with it." People like to think of motions, which is not, in my opinion, and which even in that day was not, in my opinion, the right way to think about stationary states. Nothing moves, and this is what I think I digested much earlier than most people; in a stationary state nothing moves, but this is what they did not want to accept. They said, "Well, you see something going around," when actually you don't. For instance, my shells did not move, and it was evident to me that nothing moves' (1963: transcript 2). Just to bang the point home: the idea of orbits, in the sense of something moving around, may have been a useful heuristic device in the context of the old quantum

elementary particle physics. However, the crucial point is that due to the publication date of Russell's great work, these developments in the application of group theory to quantum mechanics came too late, and have been overlooked by modern-day structuralists, with their eye only on Russell and (parts of) Poincaré, as well as by their critics. However, they were explicitly incorporated into the explicitly structuralist accounts of physics of Eddington and Cassirer whose views have, until recently, been overshadowed.

4.5 Eddington's Subjective Structuralism[39]

The basis of Eddington's structuralism in his understanding of relativity theory is well known (see Cometto 2009; Kilmister 1994; Ryckman 2005: chs 5 and 7): rejecting the usual foundation of clocks and rods as inappropriate for a structuralist reconstruction of the physical world in what he called 'strict analytical development',[40] he began with four-dimensional point events and the intervals between them and by a complicated analysis obtained Einstein's field equations relating space-time/gravity on the one side, with matter/energy on the other.[41] Not only the structuralist but also the subjectivist aspects of Eddington's position are revealed in this analysis: by reducing substantial matter to the 'unevenness' in the gravitational field, non-structural substance was eliminated from our ontology in favour of relational structures. However, this unevenness is but one of the many possible relations that could hold between the point events of the world and in his 1920 contribution to the International Congress of Philosophy, Eddington draws an analogy with the construction of constellations out of the distribution of the stars, with the distinction between substance and 'emptiness' arising from the role of the mind in recognizing certain kinds of patterns (1920: 420).

As he was later to express it, what this amounts to is 'a selection from the patterns that weave themselves' (1928: 241).[42] Here Eddington seems to be acknowledging the

theory of Bohr–Sommerfeld, but it founders on the very notion of a stationary state which, of course, was Bohr's crucial innovation in the first place. It was not until the 'second' quantum revolution of 1925–1927 that this notion came to be formally explicated in terms of the new quantum mechanics. Following this, some alternative understanding had to be obtained, according to which 'nothing moves' and as Wigner makes clear, for him this was to be found in group theory.

[39] This is a summary of French 2003a.

[40] This reflects an important issue that surfaced repeatedly throughout the development of Eddington's programme and that has obvious significance for structuralism in general: when one is engaged in an ontological construction like this, on what basis should one begin? As a structuralist, Eddington certainly did not want to begin with a foundation that presupposed the very material that his structuralist programme aimed to eliminate (and as we shall shortly see, it was substance that he had in his sights).

[41] This glosses over a complex 'cycle of reasoning' (Ryckman 2005: 7.5.3) by which Eddington sought to provide an explanation of gravitation. This cycle can be usefully compared with Cassirer's non-hierarchical arrangement of laws, symmetries, and measurements, touched on in the next chapter.

[42] This articulation of structure in terms of 'patterns' is one that recurs through various forms of structuralism (see, as a sample, Resnick 1997; Ladyman, Ross, et al. 2007; Wallace 2003). However useful

fundamental intuition underlying Newman's argument and we can already see how he might have responded to it by appealing to the subjectivist element in his philosophy. However, as we shall see, when he is brought face-to-face with Newman's criticism 12 years later, he adopts a different tack.

By the time of his 1927 Gifford lectures (1928), Eddington had adopted an important understanding of his structural 'building material', one that was insufficiently appreciated at the time (it fails to register in the early criticisms of Braithwaite (1929) and Heath (1928),[43] for example). As indicated previously, the world structure, for Eddington, consists of events, with intervals holding between them. If this structure is understood in relational terms, then the point events are the relata and the intervals are the relations and it would seem as a matter of conceptual necessity that the relata be taken as metaphysically prior to the relations, since according to the usual understanding of these matters, we can have relata without relations but not vice versa. On a not-so-usual understanding (but one that features prominently in consideration of OSR, as we have seen and shall articulate further in subsequent chapters), we might regard the relata as metaphysically derivative in the sense of their being constituted from the structure by some process of identification (Heath 1928 suggests such an approach; see also Mertz 1996). However, Eddington himself insisted that the relata and the relations come together as a package:

> The relations unite the relata; the relata are the meeting points of the relations. The one is unthinkable apart from the other. I do not think that a more general starting-point of structure could be conceived. (1928: 230–1)

As we shall see, this remark holds the key to understanding Eddington's structuralism and in this respect it bears a striking similarity to certain forms of 'moderate' structural realism that have recently been proposed (Esfeld and Lam 2008, 2009, 2010).

However, there was still a problem, namely the fundamental 'lumpiness' or particularity of matter as expressed by quantum theory that Russell had failed to adequately incorporate into his structuralism. Eddington was quite explicit that in order to understand how it is that the same quality that is chosen by the mind as that

this may seem as a lay-friendly attempt to convey the core insight of structuralism—as it certainly was for Eddington—this book seeks to go beyond it to a more detailed metaphysical articulation. Some commentators have also dismissed the 'weaving themselves' feature but this is no more than an expression of some form of realism: the patterns themselves are not woven by us—they are 'out there' in the world. Of course, for Eddington, the non-realist, subjectivist aspect enters in the choice of one such pattern from all those that are apparently available. I shall argue that we can avoid having to include such an aspect within OSR by eliminating the element of choice and taking the 'extra' patterns that appear to be available as so much 'surplus' structure.

[43] Heath usefully contrasts Russell's and Eddington's forms of structuralism and, interestingly, also touches on the impact of quantum mechanics in suggesting that the development of matrix mechanics might be regarded as an example of the replacement of substance by (law-like) function.

which we call matter[44] is also singled out by Nature for the property of atomicity, we must understand how relativity theory and quantum physics can be related. I won't go into the details here of the bridge he built between the two, based on his wave-tensor calculus, the manipulation of which appeared to give—*analytically*—the values of certain fundamental physical constants, nor will I spell out the role of quantum non-individuality in this analysis (see Kilmister 1994). What I do want to do is describe Eddington's non-standard understanding of this non-individuality and how it was absorbed into his group-theoretic structuralism (for further details see French 2003).

4.6 Scribbling on the Blank Sheet

Eddington's understanding of the implications of quantum mechanics for the notion of objects as individuals went beyond that of other physicists in that he not only took all particles of the same kind—electrons, for example—as being absolutely indistinguishable (in the sense of possessing all their state-independent or intrinsic properties such as charge, rest mass, spin... in common) but regarded particles of *different* kinds similarly; that is, he took protons to be absolutely indistinguishable from electrons, even though they apparently possess different state-independent or intrinsic properties (such as and most obviously mass and charge). This might sound bizarre but it is important to appreciate the nature of Eddington's programme on this point: he is seeking to 'analytically reconstruct'—or less contentiously, perhaps, *represent*—the world in entirely structuralist terms and thus cannot admit, at the most basic level, any features that might be deemed as non-structural. Any such features must be shown to arise or be derived from the fundamental structuralist basis. In the radical nature of its stance, Eddington's approach here offers a nice contrast to the 'Poincaré Manoeuvre' sketched in section 4.2.[45]

Thus *the* fundamental epistemological principle underpinning Eddington's work is that of the 'Principle of the Blank Sheet': in order to get the analytic reconstruction of the world going, we must first formulate some kind of background in terms of which physical phenomena can then be distinguished (Eddington 1936: 32). Precisely such a blank sheet is provided by the intrinsically indistinguishable, non-individual particles of quantum theory *and* the framework of space-time described by General Relativity (1936: 33 and 56) which then allow the relevant physical differences to be introduced openly rather than smuggled in via the initial assumptions.

[44] And the crucial feature here is that of *permanence*, expressed via 'Hamiltonian derivatives', which are a kind of generalized differential quotient, obtained by considering the variation of the action integral with respect to small changes in the fundamental field variables; see Ryckman (2005: 7.5.2).

[45] Having said that, I don't actually think one has to adopt such a thorough-going form of structuralism in order to deal with these concerns.

From this point of view, protons and electrons begin life, as it were, as completely indistinguishable units, to which various attributes—intrinsic properties—are added as the analysis proceeds:

> The Principle of the Blank Sheet requires that at the start we should recognise no intrinsic distribution between the particles which we contemplate, in order that we may trace to their very source the origin of those distinctions which we recognise in practical observation. The fundamental dynamics is the dynamics of indistinguishable particles; the dynamics of distinguishable particles is a practical application to be used when we do not wish to analyse the phenomena so deeply. (1936: 287)

As far as Eddington is concerned, such a unit cannot be taken as separate or disassociated from the system of analysis of which it is a part. The conceptual bundling together of relata and relations can then be given a mathematical gloss: 'As a structural concept the part is a symbol having no properties except as a constituent of the group-structure of a set of parts' (1939: 145).

Moving away from quantum physics, however, to the 'everyday' level of macroscopic objects, our understanding of these objects appears to represent them as more than merely group-theoretic elements. Eddington expresses this understanding in terms of 'general' concepts, from which structural concepts are obtained by eliminating everything that is not essential to the role the concept plays in a group-structure. If the structural concept becomes a mere element, denoted by a mathematical symbol, then a general concept 'is our conception of what the symbol represents in our ordinary non-mathematical form of thought' (1939: 144). However, such concepts may be no more than forms of 'self-deception' which persuade us that 'we have an apprehension of something which we cannot apprehend' (1939: 144). Thus, for example, we have a general concept of an object as an individual, which is so ingrained as a form of thought (Eddington refers to it as a 'legend of individuality') that we export it from the everyday to the quantum realm and are persuaded that we have an apprehension of that which we cannot apprehend.[46] In fact, all that we *can* apprehend is the relevant group-theoretic structure. This is a fundamentally crucial point: it is such 'legends' or general concepts that bedevil our attempts to arrive at an appropriate conception of the world that modern physics presents to us and lead to the kind of metaphysics that Ladyman and Ross excoriate (Ladyman, Ross, et al. 2007: ch. 1).

Even more radically, *existence* itself was given a structural interpretation (1946: 266) and so every metaphysical feature of the particles, as physical objects, was subsumed within the group-theoretic structure. Let me just explain a little what

[46] Interestingly, given recent discussions in metaphysics on whether there exists a 'fundamental level', Eddington argued that the 'legend of individuality encourages the view that the process of analysis has a terminus (in the individuals) but if there are none such then there is no reason to suppose that the process will ever have to stop for metaphysical reasons. We may decide to stop once we have achieved our analytical aims but that is another matter entirely' (1939: 144).

Eddington meant by this as, again, it offers something that we can bring forward to current discussions, as we shall see in Chapter 7.

Eddington felt that statements such as 'Tables exist' were nothing but 'half-finished' sentences which require completion in structuralist terms. Thus atoms and electrons, for example, 'exist', in this sense, in the physical universe; indeed they are analysed as structural parts of it. But what about the physical universe itself, does that exist? To say that it does would result in another half-finished sentence (for what further structure could the physical structure be a part of?). Indeed, Eddington saw it as an advantage of his approach that this question never arises: having described the nature of physical knowledge, understood itself as a description of the physical universe, nothing further would be added to our knowledge of it if one were to say 'and the physical universe exists'. In this manner he repudiated 'any metaphysical concept of "real existence"' (1939: 162) and introduced in its place a 'structural concept' of existence (see also 1946: 266). The structure here is simple, indeed the simplest possible, consisting of only two values: existence and non-existence (of course). This can be represented mathematically, in terms of two eigenvalues, 1 and 0, and hence, '[t]he structural concept of existence is represented by an idempotent symbol' (1939: 162). In this representation and in further work, Eddington comes close to the occupation number interpretation of quantum field theory (see French 2003a: 250–1).

As we can see, Eddington's structuralism really was all-embracing and it should come as no surprise that he had a dismissive response to the obstacle thrown up against Russell's account, namely the Newman problem.

4.7 The Battle with Braithwaite

What we have, then, is the following picture: the objects of physics—elementary particles—and the structure—represented by group theory—come as a package. The apparent individuality of the particles as objects, ordinarily conceived, is nothing more than a 'legend' which results when our ordinary frameworks of thought are transformed by the mathematics relevant to quantum theory. This 'legend' is exposed, or demystified one might say, by the Principle of the Blank Sheet which dictates that the so-called 'intrinsic' properties of particles, such as mass, charge, spin, etc., are merely aspects of structure.

However, a fundamental dichotomy between structure and content was discerned as underlying Eddington's position.[47] Thus, Braithwaite argued, the set of elements of a group do not form a group 'in themselves', but only with respect to a 'given mode of combination' (Braithwaite 1940). So, for example, a set of numbers do not form a group on their own, but only under a specified operation, such as addition or

[47] A dichotomy that has also been erected as the focus of criticism in recent discussions of structural realism, as we have seen.

multiplication. It is only by specifying the group relation, or mode of combination, that we actually have a group to begin with. But such a specification introduces a non-structural element into our structuralism, because we have to have some ground—which clearly cannot be structural itself—for selecting one mode of combination over another. Consider, for example, the rotation group—one of the most important in physics—for which the combining relation consists of performing successive rotations. Braithwaite invites us to consider a *different* mode of combination, such as, for example, that of expressing the rotations by symbols written down on the same chalkboard (1940: 462). With respect to *that* mode of combination, the rotations do not form a group at all. Hence, we must specify always the relevant mode of combination, but how can we do that except by appealing to something that is non-group-theoretical and hence non-structural? Thus Braithwaite writes,

> To say that two sets of things have the same group-structure would be to say nothing of interest unless the modes of combination of both the groups had been specified. The fact that structure depends upon content is one reason why the structure–content dichotomy of knowledge is untenable. (1940: 463)

In other words, the group-structure is only given once the relevant transformations have been specified (i.e. whether we're talking about rotations or permutations, for example), but to do this is to supply *content* and so we no longer have pure structure.

Furthermore, in a footnote to the passage just quoted, Braithwaite refers to Newman's argument and insists that,

> his [Newman's] strictures are applicable to Eddington's group-structure. If Newman's conclusive criticism had received proper attention from philosophers, less nonsense would have been written during the last twelve years on the epistemological virtue of pure structure. (Braithwaite 1940: 463)

Tackling Braithwaite's argument head-on, Eddington (1941) pointed out, first of all, that group theory enters physics as a way of expressing the relationships between relations and that whatever the nature of the entities, the use of group theory allows us to abstract away the 'pattern' or structure of relations between them. What the group-structure represents, then, is the 'pattern of interweaving' or 'interrelatedness of relations' (1939: 137–40), such as is represented by rotations acting on rotations and expressed in the associated group multiplication table. From this perspective we lose the distinction between the nature of the element and the nature of the combining relation which—according to Braithwaite—makes it an element of the group: '*The element is what it is because of its relation to the group structure*' (Eddington 1941: 269; his emphasis). We recall again Eddington's view that the relata and relations come as a package and the more general point that the unit cannot be disassociated from the system of analysis of which it is a part. Braithwaite's error is to conceptually separate out the relations from the group elements; indeed, this has the effect of rendering the latter 'impotent'. Eddington insists:

I am rescuing out of the mathematical formalism what is for physical purposes the most essential feature of the group conception of structure, namely, that primarily the elements of a group (or ring or algebra) are defined solely by their role in that group (or ring or algebra). Therefore when Braithwaite argues that it is possible to regard the elements of a group in such a way that they are not elements of a group, I answer that there is *no* other way of regarding them. Unless we import qualities not inherent in them by definition (by adopting a special realisation or representation of the group [and as he notes he means this in the non-technical sense] there is nothing to lay hold of that *could* be regarded from another point of view. (1941: 269)

As he notes, when one considers the *representation* of the rotation group, the relevant 'combining relation'—that of performing successive rotations—can be stated explicitly. Braithwaite then invited us to consider an alternative 'mode of combination' with respect to which the rotations would not form a group. But as Eddington points out, this is bizarre: if the combining relations were different, we would no longer be talking about *that* group representation. Indeed, he finds Braithwaite's example of expressing the rotations by symbols on a chalkboard unilluminating, since the act of 'laying on chalk' is neither a rotation nor a combination of rotations. How, then, are we to understand it? Eddington's answer is that this unconnected writing down of two symbols is not intended to be symbolic of anything in the physical world; it is merely a 'memorandum of the content of the writer's mind' (1941: 270). Having formed a mental concept of a rotation, holding another, disconnected, in thought is not indicative of introducing a new mode of combination but simply of holding two such concepts as possible alternatives.[48]

He went on to illustrate the difference between his view and Braithwaite's in terms of how the 'symbolic language' of mathematics should be understood (1941: 270). Consider an abstract group with elements a, b, c... whose structure is represented by equations like $c = ab$. Braithwaite would extract and make explicit the 'combining relation' by rewriting this equation as $g = a.b$ (in extracting the relation, the relata have been changed and must now be represented differently), where there is now an extra symbol, '.', expressing the mode of combination. This opens the door to the possibility of introducing an alternative symbol, say, yielding a combination $a{:}b$ not equal to g, so the group structure cannot apply to a, b, g intrinsically. Hence, Eddington writes, Braithwaite's conclusion would be,

The elements a, b, g do not form a group apart from their combining relation; therefore we can have no structural knowledge of things like a, b, g—so that's the end of structuralism. (1941: 270)

[48] It is via such holding of 'conceptual alternatives', according to Eddington, that probability is introduced into physics, but I won't consider that further here (although it relates to the discussion in Chapter 11).

Eddington observes the same process but draws precisely the opposite conclusion:

> The elements *a*, *b*, *g* do not form a group apart from their combining relation; therefore our structural knowledge is about things like *a.*, *b.*, *g.*—so that's the beginning of structuralism (1941: 270)

The point is, just because we can represent the combining relation in the symbolic language of mathematics does not mean that this relation is actually somehow 'detachable' from the relatum when the symbols are used to represent an element of our knowledge of the external world. To suggest that it can be is to fall prey to a '*suggestio falsi*' which deludes us into detaching a 'meaningless' *a* from its mode of combination; and here Eddington draws a comparison with the way in which ordinary language deludes us into detaching a 'meaningless sensum' from its mode of combination in sensation (1941: 271).

With regard to Braithwaite's conclusion that the structure–content dichotomy is untenable, Eddington insisted that this is precisely to miss the point, for there simply is no non-group-theoretic *content* to 'lay hold of'.[49] Thus he agreed that there is no structure–content dichotomy, not because structure depends on content but rather because it is content—as represented in this case by Braithwaite's understanding of the elements—that depends on and can be eliminated in favour of structure! These differences between Braithwaite and Eddington resonate down to the modern context.

This brings us, finally, to the Newman argument, and Eddington took Braithwaite's deployment of it as evidence that he hadn't in fact grasped the core of Eddington's structuralism. That this is different from Russell's should be clear:[50]

> Russell, in his pioneer development of structuralism, did not get so far as the concept of group-structure. He had glimpsed the idea of a purely abstract structure; but since he did not concern himself with the technical problem of describing it, he had no defence against Newman's criticisms. Russell's vague conception of structure was a pattern of entities, or at most a pattern of relations; but the elements of group theory make it clear that pure structure is only reached by considering a pattern of interweaving, i.e. a pattern of interrelatedness of relations. (Eddington 1941: 278)

Consider, again, that 'pattern of interrelatedness' as manifested in the multiplication table associated with the rotation group (1941: 278). The information encoded in such a table is not trivial at all and hence Eddington concluded that there is no foundation to Braithwaite's contention that the Newman objection applies in this case. Indeed, he accused Braithwaite of having failed to grasp 'the main idea' of the kind of structuralism he was advocating.

The manner in which such structural information is not trivial is revealed by the example of spin where the information encoded, as indicated previously, in the relevant structure gives all the information we can get (1941: 279). At this point

[49] A point that can also be made against Psillos' distinction between 'content' or 'nature' and 'structure'.
[50] See also Ryckman 2005: 7.6.1.

Eddington deployed a version of the Poincaré Manoeuvre which, we recall, amounts to assuming certain non-structural elements in order to be able to articulate the structure in the first place, only to discard or—perhaps better—reconceptualize these elements once the structure has been constructed. Thus, Eddington acknowledged that the components of spin can be specified in a set of mutually orthogonal planes (corresponding to spin in the x-, y-, and z-directions) and also that this represents non-trivial knowledge. Braithwaite would object that this knowledge is non-structural because we are acquainted with such orthogonal planes in the 'external' world. However, taking the set of operations represented by rotations through 90° in each of the planes, we obtain a group-multiplication table which Eddington understood as defining the relevant structure and now, he insisted, '[w]e need... trouble no further about the planes' (1941: 279). In other words, we initially associate the components of spin with the planes but this is just a kind of heuristic move (we could equally well have associated them with unit rotations in the plane) that takes us to the group-multiplication table, which in turn represents what is important, namely the structure. The information encoded in the latter is definitely non-trivial, even if 'reticent', since it conflicts with other statements, some plausible, but the apparent non-structural knowledge acquired by our acquaintance with the planes is in fact ungrounded. The appearance of a non-structural component is illusory, deriving from the heuristic role played by certain objects.

As suggested in section 4.5, much of Eddington's structuralism can be lifted free of his subjectivism and deployed in a realist context. Broad, for example, certainly insisted that the two could be separated: 'I do not think there is much connection between the "selective subjectivism" and the "structuralism" of Eddington's theory. Of course both of them may be true. But the structuralism might be true and important, so far as I can see, even if the selective subjectivism were false or greatly exaggerated' (1940: 312). It is not so clear whether such a clean separation can be achieved in Cassirer's case but nevertheless I believe that we can also extract certain features of his account and take them forward into the current context.

4.8 Cassirer's Kantianism[51]

Neglected for many years (at least by those in the Anglo-American 'analytic' tradition), Cassirer's neo-Kantian philosophy has become the subject of renewed interest in recent years (Friedman 2000; 2004). However, although Cassirer's philosophy in general and its application to General Relativity in particular have been quite widely discussed (see Ryckman 1999 and 2005), his analysis of quantum theory in *Determinism and Indeterminism in Modern Physics* (Cassirer 1936) has not received the

[51] The following is taken from Cei and French 2009. I am hugely grateful to Angelo Cei for his help in understanding Cassirer's philosophy.

attention it deserves.[52] Originally published (in German) in 1936, well after the quantum revolution had evolved into 'normal' science, and then republished (in English) 20 years later, this book explicitly incorporates the implications of quantum mechanics with regard to the individuality of objects.[53] Indeed, Cassirer argues that these are the principal implications of the newly established quantum theory that must be addressed by philosophers, rather than those concerning causality and determinism, and it is by focusing on the former that one is able to defend a neo-Kantian stance against what many had taken to be the devastating impact of quantum physics. As far as Cassirer is concerned, it is the notion of a substantival object that must be given up in the face of this impact, rather than the principle of causality, and a broadly structuralist—if Kantian—understanding of objectivity adopted. Here I shall sketch Cassirer's view, emphasizing the structuralist elements, of course, and highlighting the features that I shall draw upon in later chapters.

4.9 From Kant to neo-Kantianism

Cassirer's form of neo-Kantianism evolves from the Marburg School's interpretation of Kant,[54] according to which the fundamental principles of theoretical natural science express the universal *patterns* by means of which thought orders the manifold of phenomena. There are three features of this interpretation that are crucial for understanding this version of transcendental idealism:

a) Science and the objectivity associated with it are to be understood as *facts*. Such facts are the explananda of a philosophical theory of knowledge whose questions are how we have knowledge of nature and on what grounds we can maintain that such knowledge is objective. From this perspective, foundational issues in science have primarily an epistemological dimension.
b) In such a picture thought plays a 'constructive' role and broadly speaking objectivity is to be understood as emerging from this constructive activity.
c) The Kantian notion of pure intuition as distinct from understanding, together with the relative doctrine of mathematics as resulting from the insertion of the logical forms of the categories into the pure intuition of space and time, has to be rejected since it is denied by the development of modern mathematics.

(a) and (b) suggest a relativized view of the a priori in the scientific context. However, although different a priori principles will be instantiated in different theoretical frameworks in order to underpin the universal unity and objectivity that those

[52] It is not considered at all in Friedman's otherwise excellent encyclopaedia article, for example (2004). A sketch can be found in Itzkoff 1997: 83–98.

[53] In his introduction to the English edition Margenau presented it as 'ahead of its day; its thesis was revolutionary and radical, not, like so many, philosophical commentaries, a wordy echo of the scientists' own pronouncements' (Cassirer 1936: x).

[54] For Cassirer's relationship with Cohen and the Marburg School see Friedman (2000).

frameworks enjoy (Ryckman, 1999), Cassirer maintained that the 'historical-developmental' sequence of the structures underlying these different frameworks converges (Friedman 2004). Here we can identify a particular commonality with the views of Poincaré and modern structural realists, of course. In this context, as we will see, Cassirer's analysis of quantum mechanics highlighted precisely the kind of assumptions that allow for the construction of objectivity in the quantum domain.

The epistemological framework in which Cassirer situates quantum mechanics is grounded in his peculiar appreciation of the significance of (c). Historically, the rejection of the idea of pure intuition dates back to the crisis of the understanding of mathematics as based on intuition springing from the rise of non-Euclidean geometry (Ryckman 1991; Friedman 2000). The role played by developments in the foundation of mathematics of the late 19th century in shaping Cassirer's approach is twofold. On the one hand, the nature of mathematical concepts is logical and formal and in this sense such concepts play the same role with respect to natural knowledge as that played by the categories in the Kantian framework. They structure the manifold of experience, thus allowing for our knowledge of it. On the other hand, these developments in the foundations of mathematics underpin the revised notion of the synthetic a priori employed in the analysis of quantum theory.

Now, according to Kant, it is within the framework offered by the pure intuition of space and time that the pure logic of understanding encounters the manifold of perception.[55] Neo-Kantianism thus has to explain how this synthesis takes place if there is no pure intuition to act as the general 'theatre'. Cassirer's answer relies on the notion of *Zuordnung* or *functional coordination* (Cassirer 1907a).[56] Such a notion is taken as primitive and fundamental and 'has no other meaning than that of relation and mutual coordination of one thing to another' (Ryckman 1991: 63). Cassirer used this notion to effectively mimic the Kantian understanding of the aforementioned synthesis without making use of the idea of pure intuition:

> [these] same basic syntheses upon which mathematics and logic rest, also govern the scientific structure of empirical knowledge and first enable us, by a fixed lawful ordering of phenomena to speak of its objective significance. (Cassirer 1907: 45; quoted in English in Ryckman 1991: 65)

This notion of functional coordination is modelled on that of function in analysis. According to Cassirer its key role in allowing us to form the fundamental concepts of science has to do with the fact that a function instantiates a general rule or law that relates all the members of the series and that law, rather than being inducible by

[55] We also recall that this is also the core of Kant's explanation of the mathematical nature of physics since the schematization of categories in the pure intuition of time determines the conditions of possibility of arithmetic and the schematization of categories in the pure intuition of space yields geometry.

[56] Ryckman (1991) explores the extent to which the notion of *coordination* was in the early 20th century the focus of a wide variety of analyses of science and identifies in it a further element of commonality between neo-Kantianism and Logical Empiricism.

enumeration of each of the members, can be seen as the fundamental form of each of them (Cassirer 1953). Curiously, group theory does not explicitly feature in his analysis of quantum mechanics, and here we have an interesting contrast with Eddington.[57] Nevertheless, a similar shift away from an object-oriented stance was crucial for Cassirer's account.

4.10 Space-time, Structures, and Group Theory

The analysis of the concept of object is a central theme that runs through Cassirer's writings on physics (Ihmig 1999). And given his neo-Kantianism, the fundamental perspective from which this analysis should proceed is, of course, epistemological:

> epistemological reflection leads us everywhere to the insight that what the various sciences call the "object" is nothing in itself, fixed once for all, but that it is first determined by some standpoint of knowledge. (Cassirer 1953: 356)

As with Poincaré, Cassirer's interest in this issue can be traced back to his reflections on the nature of space and the influence of Klein's Erlanger programme, with its emphasis on the role of group theory. What this yields, we recall, is a structural conception of geometrical objects that shifts the focus from individual geometrical figures, grasped intuitively, to the relevant geometrical transformations and the associated laws.

This shift is manifested in Cassirer's neo-Kantian assertion of 'the priority of the concept of law over the concept of object'. This in turn forms an integral component of Cassirer's interpretation of the Kantian understanding of objectivity:

> For objectivity itself - following the critical analysis and interpretation of this concept - is only another label for the validity of certain connective relations that have to be ascertained separately and examined in terms of their structure. The tasks of the criticism of knowledge ("Erkenntniskritik") is to work backwards from the unity of the general object concept to the manifold of the necessary and sufficient conditions that constitute it. In this sense, that which knowledge calls its "object" breaks down into a web of relations that are held together in themselves through the highest rules and principles. (Cassirer 1913, trans. in Ihmig 1999: 522)

These 'highest rules and principles' are the symmetry principles of physics which represent that which is invariant in the web of relations itself. And these principles, in turn, are represented group-theoretically; thus the relevant group effectively lays down the general conditions in terms of which something can be viewed as an object. We shall return to the analysis of such principles shortly but again, this idea of symmetry as underpinning a structuralist conception of 'object' is a feature of

[57] Interestingly, Cassirer did deploy group theory in his analysis of Gestalt psychology; see Cassirer 1938 and for discussion, Cei and French 2009.

Cassirer's account that can be brought forward into the modern debate, as we shall see in later chapters.

Cassirer's 'application' of this framework to the foundations of relativity theory is well known (Ihmig 1999: 524–8). What it does is restore the *unity* of the concept of object which is apparently undermined by the Lorentz transformations of Special Relativity. From the structuralist perspective, this unity, apparently lost at one level, is 'reinstated on a higher level' (Ihmig 1999: 525) via the 'lawful unity' of inertial systems offered by the Lorentz transformations. The process of abstraction from a substantivalist conception of objects to a structuralist one is further supported by the development of the General Theory of Relativity. Here the role of the principle of general covariance is crucial. According to Ryckman, Cassirer viewed general covariance as a principle of objectivity that offers a 'deanthropomorphized' conception of a physical object (Ryckman 1999), a view which, he (Ryckman) claims, meshed with Einstein's own. As the requirement that the laws of nature be formulated so that they remain valid in any frame of reference, general covariance 'is a further manifestation of the guiding methodological principle of "synthetic unity" necessary to the concept of the object of physical knowledge' (Ihmig 1999: 604). Regarded as a synthetic requirement, general covariance comes to be seen as both a formal restriction and a heuristic guide for the discovery of general laws of nature (Ihmig 1999: 604). Physical objectivity—apparently lost by space and time themselves—re-emerges in deanthropomorphized form in terms of the functional forms of connection and coexistence:

> With the demand that laws of nature be generally covariant, physics has completed the transposition of the substantial into the functional - it is no longer the existence of particular entities, definite permanencies propagating in space and time, that form "the ultimate stratum of objectivity" but rather "the invariance of relations between magnitudes". (Ihmig 1999: 606, citing Cassirer 1957: 467)

What we are left with, then, is an understanding of the objects of a theory as defined by those transformations that leave the relevant physical magnitudes invariant. Thus Cassirer saw General Relativity as a natural outcome of the structuralist tendency and, far from undermining Kantian philosophy, offering further support to it in its neo-Kantian incarnation.

4.11 Quantum Mechanics, Causality, and Objects

Shifting now to Cassirer's analysis of the other major revolution of the 20th century, namely quantum mechanics, as I said, he can be characterized as attempting to protect Kantian philosophy from the impact of quantum theory by demonstrating how a neo-Kantian understanding of causality can be preserved in this new context. In a nutshell, this understanding takes causality to be a general, 'transcendental' principle that refers not to objects, of course, but to our cognition of them (1936/1956: 58). As such, it is a

> guide-line which leads us from cognition to cognition and thus only indirectly from event to event, a proposition which allows us to reduce individual statements to general and universal ones and to represent the former by the latter. (1936/1956: 65)

And from this standpoint, the concepts of chance and causality do not stand in opposition, but rather 'side by side' (1936/1956: 104), in a 'complementary relationship' (1936/1956: 103) which is as it must be if we are to determine an event as completely as possible. In classical physics the relationship is represented by that between 'the course of an event' and knowledge of its initial conditions, or more generally, by that between 'nomological' laws and 'ontological' laws which 'nowhere contradict each other' but, rather, 'interweave', giving rise to the universal form of 'order according to law' (1936/1956: 105). Thus, that which was taken to be *constructive* is now elevated to the status of a *regulative* principle, as in so far as 'the law of causality belongs... to the modal principles, it is a postulate of empirical thought' (quoted in Rudolph 1994: 241).

Thus the challenge posed by quantum physics can be met as long as we cleave to the essential idea that causality expresses 'something about the structure of empirical knowledge' (Rudolph 1994: 114). In particular, quantum mechanics does not dispense with conformity to law, even if 'law' must now be understood as 'statistical' rather than 'dynamical', as in the classical case. The challenge is to our characterization of 'the physical concept of reality' (Rudolph 1994: 128) and in particular, it is the classical concept of object which is undermined.

To get a grip on Cassirer's understanding of laws, and the role of causality, we need to note his central distinction between three 'basic' types of statements in physics: statements of the results of measurements (1936: ch. 3); statements of laws (1936: ch. 4); and statements of principles (1936: ch. 5). The first represent 'that decisive transformation' (Rudolph 1994: 31) from immediate perceptual data to experimental observation, where the latter must be understood as a determination into which concepts of measure and number enter.[58]

Statements of laws effectively join the particular to the whole and they are able to do this through the mathematical concept of function. The move from statements of measurement to statements of laws should be understood as a 'characteristic transformation' from a 'here-thus' to an 'if-then' (Rudolph 1994: 41)[59] and the hypothetical judgements embodied in the latter cannot be regarded as mere summaries of individual facts since they pertain to classes of magnitudes which typically consist of

[58] This transformation is highly complex and here we may perhaps see a 'foreshadowing' of Suppes' characterization of the 'conceptual grinder' which takes us from sense data to data models.

[59] Cf. Weyl on the here-thus and the role of the ego (Weyl 1963).

infinitely many elements.[60] What a statement of law represents is an 'abrogation' of the space-time realm in which individual facts are situated and this 'change of dimension' cannot be captured as mere induction. What, then, grants the passage from the particular to the general? It is here that causality plays its role.

We recall that Hume famously argued that there is nothing to causation over and above the representation of events as successive in time and as constantly conjoined (this is seen as the ancestor of the 'Regularity' view of causality that we shall consider in Chapter 8). Kant responded that the very possibility of such representation implies the working of a rule of ordering that makes possible that succession. Now, temporal ordering is certainly not something we grasp by perception so it must necessarily come from somewhere else, namely the understanding. Hence it is possible to attach to causality some form of a priori necessity. Cassirer in turn abandoned that standpoint not least because of his rejection of a role for intuition: he did not need to seek any principle to ground permanence, succession, or coexistence in time because in his view there is no role for the intuition of space and time in modern science. Instead, interpreted as a general principle, what causality does for Cassirer is allow for the universal application of the idea of functional coordination according to a law.

The similarity between Cassirer's consideration of the mathematical aspect of laws and more recent structuralist discussions is worth noting here. Thus, he argues that once placed in this form, phenomena are effectively established as 'enduring thoughts' (Rudolph 1994: 38), in the sense that their duration extends far beyond their original representation. As an example, he gives Fourier's theory of heat which was developed in the context of a view of heat as a fluid but whose mathematical description—in terms of which the phenomena were represented as the results of 'purely geometrical relations'—came to be seen as independent from these particular hypothetical presuppositions. It is this separation of the fundamental structure, as represented by the mathematical equations, from the underlying metaphysical commitments—which may of course play a crucial heuristic role—that was noted by Poincaré, as we have seen. Even more interestingly, perhaps, Cassirer goes on to point out how Fourier's formulae were subsequently resurrected by Heisenberg in the development of quantum mechanics. We recall from Chapter 1 that Saunders also uses this example to illustrate the 'heuristic plasticity' of such formulae (1993), a feature that Cassirer calls their 'indwelling sagacity' (Spürkraft). It is by means of this plastic mathematics that fundamental structural aspects of classical dynamics are isolated, become entrenched, and are thereby preserved in subsequent developments. In particular, as Saunders notes, certain of these features (those which are group-theoretic in particular), provide 'over-arching abstract frameworks ... within which

[60] The relationship between these two kinds of statements is certainly not inductive. Indeed, Cassirer viewed the problem of induction as the 'chief stumbling block' for the philosophy of science in general (Rudolph 1994: 39).

one dynamical structure may be embedded in another' (1993: 308). Both Cassirer and Saunders see this feature as indicative of the significant independence of the relationships represented by the equations and formulae, from the hypothetical/metaphysical presuppositions which led to their elaboration in the first place.

Moving on and upwards, as it were, statements of *principle*, seen as 'statements of third order', arise when one begins to consider how the laws themselves are interrelated. As a typical example, Cassirer considers the principle of least action and notes that as it was developed and made more precise through history, the metaphysical basis for it was increasingly lost from view (1936/1956: 48). The price for universality is the apparent loss of the subject of the principle (Cassirer nicely refers to its 'iridescent indeterminateness' (1936/1956: 51)), but rather than seeing this as a defect, Cassirer insists that it points to the real import and methodological character of such principles in general: they function as heuristic rules for seeking and finding laws (1936/1956: 52). And they do this by presupposing 'certain common determinations' which hold for all natural phenomena, and then effectively consider what, in a particular domain, corresponds to these determinations. Thus their power and value lie in this 'capacity for "synopsis"' (1936/1956: 52), which affords an overview of more than one physical domain. Unlike the laws themselves, the principles do not refer directly to phenomena, but to 'the form of laws according to which we order these phenomena' (1936/1956: 52). Symmetry principles can thus be placed here: they refer to the form of laws and play a heuristic role in discovering them (see Post 1971 for a nice account of this role and we shall return to discuss it in Chapter 10).

Putting it a little crudely perhaps, 'statements of measurements are individual, statements of laws general, and statements of principle universal' (1936/1956: 52). However, Cassirer emphasizes that the relationships between them should not be characterized in terms of any kind of spatial metaphor, as in a simple hierarchy, since these statements all mutually condition and support one another (1936/1956: 35) in a kind of 'reciprocal interweaving and bonding' (1936/1956: 35).[61] Consider the relationship between statements of measurement and statements of laws, for example: the former, as already indicated, do not constitute some bedrock of 'facts' since, as Cassirer claims, in an early reference to theory-ladenness, 'everything significantly factual is already theory' (1936/1956: 35). Thus we should not see these statements as forming the structure of a pyramid; this would suggest that the top 'layers' could somehow be removed without affecting the bottom, but such a suggestion is simply untenable since the truth of all such statements at whatever 'level' is due to their mutual interconnection. Rather than a pyramid, Cassirer likens this structure to a Parmenidean 'well-rounded' sphere, wherein the various elements can be logically distinguished, even though they cannot be ascribed any kind of independent existence. Significantly, Cassirer insists that within such a structure

[61] This might be compared with Eddington's 'cycle of reasoning' relating the laws and measurements in General Relativity.

there is 'no proper substantial carrier, nothing that per se *est et per se concipitur*' (1936/1956: 35); rather there is 'only a functional coordination in which all the elements, all the determining factors of physical truth, uniformly participate' (1936/1956: 35). Likewise, from Cassirer's structuralist perspective, there are no substantial carriers of physical properties, but only functional coordinations to which our metaphysical notion of a physical object is ultimately reduced.

Indeed, it is only through the mediation of the results of measurements that the 'concepts and judgments' of physics acquire objectivity. It is at this level of statements that we find the 'feature of individuality' associated with putative objects in the sense that such statements pertain to a definite here and now. One might characterize this in terms of what has been called 'space-time individuality', in the sense that the individuality (and distinguishability) of objects is ultimately grounded via their location in space-time (see French and Krause 2006). It is precisely this that quantum mechanics undermines. To use Eddington's phrase, this level of statement yields only a 'legend of individuality'. In this sense, in which the statements of the results of measurements are the beginning and end of physics, '[w]hat physics calls an "object" is nothing ultimately but an aggregate of characteristic numbers' (1936/1956: 36). Of course, as far as Cassirer is concerned, such an aggregate is determined and informed by the other elements of the structure, namely the laws and principles. Physical knowledge must not be thought of as a mere aggregate of data, since the data are mutually conditioned and interrelated. What is important is that 'we do not need to posit objects as sundered beings-in-themselves behind these determinations' (1936/1956: 36).

The overall framework, then, is the same as in the space-time case, at least in so far as it involves a shift from things-as-substances to relations as the ground of objectivity in science; or as Cassirer put it,

> [w]e are concerned not so much with the existence of things as with the objective validity of relations; and all our knowledge of atoms can be led back to, and depends on, this validity. (Cassirer 1936: 143)

In classical mechanics objectivity rests on the spatio-temporal persistence of individual objects and here,

> "[o]bjective" denotes a being which can be recognized as the same in spite of all changes in its individual determinations, and this recognition is possible only if we posit a spatial substratum. (Cassirer 1936: 177)

As Cassirer points out, 'The entire axiomatic system of classical mechanics is based on this presupposition' (Cassirer 1936: 177). This presupposition features explicitly in Boltzmann's axioms of statistical mechanics, for example (see French and Krause 2006: ch. 2), and it forms the basis of the 'worldview' of classical (particle) physics in which we have individual objects possessing at all times well-defined properties and traversing well-defined spatio-temporal trajectories.

It is this worldview that is apparently overturned by quantum mechanics (at least under the orthodox interpretation) and in the new situation in which we find ourselves, we cannot say that the particles unambiguously possess definite properties at all times, even beyond measurement interactions, or that they travel along well-defined trajectories.[62] It is at this juncture that Cassirer asks a pair of crucial questions: 'what *are* these electrons whose path we can no longer follow? Is there any sense in ascribing to them a definite, strictly determined existence, which, however, is only incompletely accessible to us?' (Cassirer 1936: 178). In answering these questions, Cassirer makes a fundamental demand that is analogous to the tailoring of our metaphysics to epistemology that underlies OSR, namely that we take the 'conditions of accessibility' as 'conditions of the objects of experience'. If we do that, then 'there will no longer exist an empirical object that in principle can be designated as utterly inaccessible; and there may be classes of presumed objects which we will have to exclude from the domain of empirical existence because it is shown that with the empirical and theoretical means of knowledge at our disposal, they are not accessible or determinable' (Cassirer 1936: 179). Bringing this demand forward, it rules out any epistemically inaccessible objects hiding behind the structures which we can know.[63]

What is an electron then? Not, Cassirer insists, an individual object (Cassirer 1936: 180) and here he cites Born's conclusion, reached, as we saw, at the height of the quantum revolution in 1926, that from the perspective of quantum statistics, the particles cannot be identified as individuals at all (Cassirer 1936: 184). Cassirer writes,

> The impossibility of delimiting different electrons from one another, and of ascribing to each of them an independent individuality, has been brought into clear light through the evolution of the modern quantum theory, and particularly through the considerations connected with the Pauli exclusion principle. (Cassirer 1936: 184 n. 17)[64]

Of course, this is to follow the 'received view' regarding the non-classical indistinguishability of quantum particles that draws the conclusion from quantum statistics that they are non-individuals in some sense. As noted earlier, quantum statistics is in fact compatible with the view that the particles are individuals (again, in some sense; see French and Krause 2006) and it is the metaphysical underdetermination that

[62] On the standard interpretation; for a consideration of (non-) individuality in the Bohmian interpretation, see French and Krause 2006: 178–9.

[63] Of course, the advocate of ESR could always insist that her hidden objects should not be regarded as *empirical*. However, unless she wants to adopt an explicitly Kantian view, with such objects consigned to the realm of the noumena, which would certainly conflict with her realism, it is not clear how we might understand such a move. And certainly, if these hidden objects are posited in part to act as the relata of the relevant relations, on pain of falling into the same conceptual difficulties as eliminativist OSR, it is hard to see how non-empirical objects could so serve as the relata of empirical relations.

[64] And here Cassirer follows Weyl in associating the Exclusion Principle with Leibniz's Principle of Identity of Indiscernibles (see French and Krause 2006).

arises from this double compatibility that the modern-day ontic structural realist takes to force a shift from the object-oriented stance, rather than the claim of non-individuality itself.[65]

Cassirer, like Eddington, takes the claim of non-individuality itself to further support the shift away from particles as substantival 'things'. If we want to continue to talk, in everyday language, about electrons as objects—because we lack the logico-linguistic resources to do otherwise—then we can do so 'only indirectly', 'not in so far as they themselves, as individuals, are given, but so far as they are describable as "points of intersection" of certain relations' (Cassirer 1936: 184 n. 17). And this relational conception of an object is taken straight from Kant himself:

> All we know in matter is merely relations... but among these relations some are self-subsistent and permanent, and through these we are given a determinate object. (Kant, *Critique of Pure Reason* B341, CE, p. 379; in Cassirer 1956: 182)[66]

The way in which putative objects are 'given' via relations is obviously something that relates strongly to current forms of structural realism. But there is also a more subtle point that can be imported into the modern debate: namely that our everyday logico-linguistic resources prevent us from dropping all talk of objects and hence we must do so 'indirectly', as points of intersection of relations, or nodes in a structure. However, that we are so constrained logico-linguistically should not be taken to imply that we are committed to such objects, qua elements of our metaphysics. Here again we may deploy something like the Poincaré Manoeuvre: in the absence of a basket of logico-linguistic resources that is not object-oriented, we can adopt such talk on a heuristic basis, at the level of both the everyday and that of modern physics, in order that we can continue to communicate, etc., but once the relevant relations have been articulated, in the theoretical context, we can dispense with the putative objects themselves, qua elements of our metaphysical pantheon. As a consequence, that we retain either everyday or physics-based talk of (putative) objects, given the logico-linguistic resources we are lumbered with, does not imply that we cannot adopt an eliminativist attitude towards them, *qua metaphysical entities*.

As an example of these self-subsistent and permanent relations Cassirer gives the example of charge, standardly understood as an intrinsic or state-independent property of particles. However, as Cassirer points out, in an acute rebuttal of the assumption made by today's object-oriented realist, 'the constancy of a certain relation is not at all sufficient for the inference of a constant carrier' (Cassirer 1936: 182). The permanence of charge justifies our regarding the electron, say, as a 'determinate object', where the scare quotes indicate that the sense is that of a

[65] In effect, what the formal treatment of non-individuality via non-standard (quasi-) set theory does is pull apart the concepts of individuality and objecthood, allowing us to retain the latter while dropping the former (French and Krause 2006). Obviously such a device was not available to Cassirer or Eddington.

[66] The idea of self-subsistent relations also features in certain forms of structural realism and, as we can see, it has a certain historical pedigree!

putative entity prior to reconceptualization in structural terms, but it does not justify what Cassirer calls the 'substantialization and hypostasis' of the electron in the sense of an entity which is not so reconceptualized.

Charge, like the other putative intrinsic properties, features in the relevant laws of physics and according to Cassirer, what we have here is a reversal of the classic relationship between the concepts of object and law (Cassirer 1936: 131–2): instead of beginning with a 'definitely determined entity' which possesses certain intrinsic properties and which then enters into definite relations with other entities, where these relations are expressed as laws of nature, what we now begin with are the laws which express the relations in terms of which the 'entities' are constituted.[67] From the structuralist perspective, what were regarded as intrinsic properties, like charge, are now regarded as self-substantive relations, and the putative underlying entity 'constitutes no longer the self-evident starting point but the final goal and end of the considerations: the *terminus a quo* has become a *terminus ad quem*' (Cassirer 1936: 131). Objectivity, therefore, is determinable through law, which is prior to it (Cassirer 1936: 176) and the boundaries of law mark the boundaries of objective knowledge (Cassirer 1936: 132). We shall return in later chapters to the way this perspective 'upends' the standard relationship between laws, intrinsic properties, and putative objects and there is obviously more to say about the relationship between laws and properties, for example. Here, then, is another piece of the 'hidden' history that can be uncovered and brought forward to the current debate.

Returning to quantum mechanics, the real impact of the theory for Cassirer is the way it reinforces the idea of the object as a 'terminus ad quem' by removing even the 'legend of individuality' that one might attach to the classical counterpart. Causality as a 'principle' can be retained, since it should be regarded not as a proposition pertaining to events themselves, but, rather, as 'a stipulation concerning the means through which things and events are constituted in experience' (Werkmeister 1949: 789). As such, the principle is not undermined by quantum mechanics; indeed, Cassirer insists, understood as a demand for strict functional dependence, the *essence* of causality remains untouched (Cassirer 1936: 188). At most the formulation of the principle must be corrected in the quantum context, following the articulation of the indeterminacy relations: the logical form of the causality principle is that of 'If x, then y'. Logically, of course, if indeterminacy has 'crept' into x, we are not entitled to infer any indeterminacy in the y and hence the statement 'If x, then y' is not valid. All that we can say is that in order for it to be useful in the quantum domain, the values of x must be 'permissible', in the sense that they can be determined by an appropriate mode of measurement. The causal relation as such is not affected, only its domain of legitimate application, and this is now further delineated by the indeterminacy relations. Once again we can lift certain features of this account of

[67] Again one might make a comparison here with the Brading–Skiles 'law-constitutive' view of objects.

causality and bring them forward into the current context: the most that physics supports is the functional dependence noted previously, with the productive and directed aspects of causation emerging at the level where human intervention plays a role. Again, I shall return to these points.

4.12 What We Can Take from Cassirer

From this Cassirerian perspective two conclusions about the nature of theoretical physics follow straightforwardly:

a) Relations are conceptually prior to objects.
b) The locus of objectivity shifts from objects to laws and symmetries.

The putative objects of the theory emerge from the interplay of the laws and the principles of the theory itself because they encapsulate the kind of constant pattern that ties together the empirical features that in different ways we consider to be properties of the object or consequences of the dynamics that the theory ascribes to its objects. In this sense, a working theory 'generates' its own objects, and objectivity is grounded in the universality of laws and principles.

According to Cassirer, quantum mechanics does not question the ideal of a nature ordered according to accessible laws and principles; rather, it presents us with a profoundly different picture of these items. In particular quantum 'objects' appear to lack individuality as a consequence of the laws and the principles of the framework. Nonetheless this framework provides us with perfectly objective knowledge of quantum phenomena. If the lack of individuality is taken to undermine the very notion of objecthood in this context,[68] then objectivity cannot reside with objects but must be sought in the laws and symmetries of the theory.

This is not to suggest that the debate on structuralism should move in a neo-Kantian direction (although some might approve of such a move; see, for example, Massimi 2011). Rather, I would urge, Cassirer's work presents elements of interest in a more general sense for the structuralist agenda.

In particular, if the relational notions of laws and principles can be detached from the neo-Kantian background, there are interesting consequences for the idea of objectivity. From the transcendental idealist standpoint this notion is profoundly linked with the universality of laws and derives ultimately from the nature of mathematics. If the latter is regarded as a product of human thought, then objectivity must be seen as constituted rather than given, as the realist would insist. However, in following the neo-Kantian in her rejection of objects, the structuralist need not go all the way and follow her down to what she sees as the ultimate ground of objectivity. Instead, the structuralist can resituate objectivity in the laws and principles of our

[68] And again, the modern advocate of OSR does *not* take it to so undermine objecthood.

best theories, rather than the putative objects, and the structural realist can take the former as representing features of a mind-independent reality, on the basis of the standard realist arguments (such as the No Miracles Argument).

This offers a way of responding to concerns that one cannot simply strip away such claims from their relevant context. In one sense, this is absolutely right: if one understands the claim to its full extent as originally presented, then one cannot grasp its full meaning without tracing that meaning through the web of interconnections in the original context. But that kind of context dependence cuts the history away from the current debates and leaves it as little more than a museum piece. In another sense, we can lift out of the relevant context claims that, if not identical to their originals, are sufficiently similar that they can be seen as closely related and for which we do not have to express the original associated meaning, or at least not in its entirety (as expressed by the web of interconnections). Thus we can take Cassirer's claims about the relative fundamentality of laws as compared to objects, and the shift in objectivity, and relate them to the debate over the metaphysical elaboration of structural realism without having to bring with them the associated claims about the ultimate grounding of such laws in mathematics, or the way that objectivity is constituted rather than given. And we can certainly bring forward moves such as the Poincaré Manoeuvre and deploy them in the modern context.

4.13 Conclusion

Many contemporary commentators on and critics of structural realism have hung their realist hats (or not) on the relevant equations and ignored the role of symmetry and invariance in physics. By allowing this history to be obscured by Russell's shadow, they have followed Braithwaite and Russell himself in failing to grasp the core idea of a structuralism that was appropriate for, and indeed grounded on, quantum theory. This is the structuralism of Cassirer, Eddington, and Weyl and it is this structuralism, or at least the core idea, that I shall be elaborating and defending in this book.

As we shall see, in their consideration of the nature of laws in science many current metaphysicians have similarly failed by ignoring or dismissing the role of symmetry and have similarly overlooked the possibility of an appropriate metaphysics of structuralism, fit for modern physics. Again, much of this book will be concerned with the development of just such a metaphysics. Let us now pull ourselves away from the history and continue this development, beginning with a discussion of the manner in which this structure can be represented.

5

The Presentation of Objects and the Representation of Structure

5.1 Introduction: Presentation vs Representation

Having set out the motivations for OSR, as well as some of its 'hidden' history, there are two broad sets of issues that must now be tackled. The first concerns the metaphysical nature of the structure OSR posits as fundamental. Here the kinds of questions that must be answered are: what is the relationship between this structure and the putative objects that are taken to be posited by our theories? In what sense can we take these objects to be eliminable? And does the notion of a structure without such objects make metaphysical sense? These will be addressed in Chapters 7–10. Another set of issues concerns the representation of that structure.[1] Here some of the questions are: what is the most appropriate representation for our philosophical purposes? What are the consequences of such a representation? Can we be pluralists about our mode of representation? These will be the focus of this chapter and the next one.

Unfortunately these two sets of issues have sometimes become confused in discussions about structural realism in general and OSR in particular. My intention in this chapter is to help clarify the situation by drawing on the distinction between *presentation* and *representation*, articulated in terms of the *presentation* of putative objects via the relevant 'shared structure' that our theories make available and the *representation* of such objects (as features of the world) by those theories (cf. Brading and Landry 2006). The obvious question then is: how is this (shared) structure itself represented? In addressing that question I shall again draw on the episode sketched in the previous chapter, which illustrates the presentation of group-theoretic structure within the framework of quantum physics, where it played both a foundational and an idealizational role and shall indicate how those roles can be represented within the set-theoretic framework of the 'partial structures' version of the semantic approach to theories.

[1] So, whereas the first set of issues might broadly be classed as metaphysical, this set falls within the philosophy of science. In particular, as the structure of the world is presented by, or within, theories which are then represented in certain ways at the level of the philosophy of science, I need to say something about the latter and how these ways impact on our understanding of the former.

It has been argued that the latter mode of representation is in fact surplus to requirements and that the relevant episodes can be understood from a 'minimalist' standpoint, structurally speaking (Brading and Landry 2006). However, I believe this rests on a blurring of the distinction between the role played by group structure at the 'object' level of scientific practice, and the role played at the 'meta'-level of the philosophy of science by the semantic approach. In both cases group theory and set-theory, respectively, are used as representational devices by physicists and philosophers, also respectively, but from the perspective of the meta-level, group theory also functions as the mode by which the relevant putative objects are presented to us. As far as the advocate of OSR is concerned this presentation then affords the means by which these objects can be metaphysically reconceptualized in structural terms, along the lines discussed in Chapter 7.

I shall also consider other representational devices, including that of the Ramsey sentence, in terms of which the so-called 'Newman objection' to structural realism (already introduced in Chapter 4) is typically presented. I shall suggest that this is an objection that has more bark than bite and that it can be dismissed on the basis of a more nuanced consideration of the kinds of relations that OSR seeks to capture (Melia and Saatsi 2006). These are precisely the kinds of relations that will feature in my answers to the first set of questions just given. Finally, I shall briefly explore the possibility of a pluralist account of the representation of structure at this level.

5.2 Modes of Representation: Partial Structures

The so-called 'semantic' or 'model-theoretic' approach is now perhaps the most widely adopted framework within the philosophy of science for the representation of scientific theories.[2] Since the best of these theories—according to the realist—represent the world, broadly speaking, we effectively have two levels of representation: at the level of science, we have the representational relationship between theories and the world; and at the level of the philosophy of science, we have the representation of theories themselves. An obvious move, then, is to suggest that both levels can be accommodated by the same mode of representation, namely the semantic approach (see, for example, Bueno and French 2012) and further, that the structures proffered by our best theories as aspects of the structure of the world can likewise be best represented using this framework. The form of this approach that I favour is the so-called 'partial structures' approach, which offers a certain formal flexibility that allows us to capture the various relations between theories and between theories and 'the world' in a clear and felicitous way.

The details have been given many times, but the central idea is to extend the usual notion of structure, through the device of a family of partial relations, in order to

[2] Useful presentations of this approach can be found in van Fraassen 1980; Suppe 1989; and da Costa and French 2003.

model the partialness of information we have about a certain domain (see da Costa and French 2003). Thus, when investigating a certain domain of knowledge Δ (say, elementary particle physics), we formulate a conceptual framework that helps us in systematizing the information we obtain about Δ. This domain is represented by a set D of objects (which includes observable elements, such as configurations in a Wilson chamber and spectral lines, and unobservable (putative) objects, such as quarks). D is studied by the examination of the relations holding among its elements. However, it often happens that, given a relation R defined over Δ, we do not know whether R relates all of the objects of D (or n-tuples thereof). This is part and parcel of the 'incompleteness' of our information about Δ, and is formally accommodated by the concept of *partial relation*.

The latter can be characterized as follows. Let D be a non-empty set. An n-place *partial relation* R over D is a triple $\langle R_1,R_2,R_3\rangle$, where R_1, R_2, and R_3 are mutually disjoint sets, with $R_1 \cup R_2 \cup R_3 = D^n$, and such that: R_1 is the set of n-tuples that (we know that) belong to R, R_2 is the set of n-tuples that (we know that) do not belong to R, and R_3 is the set of n-tuples for which it is not known whether they belong or not to R. (Note that if R_3 is empty, R is a usual n-place relation that can be identified with R_1.) A *partial structure* A is then an ordered pair $\langle D,R_i\rangle_{i\in I}$, where D is a non-empty set, and $(R_i)_{i\in I}$ is a family of partial relations defined over D.

With these concepts in hand, the notions of partial isomorphism and partial homomorphism can be defined. Consider the question: what is the relationship between the various partial structures articulated in a given domain? Since we are dealing with partial structures, a second level of partiality emerges: typically, we can only establish *partial* relationships between the (partial) structures at our disposal. This means that the usual requirement of introducing an isomorphism between theoretical and empirical structures cannot be met. Relationships weaker than full isomorphism, full homomorphism, etc., have to be introduced, otherwise scientific practice—where partiality of information appears to be ubiquitous—cannot be properly accommodated (for details, see Bueno 1997; French 1997; and French and Ladyman 1997).

Appropriate notions of partial isomorphism and partial homomorphism can then be introduced as follows (Bueno 1997; Bueno, French, and Ladyman 2002):

> Let $S = \langle D, R_i\rangle_{i\in I}$ and $S' = \langle D', R'_i\rangle_{i\in I}$ be partial structures. So, each R_i is of the form $\langle R_1,R_2,R_3\rangle$, and each R'_i of the form $\langle R'_1,R'_2,R'_3\rangle$.
>
> A partial function $f: D \to D'$ is then a *partial isomorphism* between S and S' if (i) f is bijective, and (ii) for every x and $y \in D$, $R_1xy \leftrightarrow R'_1f(x)f(y)$ and $R_2xy \leftrightarrow R'_2f(x)f(y)$. So, when R_3 and R'_3 are empty (that is, when we are considering total structures), we have the standard notion of isomorphism.
>
> Moreover, a partial function $f: D \to D'$ is said to be a *partial homomorphism* from S to S' if for every x and y in D, $R_1xy \to R'_1f(x)f(y)$ and $R_2xy \to R'_2f(x)f(y)$. Again, if R_3 and R'_3 are empty, we obtain the standard notion of homomorphism as a particular case.

This provides an appropriate representation of both scientific theories and models, particularly with regard to their open-ended nature and the manner in which they can be further developed (da Costa and French 2003). Furthermore, appropriately extended to include partial isomorphisms holding both 'horizontally' as it were, and 'vertically', it can also capture the relationships between theories and between them and data models (Bueno 1997 and 2000; da Costa and French 2003). Further extended again to include partial homomorphisms, it can also capture the relationship between such theories and the mathematics in which they are 'framed' (Bueno, French, and Ladyman 2002; see also Bueno and French 2012); in particular and with regard to that last point, this approach can capture what Redhead famously called the 'surplus structure' of mathematics, which has played an important heuristic role in scientific developments and which I shall draw upon in subsequent chapters (Redhead 1975; French 1999).

In his now-classic paper introducing OSR, Ladyman identified the partial structures framework as the appropriate mode of representation for this form, since it wears the relevant structural commitments on its sleeve, as it were (Ladyman 1998; see also Ladyman, Ross, et al. 2007). In particular, and in addition to the usual arguments that can be given in favour of the semantic approach, the significance for OSR of responding to theory change through the history of science—as discussed in Chapter 1—provides further support for adopting an approach, such as that here, with its associated partial isomorphisms, as a way of capturing the relevant features of such change. I have emphasized the role of this framework as a 'mode of representation' at the level of the philosophy of science and it is important to reiterate that its adoption does not entail that either theories or the structures they posit as 'out there' in the world should be regarded as inherently set-theoretic in any way.[3]

5.3 Modes of Representation: Shared Structure

Brading and Landry argue that the framework introduced here is (meta-) methodologically unnecessary (Brading and Landry 2006; see also Landry 2012). In its place they offer a form of 'methodological minimal scientific structuralism' that rejects these kinds of unitary frameworks at the level of the philosophy of science, arguing that all that we need is an appropriate grasp of the relevant 'shared structure' at the level of scientific practice. In particular, with regard to the ontological claims of structural realism and indeed, of realism in general, they state:

> What we call minimal structuralism is committed only to the claim that the kinds of objects that a theory talks about are presented through the shared structure of its theoretical models and that the theory applies to the phenomena just in case the theoretical models and the data models share

[3] For further discussion see da Costa and French 2003: 26ff; French 2006 and 2010b; French and Vickers 2011.

the same kind of structure. No ontological commitment—nothing about the nature, individuality, or modality of particular objects—is entailed. (Brading and Landry 2006: 577)

Furthermore, they insist that,

neither the framework of the semantic view of theories nor the appeal to shared structure alone offers the scientific structuralist a quick route to representation. (Brading and Landry 2006: 580)

On this last point we can certainly agree, since both representation in particular and structuralism in general may include further elements that in turn may be regarded as non-structural in certain senses. The concern then is whether the incorporation of such elements can be taken to undermine the structuralist programme and elsewhere I have argued that in relevant cases they do not (French 2007). Thus, it is clear that certain constraints must be imposed within the structuralist framework, without which it is not meaningful to talk of representation in the first place (French and Saatsi 2006). In one sense these constraints do represent significant non-structural elements, in so far as they embody theoretical content going beyond the pure logico-mathematical structure, which is linguistically specified and thereby constrains the possible systems in the world that are taken to be represented. However, the structure that the structural realist is concerned with should not be, and never should have been, construed as 'pure' logico-mathematical structure (Brading and Landry 2006; French 2007); it was always intended to be understood as *theoretically informed* structure. Although the linguistic specification of these constraints may suggest that the structuralist account of representation is not purely structural, this theoretical content was always regarded as an inherent feature of OSR to begin with (see French and Ladyman's reply to Cao in their 2003).

Returning to minimal structuralism, a crucial question is: how do we make precise this concept of 'shared structure'? According to the partial structures approach, the answer is straightforward: 'Shared structure' can be represented by (partial) set-theoretical structures plus the associated (partial) iso/homomorphism. Landry, however, offers a more general view according to which shared structure need not be shared set-structure: the shared structure can be made appropriately precise via the notion of a morphism and the context of scientific practice determines what kind of morphism (Landry 2007).

Thus she insists that,

mathematically speaking, there is no reason for our continuing to assume that structures and/or morphisms are 'made-up' of sets. Thus, to account for the fact that two models share structure we do not have to specify what models, *qua* types of set-structures, are. It is enough to say that, in the context under consideration, there is a morphism between the two systems, qua mathematical or physical models, that makes precise the claim that they share the appropriate kind of structure. (Landry 2007: 2)

Furthermore, she writes,

> I want to distinguish between semantic accounts that consider what the concept of shared structure *is* (what the appropriate type of structure is for *formally framing the concept* of shared structure in terms of some type of morphism) and those that consider what the presence of shared structure tells us (what the appropriate kind of structure is for *characterizing the use* of shared structure in terms of some kind of morphism as determined by some context), and to place focus on the latter. (Landry 2007: 8)

The core claim, then, is that all that we need to express both the content and, crucially, structure of what we take to exist can be found at the level of the relevant mathematics and, again crucially, no meta-linguistic framework, whether syntactic or semantic, is required (Landry 2012).

5.4 Modes of Presentation: Group Theory

The view under discussion here is articulated via a study of the introduction of group theory into quantum mechanics (Landry 2007 and 2012), as sketched in Chapter 4. As I suggested there, this is interesting in a number of respects, not least because of what it reveals about the relationship between physics and mathematics and the way in which the latter came to shape, in fundamental ways, the former.

We recall that a crucial stimulus for the introduction of group theory was quantum statistics and, in particular, the connection between such statistics and the symmetry characteristics of the relevant states of the particle assemblies, arising from the non-classical indistinguishability of the particles. So, just to recap: the fundamental relationship underpinning this move is that between the irreducible representations of the group and the sub-spaces of the Hilbert space representing the states of the system, with the group 'inducing' a representation in system space (see, e.g., Weyl 1931: 185). Thus under the action of the permutation group, in particular, the Hilbert space of the system decomposes into mutually orthogonal sub-spaces corresponding to the irreducible representations of this group. These include the symmetric and anti-symmetric, corresponding to Bose–Einstein and Fermi–Dirac statistics respectively, as well as those corresponding to so-called 'parastatistics'.

As well as possessing permutation symmetry, an atom is also symmetric with regard to rotations about the nucleus (if inter-electronic interactions are ignored) and again group representations can be used to label the relevant eigenstates. Weyl's mathematical work on the complete reducibility of linear representations of semi-simple Lie algebras allowed the irreducible representations of the three-dimensional pure rotation (or orthogonal) group to be deduced as well as the so-called 'double valued representations' representing spin (see Wigner 1959: 157–70). As I have noted previously, there are two important features of this case (French 1999): first of all, behind these 'surface' relationships lie deeper, mathematical ones. Thus the reciprocity between the permutation and linear groups (Weyl 1931: 281) not only functioned as 'the guiding principle' in Weyl's work (1931: 377), but also acted as a 'bridge' within group theory. Practically it was also significant since continuous groups can be more easily handled than discrete ones. Hence the appropriate representational framework in which to situate the mathematics–science relationship in this case should incorporate families of structures on each side. The application of group theory to quantum physics crucially depends on the existence of this bridge between structures within the former.

Secondly, both group theory and quantum mechanics were in a state of flux and development at this time and the structures should be regarded as significantly open in certain respects. The partial structures programme appropriately captures this feature at the (meta-) representational level. From such a perspective, both mathematical and scientific change can be treated as on a par at the 'horizontal' level as it were, and, looking at 'vertical' relations, given the *partial* importation of mathematical structures into the physical realm in this case, partial homomorphism provides the appropriate characterization of such relations (Bueno, French, and Ladyman 2002). It is precisely by accommodating and, thereby, presenting such features that a representational framework such as that provided by partial structures proves its worth.

The value of such an approach is further exemplified by Wigner's subsequent application of group-theory to the nucleus and the development of isospin based on an analogy between atomic and nuclear structure that is both partial and dependent on certain idealizations (Mackey 1993: 254–78; French 2000b). Drawing on Heisenberg's treatment of the forces between protons and neutrons by analogy with his earlier account of the exchange forces in the ionized hydrogen molecule, Wigner (Wigner 1937: 106) took both these forces and the masses of the particles to be approximately equal which allowed him to treat them as indistinguishable (apart from their charge). They could then be conceptualized as two states of a new kind of particle, the 'nucleon'. The kinds of idealizations can be represented via partial isomorphisms holding between the partial structures (French 2000b): taking them in stages we move from protons and neutrons with non-equal forces, to a model with protons and neutrons and equal forces, to one of nucleons. Merging them together, the fundamental idealization is the shift from protons and neutrons to the nucleon and in this way the nucleus can be treated as an assembly of indistinguishable particles. By analogy with the situation in the atom this in turn suggests the introduction of a further symmetry group on the back of the analogy between representations of nucleons and representations of electron spin: the relevant decomposition of the Hilbert space is analogous to the decomposition of the corresponding Hilbert space for the spin of an electron (the relevant groups have isomorphic Lie algebras).

Within the set-theoretic representational framework, we have an isomorphism between the partial structures representing the anti-symmetrized tensor power of the direct sum of two Hilbert spaces and the direct sum of products of anti-symmetrized tensor powers. The problem of determining the interaction between the protons and neutrons is then reduced to that of considering 'particles' of the same kind, the

Hilbert space of each of which is the direct sum of the proton and neutron Hilbert spaces (Mackey 1993: 257–8). The analogy between atomic and nuclear structure thus reduces to that which holds between the relevant anti-symmetrized Hilbert spaces for a system of electrons in an atom and a system of nucleons in a nucleus. However, the analogy is multiply incomplete (Mackey 1993: 259): the proton/neutron decomposition does not depend on choosing an 'axis'; both protons and neutrons also have spin 1/2 (Wigner 1937: 107) and so the representations of the rotation group in the relevant Hilbert spaces are irreducible in the electron case but the direct sum of two equivalent irreducible representations in that of the nucleons. Thus the introduction of isospin, on the physics side, requires, on the mathematical side, the use of an appropriate symmetry group that is more complicated than in the atomic case since the corresponding Hilbert space is of higher dimension (Mackey 1993: 259). This prompted Wigner to move to the representations of the four-dimensional unitary group U(4), which yields, instead of multiplets, the 'super-multiplets' of nuclei (see Wigner 1937: 112–13).

The partial structures framework nicely captures this incomplete analogy between atomic and nuclear structure. Following Hesse's classic division of analogy into positive, negative, and neutral components (Hesse 1963), there is a positive analogy that holds between the atom with its electrons and central nucleus and the nucleus itself, with its nucleons and centre of gravity. There is a further twofold analogy between the treatment of the nuclear particles as indistinguishable and the indistinguishability of the electrons; and also between the spin of the electrons and the isospin of the nucleons. The application of the permutation group then follows on the back of the former. With regard to the latter, the positive analogy holds between the direct sum decompositions into the relevant sub-spaces. The negative analogy is likewise twofold: there is no 'axis' of isotopic spin in the nucleon case but more profoundly, the relevant Hilbert space is of a higher dimension since both protons and neutrons also have spin. Thus the deeper disanalogy between the two structures concerns the replacement of the rotation group by the four-dimensional unitary group U(4). Isospin then went on to become an important feature of elementary particle physics, as the relevant structures were extended via the neutral analogy. As is well known, it was through efforts to combine the SU(2) group of isospin and the U(1) group of strangeness or hypercharge that SU(3) was proposed as the group of the quark model. Isospin then ceased to be regarded as 'fundamental', and with the development of colour and the electroweak group, so did 'global' SU(3) (see McKenzie 2014).

And of course Wigner himself extended the group-theoretic approach to elementary particles in his crucial and important work on the association of 'elementary physical systems' with representations of the Poincaré group (see Wigner 1935; also Drake et al. 2009 provide a useful summary)[4] where he noted the 'unique

[4] It is interesting to note that the abstract of the 1935 presentation indicates that a detailed discussion of this work was supposed to appear in a joint paper with Dirac 'who first perceived this problem'. I don't

correspondence' between possible Lorentz invariant equations of quantum mechanics and these representations. Such a representation, 'though not sufficient to replace the quantum mechanical equations entirely, can replace them to a large extent' (Wigner 1939: 151). It can give the change through time of a physical quantity corresponding to a particular operator, but not the relationships holding between operators at a given time. The issue then is to determine the irreducible representations of this group (Wigner 1939). We shall return to this result at various places in our discussion but it is worth emphasizing here that it is the association of the labels of these representations with the values of the properties of the 'elementary systems', such as charge and mass, that forms the basis of the claim that such properties should be conceived of structurally. Let me illustrate what I mean with another example, that of spin.

5.5 Spin and Structural Realism

A useful summary of the history of this property that maps the intertwining of theoretical and experimental aspects can be found in Morrison (2007). The conclusion reached is that spin is a 'hybrid' notion possessing both mathematical and physical features that 'bridges' the mathematical and physical domains. And this is revealed by the fact that it essentially drops out of the mathematical formalism (of the Dirac equation, underpinned by group theory), in the sense that it is required to secure conservation of angular momentum and to yield the generators of the rotation group (Morrison 2007: 546–7).[5] Specifically, spin is just a group invariant characterizing the unitary representation of the Poincaré group associated with the wave equation. This hybrid character of the property appears to pose a challenge for realism since the latter stance requires that an appropriate physical interpretation of this property be given and the manner in which the mathematical and physical are intertwined renders such an interpretation 'otiose' (Morrison 2007: 548). Now, this is a strong claim that, if accepted, would push us

know if such a joint project was ever begun. In the 1939 paper, Wigner again acknowledges Dirac, stating that the topic of the paper was suggested by him as early as 1928 and that even then, Dirac realized the connection between representations and the equations of quantum mechanics (1939: 156). The paper is presented as the outgrowth of 'many fruitful conversations', especially during 1934/1935. Dirac also published his own work in this area, presenting more elegant derivations of Majorana's results on the classification of representations of the Lorentz group. As Wigner notes, his results provide a posteriori justifications of the work of Dirac and Majorana.

[5] Spin and quantum statistics are related via the spin-statistics theorem. Although there remains some doubt over what counts as an adequate proof of this theorem (see Sudarshan and Duck 2003), one *could* interpret it as grounding the relevant statistics in an understanding of spin, thus removing the need to appeal to symmetry as playing a fundamental explanatory role. On the other hand, Berry and Robbins' 'geometric' proof turns the grounding relation in the other direction, in so far as on their account particle permutations involve a kind of 'hidden' rotation (Berry and Robbins 1997). This important approach to the theorem still awaits philosophical analysis.

either to drop standard realism or move towards some form of Platonism (again see French and Ladyman 2003).

However, the idea cannot be that the simple combination of mathematical and physical features in the description of spin renders any interpretation otiose, since that is obviously true of many such properties in physics, nor that it is required in order to save the conservation of a quantity; rather it must be that the mathematical features are such that no purely physical interpretation is possible (see also French 2015). Since '[o]ur current understanding of spin seems to depend primarily on its group theoretical description' (Morrison 2007: 552) it is obviously the latter that is problematic. What such a description yields, as Eddington pointed out, and as we considered in Chapter 4, is not simply a pattern of entities, or even a pattern of relations, but rather a 'pattern of interrelatedness of relations' (Eddington 1941: 278). What group theory gives us, then, is the appropriate algebra of operators representing rotations acting on rotations, for which the 'pattern of interrelatedness' is manifested in the associated multiplication table. Presumably it is this that is resistant to a straightforward realist interpretation.

An obvious response would be to elaborate such an interpretation in structuralist terms but after briefly sketching the respective virtues of epistemic and ontic structural realism, Morrison concludes that it cannot help in this case, since,

> [o]n this account the structures become no less mysterious than the physical entities they have reconceptualised. To say that the mathematics is a description of the structures but that they themselves are something else leaves us in the precarious position of affirming the existence of a 'something I know not what'; structures whose natures are described in a certain way. But this was exactly the problem that ontic SR was designed to solve. (Morrison 2007: 554)

However, as far as the ontic structural realist is concerned, the supposed mysterious nature of physical entities has to do with the underdetermination of what Brading and Skiles call their 'individuality profile', as we saw in Chapter 2 (Brading and Skiles 2012); that is, we cannot tell whether they are individuals or not. The 'mystery' is resolved and the metaphysical underdetermination dissipated by reconceptualizing such entities in structural terms, rather than as objects. This 'mystery' is entirely different from that which is associated with the structures. Here it has to do with the difference between the mathematical and the physical and the claim that however we understand the former, the latter will be 'something else'. But if this is a 'mystery', it is surely one that arises for *any* form of realism since it has to do with appropriately characterizing the physical. I shall return to this issue in Chapter 8.

More fundamentally, perhaps, Morrison insists that 'adding a layer of metaphysics' cannot help clarify the nature of spin beyond what is given by the physicomathematical description provided by quantum theory. On the contrary,

[t]o reconceptualize that description in terms of a metaphysics of *sui generis* structures renders the problem more convoluted. Nor do the activities associated with experimental detection become more perspicuous when understood in terms of these unexplained structures. (Morrison 2007: 554)[6]

Here the issue is: how much metaphysics should the realist in general and structural realist in particular allow into her position? Just enough, but not too much, as I argued in Chapter 3. However, it seems odd to raise a crucial problem for realism, then when attempts are made to solve that problem through the deployment of metaphysically interpreted structure, to insist that all the important features of the nature of spin are already implicit in the very physico-mathematical description that generated the problem! Again, the task of the structural realist is not to reconceptualize in terms of *sui generis* structures, but rather to do so in terms of an account of such structures appropriately metaphysically conceived. And there is an analogy here between Morrison's insistence on remaining at the level of the physico-mathematical description and Brading and Landry's: the response in both cases is to insist right back on the significance of appropriate devices that give content to our realism and philosophy of science respectively.

As for the 'activities associated with experimental detection' emphasized by Morrison, there are two things the structural realist can say. The first is that one might hope that shifting away from a metaphysics of (individual) objects and their associated (typically monadic) properties would in fact help introduce further philosophical perspicuity into these activities. The second is that the kinds of experimental traces we usually observe (tracks in a cloud chamber, etc.) are typically taken to support the exportation into the micro-realm of an inappropriate object-oriented metaphysics. I shall come back to this point in the next chapter, where I shall indicate how the position observations underlying such traces can be brought within the group-theoretic and hence structuralist fold.

However, I completely agree that '[p]art of the difficulty with attempts to generate a physical notion of spin concerns the way the electron is pictured in the hydrogen atom as a quantum mechanical object' (Morrison 2007: 554). We are led astray by this fundamentally object-based metaphysics to view spin as rotation around an axis, as described by relations between observables. But these relations are represented by operators and as Eddington perceived, it is the algebra of these operators that describes the structural 'pattern of interrelatedness of relations'—unpacking the latter will then give us our metaphysical interpretation. This goes beyond simply acknowledging the group-theoretic nature of spin, following Wigner's account of elementary particles. It is the group-multiplication table that represents the structure in this case and the metaphysics of the latter will be shaped by the features of this table.

[6] cf. Landry 2012.

Returning to our discussion of modes of representation, the point of the previous historical interlude is to illustrate the advantages of adopting an appropriate representational framework such as that offered by partial structures. In particular, it allows us to re-describe and re-present the relevant historical elements in terms that are accessible to the philosopher, such as 'positive analogy', 'partial isomorphism', and so on. Furthermore, although this re-presentation will display the work performed by group theory itself, it is clear that the latter features at the level of scientific practice, not at the level of the representations of philosophers of science; I shall return to this point shortly.

According to Landry, on the other hand,

> what does the real work is not the framework of set theory (or even category theory); it is the group-theoretic morphisms alone that serve to tell us what the appropriate kind of structure is. (Landy 2012: 11)

More generally, she claims, it is the *use* of the concept of shared structure that determines the kind of structure and characterizes the relevant meaning and all the relevant *work* is done by the contextually defined morphisms (as we shall see, what counts as the relevant 'work' in these cases is crucial).

5.6 Set Theory as Cleaver

Thus, the fundamental question is: if it is group-theoretical structures that we are going to be realists about, in the sense already indicated in the case of spin, then where is set-structure doing any real work? Landry insists that,

> if one wants . . . to use this kind of structure as a tool to carve 'the world' into its 'natural kinds', then one cannot, in addition to claiming that group theory is 'the appropriate language', claim that all such group-theoretic kinds are set-theoretic types, *unless* one is ready to hold fast to, and provide justification for, the Bourbaki/Suppesian assumption that all scientifically useful kinds of mathematical structures *are* types of set-structures. Nor can one use this assumption to make a more robust, ontologically read, structural realist claim about the structure of 'the world', unless one wants to impose (or presume) that set theory cuts not only mathematics but indeed, Nature at its joints. (Landy 2012: 15)

In other words, there is, first of all, a tension, at the very least, between the claim that group-theoretic structure is what we should be realists about and the adoption of the set-theoretic approach by the ontic structural realist, and this tension can only be dissipated if we adopt the Bourbakian line. Furthermore, that latter response would propel us into the unsavoury position of claiming that the world is somehow set-theoretic, in an ontological sense.

Thus Landry urges that structural realism should free itself from its set-theoretic ties and adopt a minimalist form of structuralism based on this concept of shared structure, understood as that structure that is actually 'doing the work' in the relevant physical context. However, I suggest that this apparent tension is the result of conflating the different representational roles being played by the respective structures and that, furthermore, there are advantages to retaining a set-theoretic representation of theories whilst also maintaining a group-theoretic presentation of structure. In particular I think we can easily resist falling into some form of set-theoretic Platonism about the world.

5.7 Presentation of Objects and Properties via Shared Structure

Let us consider briefly how physical objects are typically presented within theories. We might approach this informally via a journal or textbook presentation of the theory concerned, which might typically set out the fundamental principles, laws, etc., together with some indication of what the theory is 'about'. Or we might adopt a more formal approach, either following the logical empiricists and reconstructing the theory in a formalized language, or, more moderately, offering an appropriate description in predicative terms (Saunders 2003a). Taking this route, the non-logical symbols of the relevant formal language are derived from the theory and interpreted in terms of physical properties, relations, and functions. As Saunders puts it,

> we may read off the predicates of an interpretation from the mathematics of the theory, and because theories are born interpreted, we have a rough and ready idea of the objects they are predicates of. But there is nothing systematic to learn from the formalism to sharpen this idea of object. (Saunders 2003a: 290–1)

Saunders' concern here is with identity and indiscernibility in quantum physics and he draws on the 'purely logical aid' of his Quinean form of Leibniz's Principle of Identity of Indiscernibles (PII), as discussed in Chapter 3, in order that quantum entities can be regarded as 'weakly discernible' and hence as objects in a 'thin' sense. This effectively hones the 'rough and ready' idea of an object in the quantum case into something more metaphysically robust (although still structural). But as he notes, within the theory itself, identity signifies only the equality or identity of mathematical expressions, not of physical objects. Furthermore, the obvious worry the structuralist may have is that during the birth process, as it were, this rough and ready idea will be shaped by metaphysical preconceptions drawn from our interactions with 'everyday', macroscopic objects and inappropriately exported into the micro-realm described by modern physics.

Quine himself, of course, famously described physical objects as irreducible 'cultural posits' that are 'conceptually imported into the situation as convenient intermediaries not by definition in terms of experience, but simply as irreducible posits comparable, epistemologically, to the gods of Homer' (1951: 44). What the logical form of the relevant re-description gives us are the values of the variables that signify what exists, but ontological relativity implies that objects are nothing more than 'mere nodes' within the global structure that can be interpreted under widely different ontological frameworks while leaving the evidential base undisturbed. Since ontology is so plastic on this view, Quine concludes that structure is what matters, not the choice of objects.[7] The very notion of an object, he insists, should be seen as a human contribution, resulting from our inherited apparatus for organizing the 'amorphous welter of neural input' (and hence one can draw connections with Cassirer's neo-Kantian view again).

We can view objects as standardly presented in the context of the associated theories, either as part of this rough and ready understanding attached to the interpretation the theory is 'born' with (conceptually imported, as Quine puts it), or extracted from an (at least moderately) formal re-description of the theory, with the help of purely logical aids such as PII. What this then underpins is the standard metaphysical picture in which we 'build up' from the bottom, as it were, beginning with objects, which 'have' (in some sense) properties, that are then related in various ways, with these relations captured and described by the laws associated with our theories. Thus as an example, a particle such as an electron, metaphysically regarded as an object, possesses the intrinsic property of charge, which 'enters' into relations with other instances of charge, these relations being then described by Coulomb's Law, say.

The structuralist offers a different 'top-down' picture in which we start with the laws and principles 'presented' (on the surface as it were) by the theory, interpret these, at least minimally, in terms of relations and properties, but then resist the temptation to take that further metaphysical step and regard these last as possessed by (metaphysically robust) objects. In particular, the structuralist insists, there is nothing in the theory itself, or in the laws and principles as they are presented, that *requires* us to posit objects qua property possessors.[8] On this view, these relations and properties are features of the fundamental structure of the world (in a way that I will elaborate in Chapter 10) and what we standardly designate as 'objects' are indeed mere nodes in this structure. In particular, elementary particles are not metaphysically robust objects under this perspective, but are reconceptualized structurally and represented by the relevant symmetry groups, as indicated previously and as we shall consider again in the next chapter. And again, we can draw on Cassirer's claim that,

[7] Of course, although Quine refers to Ramsey—to be discussed shortly—and Russell he does not have structural realism in mind here.

[8] cf. Dasgupta (2009) who argues that 'primitive individuals' are redundant to all our best physical theories, just as absolute velocity is redundant to Newtonian mechanics, and are also empirically undetectable. In his words they are metaphysical 'danglers' which can, and should, be dispensed with in favour of what he calls a 'generalist' picture, whereby we 'simply ask for an account of the fundamental structure of the world that dispenses with primitive individuals but which allows us to make sense of the whole array of possible general facts' (2009: 49). There are clear connections with the picture I am sketching here and Dasgupta's 'radical holism' might offer a general (ha!) and congenial home for various forms of structuralism.

that which knowledge calls its "object" breaks down into a web of relations that are held together in themselves through the highest rules and principles. (Cassirer 1913, trans. in Ihmig 1999: 522)

We recall that these 'highest rules and principles' are the symmetry principles that represent the invariants in the web of relations itself. These in turn are represented group-theoretically and hence the relevant group supplies the general conditions in terms of which something can be viewed as a putative 'object' (see also Falkenberg 2007; Kantorovich 2003; and Lyre 2004).

I shall return to the structuralist account of laws and symmetries and the way in which putative objects are dependent upon or constituted by them, but the point I want to emphasize here is that within this 'top-down' picture putative 'objects' enter, not as part of the birth pangs of the theory, nor as imported conceptual intermediaries, nor with the help of purely logical aids, but via the relevant symmetry groups. Brading and Landry take these to be captured by the relevant 'shared structure' and I certainly agree that this is context dependent in the sense that it is the physical context that 'reveals' that aspect of the world-structure. However, we need to be clear about what, or who, is doing the relevant work in these cases.

5.8 Doing Useful Work

So, recalling the point that it is the *use* of the concept of shared structure that determines the kind of structure and that all the relevant *work* is done by the contextually defined morphisms, let's ask: who's using and what's working?

First of all, it is obviously the physicists/mathematicians who used and continue to use group theory in the relevant physical contexts, not (partial) set-structures (except maybe implicitly, if one were to insist that all mathematics is reducible to such structures!). In particular, in the context of the quantum revolution, it was group theory, not (partial) set-structures, that was effectively doing the (physical, mathematical, and hence object-level representational) work. And as indicated, it is in terms of these group-theoretical structures that we can consider putative 'objects' (taken, from the structural perspective, as mere nodes) and the relevant properties, such as spin, as presented.

However, it is philosophers of science, of course, who use various modes of representation—such as (partial) set-structures or Ramsey sentences, as we'll shortly see—to capture the structural content of theories, or to represent theories in general, together with their interrelationships, both with each other and, heading downwards, with data structures, etc., and moving up, with the families of mathematical structures into which theories can be embedded. Furthermore, these devices enable us to formalize and sharpen notions such as models and analogies and allow us, of course, to draw on a range of resources, such as, in the case of the partial structures outlined at the beginning, partial isomorphisms and homomorphisms. These can then be considered two of the various tools that philosophers can use in this representational activity. Thus at the meta-level where philosophers of science operate, it is these modes of representation and associated devices (such as partial structures) that are doing the (meta-level representational) work.

So, I agree that the appropriate structure at the level of the physics (and hence appropriate structural ontology) is contextually determined, where the context here is understood physically, rather than, say, culturally or sociologically. However, I also insist on the need for a meta-level representational unitary framework (provided by Ramsey sentences, set-theoretic models, category theory, whatever). Granted that it is 'shared structure' (group-theoretic, in the case study of section 5.5) that does all the 'work', the work that is being done is 'physical' (!) work and while I agree that this is appropriate for physicists, philosophers are doing a different kind of work, that requires a different set of tools.[9] To insist that this form of work should be dispensed with would be a radical step too far! And of course, there are other tools available that the philosopher of science might choose. In the next section I shall consider the most well known and indicate what I find unattractive about them.

5.9 Modes of Representation: the Ramsey Sentence

Perhaps the most well-known mode of representation in this context is the Ramsey sentence (RS), obtained by replacing the theoretical terms of a theory with variables bound by existential quantifiers:

$$T(t_1, \ldots t_n, o_1, \ldots, o_m) \rightarrow (\exists x_1), \ldots (\exists x_n) T(x_1, \ldots x_n; o_1, \ldots o_m)$$[10]

Advocates of ESR have adopted this as the most appropriate representation of a theory's structural content—with the theoretical terms replaced by existentially bound variables, the ontological spotlight shifts from the former (with the concomitant notion of reference to unobservable entities) to the relationships between the latter. Furthermore, this also offers a means of representing the 'hidden natures' of ESR. Being an existential generalization of the original theory, RS can be reasonably seen as describing a class of realizers far broader than that realizing the original

[9] In Landry 2012, further areas of disagreement are identified. Landry argues that the No Miracles Argument (NMA) should be understood in a 'local' form only, 'that only considers the extent to which *a particular* scientific theory *presents* the content and structure of what we say about what exists' (2012: 48). My worry is that constructing one's realism around such local instances of NMA sails perilously close to the kind of 'patchwork' view advocated by Cartwright (1999). Furthermore, she agrees with Brading and Skiles that objects can be accommodated within this methodological structural realism, where the notion of object is understood in a law-constituted manner and all that we know of it is given by its role in the relevant shared structure; as I have already noted, on this point there is perhaps only a cigarette paper between that view and OSR, with the principal difference having to do with how these positions are motivated.

[10] Philosophically what this amounts to is contentious, with different philosophers rediscovering it throughout the recent history of structuralism and putting the technique to different uses (see Cei and French 2006).

theory. In the event that the class of realizers of RS were to be constituted by more than one n-tuple of items RS would be multiply realized. Since the available empirical evidence for each n-tuple of realizers is the same there is no way to choose any member of the class over the others and thus as far as our empirical knowledge is concerned RS can always be *multiply realizable*. Multiple realizability, then, offers a way of capturing the core idea behind ESR of the structure being *epistemically* independent from the entities whose *natures* we are not in a position to know.

This notion of multiple realizability features prominently in the history of the Ramsey sentence (see Cei and French 2006). Lewis, in one of his early discussions (Lewis 1970), used Ramseyfication in order to provide a definition of theoretical terms and argued that multiple realizability should not be admitted on realist grounds, introducing a technical modification of the Ramsey sentence in order to block it. In a later contribution, however, he rediscovered it as the bedrock of a 'thesis of humility', thereby providing a way of bringing together the issues of humility and 'hidden natures' that I discussed in Chapter 3.[11] We can see how this works as follows:

A 'realization' of a theory T is an n-tuple of entities denoted by the theoretical terms of T and which satisfies the relevant 'realization formula' of T (obtained by replacing the theoretical terms by variables). Lewis demands that the theoretical terms of a multiply realized theory be denotationless and theoretical postulates containing such terms must be regarded as false, since, Lewis argues, scientists themselves appear to proceed with the expectation that their theories will be uniquely realized.[12]

To illustrate what is going on, consider for simplicity the Ramsey sentence $(\exists x)$ $[T(x, o_1, o_2, \ldots o_m)]$ (simplified for one new term only). In the case of multiple realization, we will presumably have two 1-tuples which realize the open sentence 'T(x)'. Call these 'electron' and 'smelectron'. In what sense can these actually be distinct, given that both realize 'T(x)' and, therefore, have the same properties? (Let's assume that there are no other sentences expressing different properties that one of these realizers, but not the other, realizes; i.e. this sentence is a 'final' sentence in the appropriate sense.) The issue is, how are n-tuples to be distinguished if multiple realization is to be a possibility in the case of scientific theories?

[11] Carnap, on the other hand, welcomed multiple realizability as a tool to express the openness of scientific theories noted previously and formally accommodated it through the use of Hilbert's ϵ-operator (Cei and French 2006). For a rich and interesting comparison of the views of theoretical knowledge of Russell, Ramsey, and Carnap in terms of the Ramsey sentence, see also Demopoulos (2011). Interestingly, given Melia and Saatsi's (2006) rejection of the so-called Newman objection to structural realism that I shall outline shortly, Lewis' Ramsey sentence is designed even for intensional predication and is therefore very different in terms of content from Ramsey's and Carnap's.

[12] It is not entirely clear what grounds this argument: certainly multiple realizability is not equivalent to the underdetermination of theories by evidence but even if it were, scientists themselves may remain unmoved by arguments against the latter.

If both electrons and smelectrons are supposed to satisfy T, then either T is only provisional and not final (as, we presume, most current theories are), in which case the difference in theoretical properties of electrons and smelectrons will be reflected in the replacement of T by its successor, satisfied by one or the other, or T is the final theory, in which case prima facie there should not be any theoretical difference. In the former case, multiple realizability appears to be merely a reflection of our epistemic fallibility and it is hard to see what we should be so agitated about. If, on the other hand, it is to be understood not epistemically but ontologically, then how are we to make sense of it when all the theoretical properties of electrons and smelectrons are wrapped up in T? Can we really make sense of this notion of multiple realizability?

Here is where humility enters and the way that Lewis deploys it to address this last problem fulfils some of the *desiderata* laid down by advocates of ESR.[13] What plays a decisive metaphysical role in expressing the sense of humility is a combinatorial principle applied to the properties of the entities concerned. Here *T* is taken to be the final theory of science. The language of T is formulated as earlier in this section but now T-terms label only fundamental properties.[14] Lewis further assumes that a fundamental property referred to via a T-term always falls within a category containing at least two such properties.

Once the RS is formulated in the usual way, we have the following situation: the actual realization of T prima facie seems unique but the role-occupancy of the fundamental properties is specified by the RS which has the same empirical success as that of T and is multiply realized. This means that in the case that T could be proved to be multiply realizable there is no empirical evidence that can decide between the different possible realizations. Again we face a form of humility with regard to which two factors are crucial:

a) T and its RS have the same empirical power; thus RS can be taken to specify which role the fundamental properties have to play to account for all empirical data and this is all we need for our epistemic purposes.
b) Since it is assumed that our fundamental properties belong to classes with at least two members, the combinatorial principle allows us to conclude that the same phenomena would be observable in worlds in which fundamental properties belonging to the same category are swapped. In other words there is room to argue that on this view, even the final T is multiply realizable.

[13] The general context is again Langton's analysis of Kant's transcendental philosophy as an investigation of the limits of our knowledge, with reality affecting us via relational properties only so that intrinsic properties must be regarded as 'out of the picture'. Lewis effectively detached his understanding of humility from this analysis.

[14] With the only exception of idlers and alien properties whose consideration is not important here.

The metaphysical picture here is as follows: First of all, there is the assumption of combinatorialism: we can take apart the distinct elements of a possible situation and rearrange them. Since, according to the Humean stance (again, to be returned to in later chapters), there is no necessary connection between distinct existences, the result of such a combination will be another possibility. This underpins the Humean view of laws and, in particular, entails that the laws of nature are contingent. Secondly, we have a form of quidditism (the view that *properties* have a kind of primitive identity across possible worlds; see Black 2000). Thus, different possibilities can differ only on the permutation of fundamental properties.

This offers a further way of understanding the 'hidden' natures of ESR. As Psillos, for example, has emphasized (Psillos 1999), the properties that feature in the theory's laws will be the relevant properties of the underlying entities (such as charge, mass, etc.), and hence, as French and Ladyman (2003) have argued, what remains 'hidden' will have to be something 'over and above' these properties. However, if this 'hiddenness' is understood via multiple realizability, as indicated here, we see that there is a further possibility according to which the epistemic structural realist's 'hidden natures' are cashed out in terms of the quiddities of the relevant properties.

What about the consequences for epistemic structural realism once this perspective is embraced? First of all, as Lewis repeatedly observes, the T-terms removed in this picture are a small number—as small as the number of the intrinsic properties, which in turn entails that this view admits in the Ramsey sentence a relevant amount of non-purely structural or relational knowledge although it frames it in a relational description. Secondly, the overall picture relies on combinatorialism, which in turn pushes us to abandon a conception of laws of nature as involving necessary connections. This in turn means that any articulated set of relational properties captured by the structure of the theory also loses any character of necessity.

Now this may or may not be such a heavy cost to bear, depending on one's attitude to laws and necessity, of course. The epistemic structural realist could adopt some form of regularity view, and indeed, as we shall see in Chapter 9, forms of Humean structuralism have been elaborated. However, given the problems this view faces, adopting the understanding of hidden natures via multiple realizability places further pressure on her position.

ESR must also face a more well-known objection, due to Newman (for responses see Worrall 2007 and Zahar 2001; for further details and concerns, see Frigg and Votsis 2011). Recalling the objection and configuring it in the current context, it runs as follows: as long as the given theory is empirically adequate and has a model of the right cardinality, we can always find a system of relations definable over the relevant domain such that the Ramsey sentence is true. The claim then is that if the structural realist uses the Ramsey sentence as her chosen representational mode, her realism will be trivialized. Putting it another way: if we know only the *structure* of the world, then we actually know very little indeed.

Now the structural realist might insist that she has no intention of Ramseyfying out *all* the predicates of the theory, only the theoretical ones. Even then, it has been claimed, if the Ramsey sentence makes only true empirical predictions (and gets the cardinality of the domain right) then it will be true, and structural realism is trivialized (Ketland 2004).[15]

This is perhaps the most discussed objection to structural realism. I won't cover all the responses to it (for excellent discussions of possible responses see Ainsworth 2009 or Frigg and Votsis 2011), since I believe its force has been definitively blunted by arguments due to Melia and Saatsi (2006). First of all, they point out, the objection assumes the elimination of all predicates that apply to unobservables, but it is not at all clear that a structural realist must accept that. Consider, for example, 'mixed predicates', such as 'is a part of', or, thinking ahead to the discussion in Chapter 7, 'is composed of', or 'is dependent on'. The Newman argument only goes through if these sorts of predicates are Ramseyfied away too but if a structural realist were to accept this she would be unable to formulate claims such as 'quarks are parts of nucleons', or 'spin is dependent on the Poincaré group' which seems bizarre (Melia and Saatsi 2006). Of course, the critic might object that unless *all* predicates are Ramseyfied away, the structural realism that results is not 'pure' in some sense, but I am suspicious of such demands for purity and take them to lead to a straw position.

Secondly, Melia and Saatsi argue that Newman's objection assumes that Ramseyfication takes place in an extensional framework. However, as they point out, '[t]he properties postulated in scientific theories are typically taken to stand in certain intensional relations to various other properties' (2006: 579). Such relations include being correlated in a law-like manner with, being causally dependent on, and generally, but crucially, given what I say in Chapter 10, *being modally associated with*. The extensional framework in which Ramseyfication takes place and in the context of which the Newman argument is presented cannot accommodate these. By appealing to such relations, and incorporating appropriate modal operators into the formal representation, the argument can be stymied (Melia and Saatsi 2006).

Of course, an appropriate semantics for these operators needs to be provided but now the worry is that the standard way of doing this does not enable the structural realist to escape the charge of triviality (Yudell 2010). A blunt response would be to say so much the worse for the standard semantics as a means of capturing the relevant features of scientific language and the world.[16] Certainly it is not clear that such models provide an appropriate modal semantics for OSR. Let us consider this in a little more detail.

[15] Even if all the terms, theoretical and observable, are replaced with existentially quantified variables, it might be argued that what this yields is still worth considering, from a structuralist perspective. This is the line Hintikka takes in his suggestion that the relevant structure is now effectively represented by the relationships between the second-order quantifiers and these can be revealed by adopting his 'independence friendly' logic (Hintikka 1998; for discussion see Cei and French 2006).

[16] I am grateful to Juha Saatsi for suggesting this response.

So, the construction of the standard semantics proceeds on the basis of a couple of apparently innocuous assumptions. The first is that we begin with a fixed domain of objects and the second is that in constructing the relevant possible worlds, we assume a standard accessibility condition that states that every possible world is accessible to every other (see Yudell 2010). This condition helps establish the truth conditions for the modal operators, where these latter conditions assign truth to modal sentences on the basis of what is the case in accessible possible worlds. Now this issue of what count as the truth conditions for the relevant modal sentences will loom large in Chapter 10, as will the more general issue of the construction of possible worlds from the perspective of OSR. Just to steal the thunder from that chapter, I shall argue that the standard way of constructing such worlds in order to underpin the purported necessity of laws—namely begin with a set of objects that the given law is said to 'govern', construct the worlds that are accessible from that one using the same set of objects as the basis and consider whether the same law holds—is inappropriate in the case of OSR, where the objects are taken to be dependent upon, and hence eliminable in favour of, the laws, as features of the structure of the world. Instead, I shall argue, this structure should be regarded as inherently or primitively modal. The very basis of the construction just outlined is thus rejected and certainly from this perspective, it may well be the case that 'any interesting scientific theories will make... sophisticated demands on the modal structure of reality' (Yudell 2010: 250), contrary to what is suggested.

Of course, what this brings out is that OSR rejects the very basis of the Newman objection, namely beginning with a set of objects over which the relevant relations are defined. Now, of course the set-theoretic mode of representation that I favour is going to have to introduce such a set but it is a further issue whether this set, introduced as it is in order to construct a certain kind of representation, must be taken seriously ontologically. By appealing to Poincaré's Manoeuvre again (see Chapter 4), we can write down such a set, without having to be ontologically committed to it. That the counter-response to Melia and Saatsi's rejection of the Newman objection depends on taking the set of objects ontologically seriously is clear from the role the first assumption (stated in the previous paragraph) plays in the construction of the semantics.

How then are we to understand the modal operators that Melia and Saatsi introduce? An alternative to the standard semantics is to draw on some account of the nature of laws in order to provide an interpretation. Two obvious options are, first, the Humean account which takes laws to be those regularities picked out by our 'best' theoretical system (a view we shall look at in more detail in Chapter 9) and secondly, the Armstrong–Dretske–Tooley (ADT) account, which takes laws to have a natural necessity grounded in universals. However, in both cases we must restrict the quantifiers of the Ramsey sentence in ways that Melia and Saatsi might not be comfortable with (Yudell 2010: 250–2). Now it is not clear just how extensive their discomfort would be,[17] but more important, for our discussion here, and again foreshadowing the discussions in Chapters 9 and 10, I shall argue that the ontic structural realist should not be committed to such accounts. A further option is Lange's account of laws, according to which law statements have a certain non-nomic stability and their modal features are grounded in so-called 'primitive subjunctive facts'. In this case, it is, at the very least, not clear that there would be any resultant commitment to a restriction in the quantifiers that would cause problems for the Melia and Saatsi response.[18]

This is true also of the account I shall advocate, according to which the laws are taken to be inherently modal. In both the Humean and the ADT cases, the restrictions arise from taking the relevant set of objects to be extensions of natural predicates, given as prior, or 'genuine', universals respectively. In the OSR case, we may not have such universals (depending on our metaphysics), nor do we necessarily have a prior notion of 'natural' predicate that effectively divides up the domain. On the contrary, the structural realist views metaphysicians' now-standard invocation of 'naturalness' with regard to properties with considerable suspicion (as indeed should all realists; see McKenzie 2014). As far as she is concerned, the sense of 'natural' here needs to be grounded in the relevant physics (and here issues of 'reading off' from theories come to the fore) and once one looks closely at such grounding one can see that the properties are yielded by, for example, the relevant group representation, as in the case of spin in section 5.5 (see McKenzie again). Thus as far as OSR is concerned and as we shall discuss further, it is the laws and symmetries that are taken as ontologically prior (as manifestations of the structure of the world) and upon which the relevant 'natural' predicates are dependent.

One might worry that appealing to laws rather than natural properties will not help here because of the 'deep connections' between laws and natural kinds (Yudell 2010: 252) and allowing the latter to be dragged into the picture by the former simply generates the same problems again. But setting aside the point that this actually needs to be shown, as far as I am concerned such kinds are likewise dependent on the relevant laws and symmetries—that is, the structure. Consider, yet again, the case of those fundamental kinds of bosons and fermions, into which all known particles are divided. These can be 'read off' from quantum statistics where one can see that the distinction is grounded in the symmetry expressed in Permutation Invariance (again,

[17] Yudell locates their discomfort in their apparent rejection of an appeal to natural properties as a way of restricting the quantifiers, but Melia and Saatsi remain neutral as to which specific metaphysics of properties and laws one should adopt. Their rejection of Newman's argument is grounded in a delineation of appropriate conceptual resources that mesh with our scientific language and allow a response to what is basically a model-theoretic problem. These conceptual resources can be viewed as acting as a constraint on the relevant model-theoretic constructions and only require that certain non-trivial conceptual distinctions can be made. Again I am grateful to Juha Saatsi for helping me to be clear on this.

[18] Yudell suggests that although Lange does not draw on a prior notion of natural properties, his view 'does end up being part of a systematic picture that includes natural properties' (2010: 251). But that doesn't entail the kind of quantifier restriction that might be problematic.

symmetrized wave-functions yield bosons, anti-symmetrized, fermions) and thus the kinds come from the structure. Hence their admittance does not come with the cost of restricting the quantifiers in the way that might be problematic. And again—just to hammer home the point—the advocate of OSR does not read the set of objects ontologically, and so does not take them to be extensions of predicates, universals or whatever, or at least does not take them to be so in a serious ontological sense that, again, generates the problems previously indicated.

Thinking of the structure of the world as inherently modal also offers an obvious counter to recent dismissals of Melia and Saatsi's account on the grounds that it is not clear how to motivate the introduction of the modal operators as logical primitives (Ainsworth 2009; Frigg and Votsis 2011). But actually, one does not have to view the structure of the world as inherently modal to generate the requisite motivation: if one took the modality to reside in, or be grounded on, the relevant dispositions (a view we shall come to in Chapter 9) one might also argue that in so far as this feature goes beyond the determinate aspect of the dispositions concerned it must be represented by an operator that is primitive in the sense of not being reducible. How else could this feature be represented formally?

A similar reply can be made to the argument that 'we surely cannot accept that modal operators expressing things like "it is physically necessary that" can be taken as logical primitives, since whether or not something happens as a matter of physical necessity is an issue that must be decided empirically, not as a matter of logic' (Ainsworth 2009: 162). Introducing such operators into one's mode of representation as primitives does not imply that it is logic that is deciding whether something happens as a matter of physical necessity or not. That decision is reflected in the choice of statement to which the operator is appended and hence the worry that Melia and Saatsi have conflated physical and logical necessity can be avoided. Furthermore, if one posits modality 'in the world', rather than 'in' our theories as the Humean does, then, as already indicated, one is going to have to have some way of representing that modality within the formal framework one has chosen. If—and I think it remains a big if—one were to insist that the Ramsey sentence still remains the best such framework for the structuralist then how else is that modality going to be captured? One can't simply point to the relevant relations, since these are going to be cashed out extensionally and the Ramsey sentence, as standardly set down, is perfectly compatible with a Humean account of laws, of course. So, 'building in' modal operators seems an appropriate way to go. And these can represent physical necessity as it is grounded in dispositions, say, or as regarded as inherent in the laws, as I prefer. Either way, this seems an acceptable way of representing that necessity without implying that what is necessary is decided by logic rather than science.

In conclusion, then, *if* one were to insist on the Ramsey sentence mode of representation, the Melia and Saatsi approach is surely the way to go in order to overcome the Newman objection, particularly given the long-standing emphasis on the modal nature of the structures in structural realism (Ladyman 1998; French 2006;

Ladyman, Ross, et al. 2007). Nevertheless this discussion may also be taken to illustrate the danger of harking back to Russellian structuralism and its Newmanian nemesis and taking the former as representative of modern forms of structural realism and the latter as undermining these as well. As we saw in Chapter 4, even at the time of Russell's exchange with Newman, Eddington was developing a form of structuralism that, he insisted, could evade Newman's criticism. And the crucial point for Eddington was that, unlike Russell's 'vague conception' of structure as a pattern of entities or at best a pattern of relations, he thought of structure in terms of a 'pattern of interweaving' or a 'pattern of interrelatedness of relations', giving as an example the group algebra of operators representing rotations acting on rotations. Here, Eddington argued, the group elements are defined by their role in the group and that role will not be captured by the Ramsey representation and hence the Newman problem does not apply.

Nevertheless, let us stay with the Ramsey sentence mode of representation for the time being as it will help us to articulate a further feature of scientific realism, namely that the relationship that holds between a theory and the world can be articulated in terms of the notion of 'reference'.

5.10 Realism, Reference, and Representation

According to the standard account of scientific realism, the theoretical *and* observational terms of our best theories are taken to *refer* (Putnam 1978: 20–1; Boyd 1973). For example, the term 'electron' is taken to refer to an elementary particle that falls under the kind 'fermion', has charge *e*, (rest) mass $9.10938291(40) \times 10^{-31}$ kg, and so on. Now, let us consider the question: what *fixes* the reference of a theoretical term such as 'electron'?

There is a well-known answer given in terms of the Ramsey sentence, following Lewis (Kroon and Nola 2001), as touched on in the previous section:

the reference of theoretical term $t = (\iota x)\ [T(x, o_1, o_2, \ldots o_m)]$

(simplified for one new term only).

Thus the term *t* refers to whatever uniquely realizes the open sentence 'T(x)'; if there are no realizers, there is no reference, whereas if there are multiple realizers, the reference is deemed to be indeterminate (see again Cei and French 2006).

The next question is: how much of the theory T do we need to invoke to fix the reference of *t*? Papineau (1996) offers a plausible approach that divides T into

— T_y which contributes to the fixing of the reference of *t*;
— T_n which does not contribute to the fixing of the reference of *t*;
— T_p which might contribute to the fixing of the reference of *t*.

This nicely accommodates the imprecision that occurs in practice, and it bears an obvious comparison with the partial structures approach.

Now, what goes into each component? The answer presumably depends on the kind of theory under consideration, and also the kind of realist adopting this framework. For many theories, T_y would include the relevant causal properties and the entity realist, say, might well insist that these are all it should include. Structural realists, on the other hand, would note that in practice T_y would include general symmetry considerations, as in well-known cases from elementary particle physics (Kroon and Nola 2001). Thus, we can characterize the shift from one theory of, say, the electron, to another in terms of the relevant properties moving from T_p to T_y.

Nevertheless, problems arise. Recall the 'classic' example of the ether, given as part of the inductive base for the Pessimistic Meta-Induction discussed in Chapter 1. Insisting that this term did not refer, despite the success of the theories it featured in (such as Maxwell's theory of electromagnetism), lays the realist open to precisely the concern that if terms of past successful theories are found not to refer, then the same may happen for terms of our present successful theories, thus undermining realism. We also recall the strategy of including descriptive elements as well as causal roles in one's account of reference (Psillos 1999: 293–300). Using this strategy, Psillos argues that the term 'ether' actually refers to the electromagnetic field (1999: 296–9), where the 'core causal description' is provided by two sets of properties, one kinematical, which underpins the finite velocity of light, and one dynamical, which ensures the ether's role as a repository of potential and kinetic energy. Thus—in terms of Papineau's framework—T_y excludes the problematic mechanical properties of the ether, which are effectively shunted off into the relevant models. The worry, however, as previously noted, is that this obscures precisely that which was taken to be important in the transition from classical to relativistic physics (da Costa and French 2003: 169). But if these properties are included in T_y, then there can be no common reference with the electromagnetic field.

Now, again, the standard realist might insist that when she, as a realist, insists that the world is as our best theories say it is, that covers the relevant scientifically grounded properties only and not these metaphysical natures. But then, what is being referred to is only the relevant *cluster of properties* which are retained through theory change. Hence, reference to the ether was secured via a certain cluster of properties that also feature in reference to the electromagnetic field. In so far as these properties feature in or are the subject of the relevant laws, certain *structural* aspects of theories are retained through theory change.

Given this, one might expect reference to play at best an attenuated role in the structural realist picture. Worrall, for example, insists that he has no need for reference at all, even though his epistemic view still retains objects (albeit 'hidden' behind an epistemic veil) and characterizes structure in terms of Ramsey sentences. It is easy to see how the epistemic structural realist could appropriate the account outlined here, with Chakravartty's detection properties (Chakravartty 1998) featuring in T_y and the auxiliary properties (introduced as part of our efforts to get a theoretical grip on the entity concerned, but which may eventually be abandoned) falling in T_n

or T_P since it is possible for such a property to come to be regarded as a 'detection property'.[19]

However, as far as Ladyman is concerned, such epistemic forms of structural realism fail to address the problem of ontological discontinuity across theory change:

> The Ramsey sentence of a theory may be useful to a concept empiricist because it shows how reference to unobservables may be achieved purely by description, but this is just because the Ramsey sentence refers to exactly the same entities as the original theory. If the meta-induction is a problem about lack of continuity of reference then Ramsefying a theory does not address the problem at all. (French and Ladyman 2003: 33; cf. Ladyman 1998)

And certainly, it is difficult to know how to understand all those existential quantifiers with reference out of the picture.[20] Of course, the epistemic structural realist might respond by insisting that there is a difference between the standard realist's reading of the original theory and her own reading of the Ramseyfied version in that on the latter, what is being referred to are these 'hidden' natures and not the objects fully clothed as it were. However, here one can recall an earlier comment by Shapere:

> to say that continuity is guaranteed by the fact that we are talking about (referring to) the same "essence", where we do not or cannot know what that essence is, is merely to give a name to the bald assertion of continuity. (Shapere 1982: 21)

Thus we have a dilemma: if reference is simply to the hidden 'essence' of unobservable entities, then Shapere's point bites; if however one were to maintain that reference is to the entities as usually understood, then Ladyman's criticism applies and the problem of ontological shift rears its ugly head again.

Thus one may simply reject the Ramsey sentence as the most appropriate way of representing the structure that the realist should be committed to. However, even if it is granted that the partial structures mode of representation can accommodate the structural aspect of structural realism,[21] there is still the realist side. How can this be maintained if reference is dropped as well?

One response is to develop a distinction previously made by Suppes (see da Costa and French 2003): from the external perspective, the 'world structure' (for want of a better name) is understood to be represented via the interrelated models of the semantic approach. If one wants to talk of truth and reference, strictly speaking, one should shift to the internal perspective, in which we have propositions which are true if satisfied in the relevant model (and this must be modified of course if the

[19] Bain and Norton's structural realist account of the development of theories of the electron might be nicely couched in these terms (see Bain and Norton 2001).

[20] Similarly, Cruse and Papineau have argued for a form of 'standard' realism without reference in the context of a Ramseyfied characterization of theories, insisting that we should regard the existential claims as 'approximately true' but it is hard to know how to understand this (Cruse and Papineau 2002).

[21] Ainsworth (2010) argues that Newman-type issues arise within the semantic approach as well and hence French and Ladyman's (2003) dismissal of these issues must fail.

realist wants to appeal to a notion of approximate truth) and which contain terms which refer to either 'thin' objects, or, on the more radical form of OSR, aspects of the structure of the world, understood in ways I shall outline in subsequent chapters.

From this dual perspective, then, one can have all the representational advantages of the semantic approach whilst retaining truth and reference in a way that satisfies one's realist inclinations. However, there might be a sneaking suspicion that this is too much like having one's philosophical cake and eating it too! It might be argued, for example, that the purpose of introducing the notion of reference is to provide an appropriate connection between words and the world and hence a certain philosophical economy is achieved by having it play this role. But of course, economy comes at a price, which in this case is the aforementioned representational advantages of the semantic approach.

Alternatively, one could bite the bullet and focus on the representational side only, as already suggested here, arguing that a robust notion of representation can provide the requisite connection between theories—conceived of model-theoretically—and the world and that some understanding of 'good' and 'bad' representations can appropriately underpin the realist's epistemic attitudes (see Contessa 2011).

5.11 Models, Mediation, and Transparency

However, the following concern arises. It is now generally accepted that between the 'high-level' theoretical models and the 'low-level' data models there is a hierarchy of so-called 'mediating models' which enable the various levels of the hierarchy to be appropriately related (Morgan and Morrison 1999).[22] This in itself is not a problem for the partial structures approach, where the structures were explicitly designed to accommodate such interrelationships (as suggested originally by Suppes, for example), via the device of partial isomorphisms holding between the various levels (see, again, Bueno 1997 or da Costa and French 2003). However, it has been claimed that these mediating models may be mutually incompatible, in the sense that different such models may be applied in different ways and may thus be related to different data models. Unfortunately this creates a potential problem for the structural realist (Brading 2011: 52–7).

Consider, first of all, realism in general and the question: given the hierarchy of models, what is the realist who adopts the set-theoretic semantic approach as her meta-level mode of representation going to take as her theory? If she takes the whole hierarchy, then she is going to have to confront the issue of the mutual incompatibilities between mediating models in order to tell a consistent story about how the world is (Brading 2011: 53–4). The obvious alternative is to take just the highest-level

[22] Earlier expressions of this idea can be found in Apostel (1961: 11) and Hutten (1953–1954: 289). For a critical discussion of the supposed autonomy of such mediating models see Bueno, French, and Ladyman 2012.

theory as telling such a story, with the hierarchy understood as simply linking up this high-level theory with the relevant phenomena. As far as the content of the realist's beliefs are concerned, the intermediate levels of the hierarchy become 'transparent' (Brading 2011: 54). The issue now is to justify this 'transparency of the hierarchy'.

For the object-oriented realist this can be achieved by appealing to the relevant objects and their properties that are the subject of her claims. The kinds of objects the theory is concerned with are characterized by the high-level theory and then may be traced up and down the hierarchy as it were, ensuring that at any given level we are talking about the same kind of thing (such as 'electron', for example). But then,

> [t]he kinds of objects appearing in any model at any level of the hierarchy are labeled *as that kind* from resources *outside* the model (and for mediating and data models, from outside that level of the hierarchy altogether). Therefore, it is legitimate to point to objects in a model of the high level theory and call them electrons (say), and then trace (with further pointing) the presence of these objects (or rather, their trajectories) down through the hierarchy to the data models (and, so the realist hopes, into the world). (Brading 2011: 54)

Thus even if different mediating models ascribe incompatible properties to a given object, these models can be regarded as involving different idealizations or approximations of the same fundamental kind of object, where that fundamental kind is characterized solely by the high-level theory (Brading 2011: 55). The hierarchy thus remains transparent with regard to the content of the realist's belief. The structural realist, on the other hand, appears to face problems in justifying a similar transparency.

So, she faces the same choice in cashing out her commitments. Again, however, there are good reasons for not taking the whole hierarchy as representing these; or, better, as not representing her commitments regarding the fundamental structure of the world. There is a sense in which the hierarchy can be said to represent the structure of the world where this is taken to encompass fundamental, intermediate and, as it were, observable levels, if the mediating models can be construed as representing the structure of the intermediate levels, say. However, it would be odd, to say the least, to insist that the whole hierarchy, with its panoply of different models, represents the structure of the world at the most fundamental level.

So, let us suppose that the structural realist adopts the same understanding as the object-oriented realist and takes the high-level theory as yielding the (structural) content of her beliefs. It would appear that she can account for the transparency of the hierarchy by appealing to, say, the relevant partial isomorphisms linking the models at each level (for an explicit representation of these relationships see Bueno 1997). Now what is being traced is not the kinds of objects the realist is committed to, but the relevant 'shared structure' in terms of which the structural realist's beliefs are expressed (Brading 2011: 56). However, if the mediating models are mutually incompatible, then what we have is a proliferation of incompatible structures at the lowest levels of the hierarchy. But then, without a unique structure 'cascading' down the hierarchy, it becomes unclear in what sense the structural realist can claim that she has, at the high level, 'latched onto' *the* structure of the world (Brading 2011: 56–7).

However, I think the problem can be dissolved. First of all, we should be careful when we ascribe this mutual incompatibility of mediating models.[23] Two such models might be entirely incompatible in the sense that they share no features in common. In this case they would have to be placed in different hierarchies and thus be mediating between different theories and the relevant data models. This obviously raises no concerns for the structural realist. In order to be part of the same hierarchy but also be mutually incompatible, the models must have some features in common but in that case these features or 'parts' of each model can be related via partial isomorphisms and thus traced through the various levels of the hierarchy. In this manner, the 'transparency' is justified again.

But perhaps the problem is deeper, arising from the assumption that the structural realist must take there to be a unique structure that 'cascades' down the hierarchy. However, it is not clear how reasonable an assumption this is, since we should not expect such a cascade given the role of idealizations and approximations in relating our high-level theories to the low-level data models. If we drop the uniqueness requirement then we can still claim there is structure cascading down, or fountaining up, and that the relevant parts of this structure can be interrelated via the device of partial isomorphisms.

We can see how this works in the case of the literal fountaining of liquid helium 3 (see Bueno, French, and Ladyman 2002). Here the explanation of this phenomenon in terms of Bose–Einstein statistics can be represented within the partial structures approach, with the relationships at the bottom of the hierarchy captured via an extension of the notion of empirical adequacy, and those at the top represented via a notion of partial homomorphism, which allows us to represent the partial importation of the relevant group-theoretic structure into the physical domain.[24] In this case the relevant hierarchy can be explicitly represented within the partial structures approach. In particular,

> only some of the structural relationships embodied in the high-level theory of Bose–Einstein statistics (the 'general features') needed to be imported in order to account for the (low-level) qualitative aspects of the behaviour of liquid helium and this importation can be represented in terms of [this] framework above of partial homomorphisms holding between partial structures. (Bueno, French, and Ladyman 2002: 516)[25]

[23] And of course this incompatibility can be straightforwardly captured by the partial structures approach (da Costa and French 2003: ch. 5).

[24] I say 'partial importation' because not all of the structure of the permutation group is so imported in this case, of course—the structures corresponding to the anti-symmetric and para-symmetric representations are not, for example.

[25] The use of partial homomorphism as a representational device in this manner blurs the distinction—at the meta-level—between mathematics and physics, an issue that I shall return to in Chapter 9.

But perhaps there is a yet deeper worry behind the problem here, namely that the notion of kinds provides, for the object-oriented realist, an indication of what the theory is about, such that she can still state this, even when faced with incompatible mediating models. And without such a notion, the structural realist cannot do this, as she can only point to the various bits of structure at the appropriate level. Here too a response can be constructed. First of all, the structural realist can in effect piggyback on the object-oriented realist's use of kinds here, but where the latter insists these are kinds of objects, understood in a metaphysically robust manner, the structural realist offers her structuralist reconstrual. Thus, in the case of explaining the behaviour of liquid helium, the object-oriented realist may point to the role of bosons, as a kind, in that explanation and assert that this is what the theory is 'really' about. However, the structural realist can then point out that the relevant kind classification here has to be understood group-theoretically, and hence structurally, so there is nothing particularly object-oriented going on in the physics. She can still track these kinds, still use them to say what the theory is 'about', but when it comes to the ontological crunch she will cash out this notion of kind not in terms of sets of objects possessing certain properties but in terms of the relevant symmetry conditions.

Secondly, she can simply point to the relevant structure given at the highest level of the hierarchy and insist that that is what the theory is about and maintain that the various features or bits of this structure can be tracked up and down the hierarchy via the relevant partial iso- and homomorphisms. Indeed, she might well insist that her view has an advantage over that of the object-oriented realist in so far as she does not need to worry about ensuring a particular term has the same 'meaning' up and down the hierarchy, where this is given via reference to some object.

In these ways, then, the transparency of the hierarchy can be secured for the structural realist as well.

5.12 Modes of Representation: Morphisms

Furthermore, I think that this response holds certain advantages over the category-theoretic approach which, as I have noted, is sometimes offered as an appropriate representational framework for the structural realist (see, for example, Bain 2013; for criticism see Wuthrich and Lam 2015). As briefly indicated by da Costa and French (2003: 26), one could certainly consider representing theories in such terms but it's not clear what would be gained given the level of abstraction at which the relevant categories sit. In particular, when it comes to the issue of capturing the kinds of inter-theory relationships that motivate structural realism, it is unclear whether category theory offers a better framework than the set-theoretic one.

Now this claim might be challenged in two ways. First of all, category theory might offer a useful meta-(meta-) framework for representing the interrelationship between the two aspects of OSR arising from its twin motivations: on the one hand, we have a focus on inter-theory relationships; on the other, we have the group-theoretic representation of objects and properties. One suggestion might be that category theory could offer an appropriate way of characterizing the relationship between these two aspects via the relationship between the categories 'Set' and 'Group'. Of course, one can respond that the relationship between the laws of a theory and the symmetries is already nicely captured set-theoretically, but nevertheless, the role of internal symmetries in this context might push us towards a more general category-theoretic account.

The second would be to consider whether category theory offers a better framework for OSR because a category is characterized by its morphisms and not the relevant objects, with the latter regarded as secondary at best, or as definable in terms of, and consequently but more radically perhaps, reducible to, the morphisms going in and out. Thus category theory might offer a way of representing the shift in focus from objects to structures that is central to OSR.

Certainly the set-theoretic representation appears inelegant at best in this regard. If we recall Cantor's original formulation, and its motivation, we can see that a commitment to objects appears to lie at the heart of the origins of the theory and even if we introduce novel formulations that capture the sense in which these objects might not be individuals, that commitment remains. This is not to say that there aren't ways of handling the structural realist's 'reconceptualization' of objects within the set-theoretic framework (see French 1999 and 2006; French and Ladyman 2011). We can perform what I have called the 'Poincaré Manoeuvre' (see Chapter 4): as we recall, we begin with the standard presumption that theories are committed to objects, at least as the subjects of property instantiation; we then reconceptualize and, on the more 'radical' form of OSR, eliminate those objects in structural terms. Thus the putative objects come to be seen as merely stepping stones or heuristic devices to get us to the relevant structures. Given the initial presumption, it may seem natural to employ a set-theoretic representation, which includes the putative objects of course, but then we must insist that this be read 'semitically'; that is from right to left, so that, taking the simple formula:

$$<A, R>$$

the relations R are understood as having ontological priority over, and can be understood as constituting, the objects of the domain A.

Thus we are faced with the following situation: the set-theoretic framework nicely captures the various inter-theory and maths-theory relationships that the structuralist will be interested in but has to be manoeuvred into accommodating the shift away from objects; whereas category theory has that shift 'built in' as it were, but operates at too high a level to straightforwardly capture the inter-theory relationships, etc. In the spirit of a pluralist approach to this issue of meta-level representation one option would be to again follow a Suppesian line and suggest that when it comes to accommodating the structuralist response to the pessimistic meta-induction we adopt an 'external' characterization of the relevant interrelationships in set-theoretic terms, then shift to an 'internal' or ontological characterization through category theory in order to capture the implications of modern physics for the notion of object.

Alternatively, one can view category theory as a language in terms of which we can analyse *systems that are structured*, rather than as offering a 'meta-science' of structures (Landry 2007). As such, it presents a framework that is 'prior in definition' to any particular system without being committed to the claim that mathematics is 'about' actual or possible objects and structures; in this latter sense, then, 'it is philosophy without either metaphysics or modality' (Landry 2007). However, importing such a view into the current context is problematic (Landry 2011). As I shall argue in Chapter 8, talk of systems that are structured, in the sense that the systems are ontologically prior to the structure, is not appropriate for OSR (French 2006). In particular, the pulling back from metaphysics and modality would be highly questionable in this context (as Landry herself acknowledges). As indicated previously, it is in its presentation of putative 'objects' and their properties that group theory contributes to a metaphysics of them and category theory's contribution to such a metaphysics is attenuated by the comparatively higher level at which it operates. Thus, a category-theoretic reconceptualization of physical objects in terms of the relevant morphisms 'in and out' may sit at too high a level to capture the relevant physical particularities.[26] As for modality, again as we'll see later, there are advantages to be gained from regarding (physical) structure as modally informed.

5.13 Modes of Representation: Structure as Primitive

The following question now arises: if the set-theoretic approach is compromised by its surface-level commitment to objects, and the category-theoretic stance operates at too abstract a level, why not offer an alternative characterization that defines structure *directly*, without the prior device of elements over which hold the relations we are actually interested in, and at the appropriate level of concreteness? However, not for nothing has set theory come to be widely regarded as an appropriate foundation for most, if not all, of mathematics and its adoption in the form of the semantic approach, outlined previously in this chapter, followed Suppes' declaration that the appropriate representational framework for the philosophy of science was mathematics, not meta-mathematics. The sense of appropriateness here has to do with the mathematization of much of modern science, especially physics, of course: given that, it makes sense to use as a representational framework for the analysis and understanding of science that which sits at the foundations of mathematics.[27] So, if we're going to abandon this framework, we need to be given some other way of

[26] See also Muller 2010.

[27] In originally presenting his version of the semantic or model-theoretic approach Suppes gave group theory as one of his principal examples (the other was psychological learning theory; Suppes 1957).

representing the mathematics in physics; or, at least, if we're going to define the term 'structure' directly, then we will need to ensure that the framework constructed on the basis of that definition can accommodate this mathematics.

An attempt to offer just such a characterization has been made (Muller 2010). The motivation given is that neither the set-theoretic nor category-theoretic approaches clarify what it means to say that a given system *is* or *has* some structure, where this clarification is necessary for reference, which in turn, Muller maintains, is required by any viable form of realism.

Thus, Muller takes as his example of a system that of a helium atom in a uniform magnetic field HeB and begins with the semantic approach. From this set-theoretic perspective, the quantum mechanical structure used to describe this system is as follows:

$$S(\mathrm{HeB}) \equiv \langle \mathrm{L}^2(\mathrm{R}^3), H(\mathrm{B}_0), \psi, \mathrm{Pr}_t \rangle$$

where $L^2(R^3)$ is the relevant Hilbert space, $H(B_0)$, the Hamiltonian, ψ, the wave-function for the system, Pr_t the Born probability measure that gives the probability for observing a value for the energy of the system when in the state given by ψ (Muller 2010).

Now, the crucial question is how are we to understand the earlier claim in this context, namely that HeB is or has the structure *S*(HeB)? The 'is' here cannot be taken to be that of identity, on pain of falling prey to the accusation that OSR collapses into a form of Platonism. Alternatively, as Muller notes, we can take either 'is' or 'has' in the claim to be associated with predication, just as we would with the claims 'the tomato is red' or 'the tomato has flavour'. To avoid the problem of having to regard the helium system as set-theoretic again we must expand our set theory to include so-called 'Ur-elemente' which represent physical systems. The appropriate language now includes set-theoretic variables and 'physical-system-variables'. This is all unproblematic and standard. And as Muller shows, one can then construct a predicate that holds between the structure and the physical system, as denoted by the relevant variable within the language. The problem now is that, as Muller notes, variables do not refer (Muller 2010). Again, drawing the comparison with the humble tomato, consider the sentence 'Red (this-tomato)': this will be true and 'this-tomato' will refer if there is a red tomato on the plate in front of us. Similarly, we want to say that the relevant expression for the helium atom is true and that the relevant variable refers to a helium atom in a uniform magnetic field.

So, we need some account of reference. Muller plumps for descriptivism, on the grounds that the standard causal account is inadequate for science.[28] Briefly put, on

[28] He does not consider Psillos' hybrid causal-descriptivist account, which meets some of the objections to the standard causal view in this context. For a general discussion of the notion of reference in the quantum context which addresses some of the more well-known criticisms of the causal account, see French and Krause 2006: ch. 5.

the descriptivist view a term refers via the descriptive content associated with that term.[29] Now, if we take that content to be that given by set-theoretically reconstructed quantum mechanics, it turns out that the description applies indiscriminately, across all physical variables, which means that every single physical system counts as HeB. And it gets even worse, since one can show that for every set structure there are as many structures as there are systems, so 'every physical system is everything' (Muller 2010).[30]

Now there are various ways in which one might evade this conclusion: one might adopt an alternative account of reference, or abandon reference entirely, or drop the set-theoretic approach, or re-construe it in the manner I have indicated here, or, as Muller prefers, take structure as a primitive. Let me briefly consider each of these, since this will further help to illuminate what is at stake here.

First of all, Muller is assuming a standard extensional understanding in his analysis and as we have seen in the discussion of the Newman problem, incorporating causality and other modal notions in our framework will take us beyond this. So we might consider adopting an intensional framework[31] and some causal-based theory of reference. However, we have already touched on the most plausible form of the latter, namely the causal-descriptivist view advocated by Psillos, and found it wanting.

What about dropping reference from our framework altogether, as Worrall does, and taking structural realism to be a form of 'realism without reference'? Muller is dismissive, insisting that this '[s]mells like realism without reality' (2010: 10). However, given the well-known problems with reference in the context of quantum mechanics (French and Krause 2006: ch. 5), perhaps realists should hold their noses! Indeed, the structural realist might well feel that the whole framework in which this conception is expressed sits at odds with her stance, as expressed here. So, typically it is the theoretical terms of our theories that are taken to refer, and, of course, what they are taken to refer to are objects, whether unobservable or observable. But this is already to adopt a particular way of 'reading off' our commitments from our theories that I have suggested should be dropped. The advocate of ESR will argue that what should be read off are the relevant equations that are retained through theory change, and the defender of OSR will agree, but urge that the same attitude should be adopted towards the symmetries of the theories. Indeed, the terms themselves only have meaning because they are embedded within this nexus of laws and symmetry

[29] Of course, if that content is given by the relevant theory, then theory change raises obvious problems for this view, which was one motivation for coming up with the alternative causal account in this context.

[30] According to Muller, the same conclusion holds for Brading and Landry's approach: if we identify the 'objects' that they take to be 'presented' by a structure as the Ur-elements, then one can show that everything can be 'presented' by every structure, so all of them 'present' everything, or conversely, every structure can 'present' anything (2010: 9 n. 16). Brading and Landry would, of course, reject such an identification.

[31] This possibility is briefly canvassed in da Costa and French (2003), and meshes with Carnap's use of higher-order logic to define the relevant notion of structure.

principles and taking them to have ontological priority and thus to be the relata of a relation of reference that has as its other end point objects in the world is to put the metaphysical cart before the horse.

But how, then, is the structural realist to articulate her view of the way in which theories—and in particular those structural features to which she is giving ontological priority—latch onto the world, as Ladyman, for example, puts it? One option is to appeal to idea of theories and models as *representations* of systems in the world (French 2003b; Bueno and French 2011). Muller is again dismissive, pointing out, in effect, that one needs to appeal to the intention of the person using the relevant theory or model, or, equivalently, to the purpose to which the theory or model is being put, in order to fully articulate the notion of representation in this context. This is a familiar theme. However, as French (2003b) noted (see also Bueno and French 2011), intentions play a much reduced role in the scientific context than in the artistic, for example, and, furthermore, it can be argued that they should not be seen as constitutive of the mechanism of representation but rather as part of the relevant context that allows one to choose one of the many possible representational relationships that may hold with regard to a particular theory. Furthermore, as Muller himself notes, as far as the realist is concerned, there is only one purpose of representation: to describe the world as it is (2010). Of course, this may still leave more than one possible representation on the table but this amounts to the usual situation of underdetermination (2010) and if the situation persists one can deal with it in the ways I've indicated in Chapter 3.

Nevertheless, a problem remains: suppose we are left with one structure and we assert that the system—the helium atom in a uniform magnetic field, say—is or has that structure. Then, 'we still need to know what "structure" literally means in order to know what it is that we attribute to [the system], ... and, even more important, we need to know this for our descriptivist account of reference, which realists need in order to be realists' (2010: 15). However, it is not clear what is being asked for with the demand that we need to know what 'structure' literally means here. In one sense, it is, or should be, quite obvious in the given context what we mean when we say that the system is or has a certain structure, since that structure will be given to us, or *presented*, by the relevant theory. Thus the structure of the helium atom will be that given by (the relevant part of) quantum mechanics. The role of set theory here is to offer us a (meta-level) representation of that structure, for our purposes as philosophers of science. And of course, from this perspective, we see no need to have an account of reference—what the realist needs is some account of how theories latch onto the world and representation clearly fits the bill (at the 'object' level).

Nevertheless, there may be costs for the realist if she adopts such an account. Can she talk of the truth of representations in any way other than as a *façon de parler*, for example? It would seem not if she wants to adopt a Tarski-style account of truth, and thus she will be reduced to talking of representations as more or less 'faithful', as already indicated (Contessa 2011; Suarez 2004).

Alternatively, one could again deploy the Suppesian dual perspective: adopting the 'intrinsic perspective' and representing theories in syntactic terms, one can still avail oneself of the standard Tarski conception of truth as correspondence (da Costa and French 2003). One could also deploy reference from this perspective, although as I have said, it might seem at odds with the way I've suggested that structural realists should read off their commitments. Shifting to the extrinsic perspective one can (meta-) represent theories set-theoretically and not only characterize their interrelationships in terms of partial isomorphisms and the like, but also take them to represent systems via a similar formal mechanism. Thus we would retain truth, from within the intrinsic perspective, while replacing reference with representation, from within the extrinsic.

Muller's preferred option is to elaborate a 'direct characterization' of structure.[32] The idea is that just as the standard set-theoretic formalism takes the concept of set as a primitive, introduced via set variables, so the language of structures should introduce structure as a primitive via 'structure variables', and not be reduced to either sets or category-theoretic objects. The structural realist will take these variables to range over all the structures in physical reality, where it is science that tells us which of all the possible structures covered by our theory of structures are actually realized or instantiated. According to Muller, the structural realist can then say that those predicates in the language of the theory of structures that single out these realized structures provide literal descriptions of these structures, and on this point a descriptivist account of reference can get a grip (2010). Thus the claim that a given system 'is' or 'has' a certain structure can be articulated in the following terms: to say that a system HeB, say, is or has a structure *S* of type F is to say that F is a predicate in the language of structures such that F(*S*), where '*S*' is the relevant structure variable of the language and the predicate F also supplies a structural type-description of the system, such that we can say F(HeB).

Now I am sympathetic to such a project. It would give us a theory and language of structures directly appropriate for structuralism in general. It would mean we wouldn't have to indulge in the fancy footwork of the 'Poincaré Manoeuvre' or read the set-theoretic representation semitically. It would mean, perhaps, that we could finally and definitively respond to those critics who insist that we cannot have structures without objects. However, as it stands, it remains a promissory note.

Furthermore, it is important to appreciate that even if we were to be given such a new framework that takes structure as primitive, it would simply be one more such mode of representation. If we go back to the crucial claim that HeB *is* or *has* the structure *S*(HeB), the expression, *S*(HeB), whether understood within set theory or structure theory, stands for, or characterizes, at the meta-level, the quantum

[32] One might also consider adopting the category-theoretic framework, as already indicated. However, Muller maintains that one ends up with the same conclusion as he obtains in the set-theoretic case: there are as many category-theoretic structures as there are physical systems and the descriptivist account fails to get off the ground here as well.

mechanical structure that the theory presents to us at the object level. Thus, in one sense, as I have said, when the structural realist is asked what 'structure' literally means in this specific case, she should say 'it means *this* particular quantum mechanical structure', pointing perhaps to the relevant sequence of symbols in the textbook, or, more generally, the relevant part of the theory. If I am asked 'what is this structure in terms of which you claim the "kinds" that we denote as bosons and fermions can be articulated?', all I can do is point to the relevant features of the permutation group (and its symmetric and anti-symmetric representations). Again, this is not to slide into a naïve Platonism and say that the world is group-theoretic—the group theory itself represents that feature of the structure of the world and it is in such terms that these features are presented to us in the appropriate theoretical context. But to represent that structure, or those features of the relevant structure, for my purposes as a philosopher of science—to make ontological claims about it, to insist that it is a common feature of a particular sequence of theories, to articulate the relevant interrelationships with aspects of other theories, and so forth—I need to choose an appropriate mode of representation (appropriate that is for my purposes as a philosopher of science). As we have seen, there are a variety of such modes available, including the Ramsey sentence, category theory, and set theory under a semitic reading. Muller's theory of structures (as primitives) will certainly represent a considerable formal advance but it will remain just one such mode of representation, albeit one that may well be more convenient for the purposes just listed.

5.14 Conclusion: Presentation and Representation

Without a formal framework, set-theoretic, category-theoretic, structure-theoretic, or otherwise, that can act as an appropriate mode of representation at the meta-level, our account of episodes such as the introduction of group theory into quantum mechanics would amount to nothing more than a meta-level positivistic recitation of the 'facts' at the level of practice. Any concern that the choice of a set-theoretic representation of such an account would imply that set theory is *constitutive* of the notion of structure can be assuaged by insisting on the distinction between levels and modes of representation. To reiterate: at the level of scientific practice, group theory was introduced and used to represent physical objects, their properties, and the latter's relevant interrelationships. This is the mode by which these objects are *presented* at this level. At the level of the philosophy of science, there exists a variety of modes by which we can represent both this practice and our structural commitments. In deploying the semantic approach, or partial structures, there is no suggestion that, first of all, physicists themselves had such an approach in mind when they applied the mathematics that they did, or related the theories in the way they did;[33]

[33] Brading and Landry acknowledge that they are not implying that such a suggestion is being made.

nor should this be taken to imply the view that the world is somehow, in some Platonic sense, set-theoretical. The claim is merely that in order to appropriately represent the physicists' representation of the phenomena, the semantic approach offers a number of advantages to the philosopher of science, and in particular, for the structuralist, by 'making manifest' the relevant structures.

Furthermore, as I have said, there is certainly a degree of context dependence here in the sense that the physical context 'reveals' and hence presents that aspect of the world-structure that is represented by group theory. And I agree on the significance of 'shared structure' in this sense for the presentation of the aforementioned objects and their properties. It is certainly this shared group-theoretic structure that is doing the work for the physicists at this level and not partial structures or anything of that kind (except maybe implicitly if one accepts set-theoretic reductionism). But I disagree that this is sufficient: at the meta-level where philosophers operate, it is (partial) set-structures that are doing the work (at least in the account I have offered). Within such an account, the structure is represented set-theoretically but the putative objects are presented and reconceptualized (and hence metaphysically eliminated qua objects) via group theory and it is the particularities of the latter's representations (in the technical sense) that reveal, represent, and present to us the concrete features of the structure of the world.

In the next chapter I will consider three concerns that have been put forward with regard to this presentation and reconceptualization: that a form of underdetermination arises again with regard to the so-called 'automorphism towers' that can be generated within group theory; that group theory alone cannot capture the full extent of the structure of the world; and that objectivity cannot be captured in these group-theoretic terms. In responding to them I hope to make good on my various promises to flesh out the structuralist picture offered by OSR.

6

OSR and 'Group Structural Realism'

6.1 Introduction

In the previous chapter I argued that what does the relevant work when it comes to the physics is group theory and that what does the work when it comes to the philosophy of science can be set theory (although other representational devices are also available). It is the former that presents at the level of theory those features of the structure of the world that are associated with the fundamental symmetries and invariances that are so important in modern physics. In bringing these features and the group-theoretic presentation to the fore, OSR places itself in a long tradition whose history, although overshadowed by the likes of Russell, represents a significant intertwining of physics and philosophy, as we have seen. Indeed, this tradition has been identified with a distinct variant of structural realism, called 'Group Structural Realism' (GSR; see Roberts 2011; see also Kantorovich 2003), although the significance of group structure is so intimately bound up with OSR that I shall take the former to be an articulation of the latter.[1] However, GSR, and hence OSR, have been the subject of three important objections: first, that group theory generates further structures and there are no grounds for identifying which represents the structure of the world; secondly, that group structure does not capture the relevant dynamics; and thirdly, that the emphasis on invariance that the group-theoretic framework embodies is not sufficient to ground an appropriate account of the objectivity of science. I shall consider each of these in turn.

6.2 Concern 1: Toppling the Tower of Automorphism

The first objection concerns the structural describability of *structure* (Roberts 2011).[2] In one sense, this is unproblematic in that one can appeal to structure to describe and represent structure; indeed, given, for example, Eddington's emphasis on the relevant

[1] As Roberts illustrates, GSR can nicely account for the kinds of theory change that motivate the move to structural realism, as discussed in Chapter 1 (Roberts 2011).

[2] I'll touch on a similar concern in Chapter 8, to do with the structural describability of causality.

structure as understood in terms of the interweaving of and hence relations between relations, such a representational move lies at the heart of OSR. However, this describability function generates a further concern, namely the 'higher structures problem': 'If S is a structure, what is the status of the structure of S itself?' If by the term 'status' is meant 'metaphysical' status, rather than representational, say, the following dilemma arises:

> On the one horn, we would like to choose just one structure to be at the top of our metaphysical hierarchy. But it is unlikely that we will be able to give a well-motivated reason to choose between a structure S, and the structure of S itself. This pushes us to the other horn: we must promote the whole shebang, both S and the structure of S, to a metaphysically 'fundamental' status. But this account of metaphysics, if one can even make sense of what counts as the 'whole shebang,' leads to a much more complex hierarchy, which need not satisfy the aims of structural realism. (Roberts 2011: 57)

Take as a concrete example the automorphism group *AutS* of S; then the dilemma bites like so: either we give a reason for choosing *AutS* over S as more fundamental (or vice versa) or we swallow the 'whole shebang' but that's a big shebang, given the existence of so-called automorphism towers; that is, a succession of automorphism groups of automorphism groups that are non-trivial in the sense of generating new groups and that may only terminate in the transfinite, or even cycle.

Let us begin with horn number 1: one approach might be to adopt a variant of the point already noted that whenever we take the physical structures we're interested in and embed them into 'higher' mathematical structures, we obtain a lot of surplus structure that may or may not be heuristically very useful (Redhead 1975; see also his 2003). As I also noted in the previous chapter, this embedding can be represented set-theoretically via the notion of partial homomorphism (see Bueno, French, and Ladyman 2002; Bueno and French 2011). Of course, the situation with the automorphism towers cannot be represented in this way (since *AutS* is not a sub-structure of S), but nevertheless the core issue is the same: where do we draw the metaphysical line between those structures we take to represent the world and those that are surplus?

One option would be to appeal to mathematical considerations but the tower can be extended downwards and in different ways, and a kind of underdetermination arises (Roberts 2011). Again, this seems little more than a reiteration of the point that mathematics yields more structure than we need to represent the world, which hardly comes as a surprise and the issue remains as to how to draw the relevant line.

Alternatively, we might draw that line on physical grounds, by appealing to the objects to be represented. However, this leads to circularity if we think of the group as providing physical objects with their properties, so we can't appeal to those objects to pick out the group. Perhaps, then, one can approach the issue from a different direction: what 'picks out' the group is the relevant theoretical context via the usual justificatory moves (and thus is grounded in the appropriate empirical context). The structural realist then metaphysically reconstitutes any putative physical objects in group-theoretic terms, claiming that it is the latter that articulates the sense of structural reconceptualization or elimination of these objects. Thus rather than a circle, we have two 'arms': one justificatory and hence epistemological; and the other metaphysical and hence ontological. (As we shall see, a failure to note the justificatory side of things undermines the further concern regarding objectivity that I shall consider shortly.)

A related option is to stick with whatever is closest to the physics, by adopting some kind of 'natural physical attitude'; that is, we accept the group that is most naturally suggested by the physics (Roberts 2011: 64–5). That in effect is what Brading and Landry urge us to do (see also Lyre 2011). But of course, for the structural realist this is not enough. Leaving aside the issue of the representation of the relevant results at the meta-level of the philosophy of science, there is the question of how the physics is to be interpreted. By interpretation here I mean an appropriate *metaphysical* understanding of structure that best fits the mathematically informed physics of SU(3), say, in the sense of avoiding object-talk and reducing the level of humility involved. Given the stated aim of OSR to provide such an understanding, it is hard to see how the structuralist could be 'barred' from appealing to interpretation here. Again, the move is to take what the physics gives us, as it were, as revealing what the structure of the world is like (so, for example, the claim might be that that structure can be represented at the object level by SU(3)) and interpreting that via an appropriate metaphysical understanding of structure (again, a similar move can be found in Lyre 2011).[3]

One might, instead, seize the second horn and simply accept the whole 'tower' of structures, as it were. The obvious worry here is that this is just too 'wild' and ontologically extravagant.[4] Now of course if one were to reject the kind of hierarchical framework that the problem assumes, where there has to be a fundamental structure underpinning all the rest, then this worry might dissipate. It is important to note, however, that what we have are not *physical* structures represented by different groups all the way down (cf. Saunders 2003 b and c), but a tower of mathematical excrescences associated with the one group (e.g. SO(3)). Are all of these mathematical objects to be seen as further features of the structure of the world? That does seem ontologically inflationary. But again one can see this as a consequence of the surplus structure (understood broadly) that mathematics inevitably provides and we return to drawing the line in terms such as already presented. The fundamental point is that we have to draw the line anyway since the structure we are realists about is physical, not mathematical (a point I shall return to in Chapter 8). Nevertheless, we should not simply dismiss the tower since it in effect encodes the possibility of the groups, just as the groups encode the possibility of the relevant representations. And I shall draw on the latter (in Chapter 10) as a way of understanding how the structure of the world

[3] Where that understanding might be obtained via the 'Viking Approach' introduced in Chapter 3.

[4] Although as Roberts has noted, there is a sense in which the automorphism group does empirical work, so why not be realist about it?

can be said to be 'modally informed'. In the end, then, all I can do is wriggle between the two horns of Roberts' dilemma: we should take the physics as revealing what the structure of the actual world is like, and in this respect we draw a context-based line through the group-theoretic edifice that allows us to say 'this is the structure of the world';[5] but then in articulating the metaphysical nature of that structure and in particular the way it is modally informed, we find ourselves having to ascend the tower, although not, given the distinction I'm trying to draw, in a way that undermines our claims about what the structure of this, the actual, world is.[6]

6.3 Concern 2: From Group Structure to Dynamical Structure

The second concern is that a group-theoretic conception of structure does not give us enough, since it does not capture the relevant *dynamics*. Now in a sense this is tilting at a straw person, since the claim is not that all there is to structure is group theory—on the contrary, laws *plus* the kinds of symmetries that group theory so beautifully captures make up what the advocate of OSR insists is 'the structure of the world'.

Nevertheless let us take the example of Yang–Mills theories (Bain preprint), which are gauge theories that play a prominent role in the construction of the Standard Model in elementary particle physics.[7] Gauge theories in general are theories for which the Lagrangian (see Chapter 2) is invariant under a continuous (Lie) group of local transformations that hold between possible gauges, or redundant degrees of freedom in the Lagrangian (as we'll see, this redundancy has been taken to undermine the physical significance of gauge symmetry). The group generators of the associated Lie algebra yield the corresponding (vector) gauge field, which, when included in the Lagrangian, ensure invariance under the relevant transformations. When these fields are quantized, the resulting quanta are called gauge bosons. So, for quantum electrodynamics, the symmetry group is the U(1) group, the gauge field is the electromagnetic field, and the gauge boson is the photon. Indeed, it is sometimes said that the photon 'drops out' as a result of the requirement of gauge invariance. The Standard Model is based on a gauge theory[8] that has the symmetry group U(1) x SU(2) x SU(3) yielding twelve gauge bosons: the photon of the electromagnetic force, three bosons for the weak nuclear force, and eight gluons associated with the strong nuclear force (and the associated theory of quantum chromodynamics).

[5] And as emphasized in the previous chapter, the context here is grounded in the physics.

[6] The question has been raised whether I am proposing a metaphysics-driven account of physics or a physics-driven account of metaphysics. The answer, of course, is neither. What I am proposing is an understanding of the structure of the world, based on physics but informed by metaphysics, along the lines articulated in Chapter 3.

[7] For a useful introduction, see Jaffe and Witten (undated).

[8] Unlike quantum electrodynamics this is non-Abelian: the symmetry group is non-commutative.

Now in the twistor formulation of anti-self dual Yang–Mills theories, the relevant partial differential dynamical equations 'evaporate' into certain global (holomorphic) geometric structures (Bain preprint). The point is: the equations are not the only way of encoding the dynamics. Furthermore, with such examples it becomes unclear what group should be taken as fundamental and hence which set of invariants should be understood as constituting the structure of the world (Bain preprint). On the space-time formulation of Yang–Mills theories, the relevant groups are the Poincaré group and the relevant gauge group (or 'local' symmetry group).[9] How we should understand the latter is a matter of contention: on the one hand, as just noted, gauge invariance plays a hugely significant role in modern physics; but on the other, gauge freedom may be regarded as a mere ambiguity in our representation, with no physical significance (see, for example, Redhead 2003).[10]

Let us return to the example of the Dirac equation, the U(1) group and the photon, since it quite beautifully exemplifies a number of issues we have discussed so far (here I shall follow Martin 2003: 42–3). The equation describes a free field of electrically charged matter and is the Euler–Lagrange equation for the relevant Lagrangian (here we recall Curiel's argument for the significance of the latter, discussed in Chapter 2). The corresponding action[11] is then invariant under U(1) and Noether's first theorem then implies conservation of the current. Now, if the symmetry is global, then choosing the gauge at one point effectively fixes it for all other points. One of the crucial innovations in the history of gauge theories was to lift this requirement and allow the gauge invariance to be local (I shall shortly return to the justification for this). Doing this in the electromagnetic case requires the introduction of another field—the gauge potential—which couples with the matter field and can be understood as representing the electromagnetic potential. The free field Lagrangian must now be replaced with its interaction counterpart, which is invariant under the local phase transformations and a kinetic term must be added for the gauge potential. This 'imbues the field with its own existence' (Martin 2003: 43) and yields the Lagrangian for the fully interacting theory. Varying the corresponding action with respect to the

[9] For a useful introduction to the relevant history and the central philosophical issues, see Martin (2003) and, as presented in a broadly structuralist context, Cao (1997) (see also his 2010). In terms of the kinds of considerations I crudely sketch in Chapter 4, this history has been explored further by Ryckman (2003b), who, as I have noted, illuminates the philosophical roots of Weyl's introduction of the gauge principle in Husserl's phenomenology.

[10] Martin represents this tension as holding between what he calls 'the profundity of gauge' and the 'redundancy of gauge' (2003: 52) and his exploration of its origins nicely brings out certain features of how symmetries are regarded that relate to my considerations here.

[11] The action is the integral of the Lagrangian and minimizing the action yields the trajectory of the system. The Principle of Least Action that encapsulates this yields the classical equations of motion. We recall Cassirer's emphasis on this Principle helping to yield the relevant laws. Interestingly, given my discussion to come in Chapter 9, Katzav has argued that dispositionalism is incompatible with the Principle of Least Action, because the latter demonstrates that the equations of motion are not made true by the intrinsic properties of the given particle, contrary to the central claim of the dispositionalist approach to laws (Katzav 2004); see also Ellis (2005) and Katzav (2005) for further discussion.

gauge potential yields the inhomogenous coupled equations of motion for the electromagnetic field. And, further, requiring local gauge invariance (specifically, requiring that the mass term for the gauge field be gauge invariant) implies that the 'carrier' particle, that is, the photon, be massless: '[l]ocal gauge invariance thus necessitates a massless photon' (Martin 2003).

From the structuralist perspective we can see how the symmetry, associated with the relevant law (as represented by Dirac's equation), imposes certain requirements that yield the particle; or, more specifically, given the assumption that a 'force carrier' particle be associated with the relevant field, the symmetry yields the crucial (some might say essential) property of the particle, namely that, in this case, it be massless. From this perspective, reading the theory as positing an ontology of objects, with certain properties (zero mass, for example), the interrelationships between which are encapsulated in the relevant laws and symmetries, and taking *that* ontology as fundamental, just seems perverse.[12]

Following the successful formulation of a renormalizable quantum electrodynamics (QED) but in the face of the 'hadron zoo' of elementary particle physics in the 1950s and early 1960s, many physicists abandoned the appeal to symmetries and group theory (Martin 2003: 38). Others persisted, however, and Yang and Mills, in particular, made an important advance in the application of local gauge invariance to nuclear interactions, drawing on Heisenberg's consideration of the similarities between the proton and neutron that underpinned the introduction of isospin and SU(2) symmetry that I briefly sketched in Chapter 5.

We recall the underlying 'idealization' of regarding the proton and neutron as two states of the same particle, namely the nucleon, transformed into one another via the transformations of the SU(2) 'internal' symmetry group. As in the case of QED, Yang and Mills lifted the restriction imposed by regarding SU(2) as a global symmetry and recast isotopic spin in terms of a local gauge invariance. This yielded a SU(2) gauge field, which, given the nature of isotopic spin, is self-interacting and carries its own 'charge' (unlike the electromagnetic case). The relevant equations are thus non-linear and the group is non-Abelian (i.e. non-commutative). Quantizing the gauge field led to an immediate problem, however: it could not be massless, as in the electromagnetic case, since the interaction would then be long range, contrary to the known short range of the nuclear force (see Martin 2003: 39–40).

The solution lay in the idea of spontaneous symmetry breaking,[13] whereby 'given a symmetry of the equations of motion, solutions exist which are not invariant under the action of this symmetry *without the introduction of any term explicitly breaking*

[12] We recall Pashby's claim (2012) that there are structural discontinuities here, associated not with the relevant symmetries but with the laws, as we shift from quantum mechanics to quantum field theory. I have indicated how the advocate of OSR might respond to these in Chapter 2, note 9.

[13] For the history of this idea, see Cao 1997 and 1999; for broadly philosophical considerations of its impact see Castellani 2003 and Brading and Castellani 2008.

the symmetry' (Castellani 2003: 327). Now, when we have a global continuous symmetry—as with global gauge invariance—spontaneous symmetry breaking yields massless bosons (so-called Goldstone bosons). If the symmetry is then taken to be local and the Higgs mechanism applied, these massless bosons acquire mass (it is sometimes said that the Goldstone bosons are 'eaten up') and the short range of the force thereby underpinned.[14] Furthermore, it was shown that this mechanism also underpins the renormalizability of these theories. So, returning to our brief history (Martin 2003: 40), Weinberg and Salam appealed to spontaneous symmetry breaking to develop their electroweak unified theory of electromagnetism and the weak nuclear force. Here the relevant gauge group is SU(2) x U(1) and the particles that 'drop out' are the massless photon, associated with the unbroken U(1) sub-group and three massive bosons corresponding to the broken part. The prediction and subsequent discovery of these latter particles (with the required properties) and the associated weak neutral current have been taken to represent a major success for gauge invariant theories.

The extension of this kind of theory to the strong nuclear force hinged on the realization that non-Abelian gauge theories display 'asymptotic freedom', thus explaining why nucleons behave as if their constituent particles are free under certain circumstances (for more details see Cao 2010). Here the unbroken gauge group is SU(3), describing the quark colour multiplets and yielding eight massless gluon fields (which also carry colour and are thus self-interacting). This underpins quantum chromodynamics—also hugely successful—and together with the developments already outlined here, led to the construction of the so-called 'Standard Model', based on the gauge group SU(3) x SU(2) x U(1).

So one can begin to see, I hope, that heuristically, at least, gauge invariance has been enormously successful. How is it, then, that it can be regarded as 'redundant'?[15]

First of all, one might follow tradition in drawing a sharp distinction between the contexts of discovery and justification and insist that the heuristic value of these symmetries is confined to the former and speaks not at all to the issue of understanding theories, which has to do with the latter. On this view, then, 'we should... count ourselves amazingly fortunate that the "right" theories just happened to have such a nice structure, i.e. that seen in the theories' tight group-theoretic structure which accompanies the characteristic symmetry/invariance' (Martin 2003: 41).

[14] As Castellani notes (2003: 322), rather than conceiving of the relevant symmetry as 'broken', in some ontological sense (whatever that may be!), the situation is better understood as one where the relevant phenomena is characterized by a symmetry that is 'lower' than the 'unbroken' symmetry. This means that the group characterizing the latter is broken into one of its subgroups and so the process can be described in terms of relations between transformation groups. As I said, I shall return to this later but this brief comment will perhaps assuage the concerns of those who might think that the idea of 'breaking' symmetries presents a further obstacle to the structuralist picture I am drawing here.

[15] Ismael and van Fraassen argue that symmetries in general act as 'beacons of redundancy' (2003: 391).

Now leaving aside the issue as to whether a sharp discovery–justification distinction can be enforced (answer: no it can't; see da Costa and French 2003: ch. 6[16]), the realist is going to have to take the success of such principles as more than merely 'pragmatic' and the associated value as going beyond the 'heuristic'. And this will be for the same reasons as in the object-oriented case of electrons and the like; that is, having to do with the No Miracles Argument (NMA), construed along the lines of 'it would be a miracle if gauge field theories were so successful and they did not describe the way the world is' (or as Ladyman puts it, did not latch onto the world). Indeed, Martin's statement can be read as an assertion of the 'it's just a miracle' anti-realist counter-position. A crucial difference from the standard use of NMA of course is that as far as structural realism is concerned it should motivate commitment not just to the relevant entities (particles, fields, etc.) nor just to the corresponding equations but to the group-theoretic structure that captures the relevant symmetry/invariance.[17]

Nevertheless, some caution must be exercised in simply, or straightforwardly, reading off our realist commitments from gauge field theory, under the demand of the NMA. It is certainly not the case that the imposition of local gauge invariance either uniquely dictates the form of the interacting theory, or dictates the origin of a new physical gauge field (Martin 2003: 45). Various other factors come into play, from the imposition of some form of simplicity requirement[18] to the addition (more or less 'by hand') of the kinetic term to the Lagrangian that, in a sense, 'gives physical life' to the field.[19]

Of course, much the same can be said of any major theory in physics and if concerns cluster around these factors within the realism–anti-realism debate then they will hold just as much for other forms of realism as for structuralism.[20] One can take gauge invariance as privileged simply by virtue of being regarded as a kind of axiom of the relevant theory. However, there are other ways of setting up the relevant

[16] Nevertheless, one should be careful not to take the heuristic role of symmetries as implying a certain form of relationship between such symmetries and the associated laws. I shall come back to this in Chapter 10 where I shall argue that this heuristic role should not be taken to mean that symmetries must be regarded as requirements imposed on laws rather than as by-products of them.

[17] As for the all-important novel predictions, in addition to the much discussed Ω^- particle, we also have the Z^0 and $W^{\pm}$ bosons, as well as the electroweak current and other predictive successes of the Standard Model. A critic might insist that only those novel predictions that follow directly from the group-theoretic structure, such as Ω^-, should count but this is to suggest too restrictive a view of how and to what one should attribute success. Even if we adopt the view that such attribution should follow the lines of explanatory connection (Saatsi 2007; see Chapter 3), it is hard to see what motivation there is for stopping before one reaches the invariance.

[18] And we all know how hard it is to capture that formally!

[19] Indeed, the latter point might be taken as helping to explain how it may appear that one gets more physics out of the gauge argument than one puts in—a suggestion that obviously bears on Redhead's concerns as previously indicated (Martin 2003).

[20] Of course, in terms of tracing the line of explanatory connection to the relevant success-inducing elements, it may well be that this line does not always run straight, as it were, but has to depend upon such factors in connecting the relevant elements with the requisite empirical sub-structures. This is an issue more to do with justification in general than with realism or structural realism in particular.

formalization such that gauge invariance emerges as an 'output': imposing renormalizability in a certain form also gives rise to Yang–Mills gauge theories, for example. More interestingly, perhaps, according to the effective field theory programme such theories are just low-energy approximations to, or residues of, a more fundamental underlying theory. The ontological significance of gauge invariance might then be further undermined.

Now, first of all, it is no part of structural realism that gauge invariance should be taken as an *axiom* of the relevant theory. Indeed, from the perspective of the semantic approach to theories adopted here, the idea that such principles that arise in scientific practice can or should be taken as axioms in the traditional Euclidean sense is deeply problematic (see van Fraassen 1980; da Costa and French 2003). Again, the motivation for adopting a realist stance towards such principles has to do with the success (particularly predictive success, given the NMA) of the theories with which they are associated. And of course, as has just been acknowledged, the nature of that association may be less straightforward than a simple deductive schema might suggest but that does not undermine the general strategy of how we should 'read off' our realist commitments that I have outlined here. As for the impact of effective field theory: if it were to be generally accepted that our current theories are in fact nothing more than such residues of a more fundamental one (perhaps articulated along string-theoretic lines) then as long as the structural realist can point to the relevant commonalities between the former and latter, I see no real problem here.[21]

Returning to the question of how gauge invariance might be regarded as 'redundant', one can begin to get a grip on this by taking gauge in its most primitive sense as involving the association of physical magnitudes with mathematical entities such as numbers (Redhead 2003). And just as one can associate the hardness of various minerals with a scale from 1 to 10, so one could just as well associate it with a scale from 10 to 20 or 11 to 21 or whatever. Expressing this in terms of the semantic approach discussed in Chapter 5, what we have is a homomorphism that is established between a physical structure P and a mathematical one M, where the latter acts as a gauge for the former (Redhead 2003: 125–9). The conventionality of gauge is built into the concept, so the question now is how it can have any physical significance.

Presented in terms of a model-theoretic meta-level representation of symmetry (Redhead 1975; 2003: 127–8) the so-called 'gauge freedom' can be understood as an ambiguity of that representation. If we take a single mathematical structure M and two distinct isomorphisms x: $P \rightarrow M$ and y: $P \rightarrow M$, then the mappings $y^{-1}.x$: $P \rightarrow P$ and $y.x^{-1}$: $M \rightarrow M$ are automorphisms of P and M respectively[22] and, of course, are

[21] Nevertheless, if accepted, the effective field theory would have implications for our understanding of the notion of fundamentality in this context, as Martin notes (2003: 47); see McKenzie (2012) for further discussion.

[22] These correspond to what are called 'active' and 'passive' symmetries of P respectively, where the former is taken to be physically meaningful and the latter to be merely a trivial change in representation. I'll come back to this distinction.

in 1–1 correspondence since *P* and *M* are isomorphically related. Since they share the same abstract structure, the structural properties of the former represented by the relevant symmetries can be simply read off from the corresponding properties of the latter (Redhead 2003: 127). Moving to the case where *M* is a sub-structure of a larger mathematical structure *M'*, the relative complement of *M* in *M'* in this case represents the 'surplus structure' in the representation of *P* by means of *M'* (Redhead 1975; 2003: 128). As Redhead puts it, '[c]onsidered as a structure rather than just as a set of elements, the surplus structure involves *both relations among the surplus elements and relations between these elements and elements of M'* (Redhead 2003: 128; my emphasis).

However, if the structure is genuinely surplus in the sense of bringing new mathematical resources into play, then the physical theory should be regarded as embedded into a whole family of such structures (Bueno 1997). Within the partial structures approach, as outlined previously, with the third component in the family of partial relations—namely R_3—left open, there is structural 'space' to accommodate this. But of course, this surplus mathematical structure cannot be represented simply in terms of more n-tuples of objects in the relevant domain. Instead, it must be represented in terms of a family of structures $(S^K{}_i)_{i\in I}$, associated with a given structure K (Bueno 1997). We can then represent how a given structure can be extended by the addition of new elements to its domain, or the addition of new relations and functions defined over these elements. Each $(S^K{}_i)_{i\in}$ represents such an extension and the whole family of such extensions represents the surplus structure.[23] In terms of this framework what we have is the partial importation of the relevant mathematical structures into the physical domain. This partial importation can in turn be represented by a partial homomorphism holding between the structures $(S^K{}_i)_{i\in}$ characterizing the mathematical surplus structure, and the structures of the physical theory under consideration. This effectively allows the carrying over of relevant structural features from the mathematical level—captured by the R_1 and R_2 components of the relevant partial structure—to that of the physical theory. The heuristic fertility of the application of mathematics rests on the surplus, in the sense that more structure from the family can be imported if required; it is this crucial aspect that is captured by the openness of partial structures.[24]

Redhead was perhaps the first to emphasize the heuristic fertility of this surplus structure, giving the example of Dirac's famous hole theory of what came to be identified as the positron, where the surplus structure allowed a physical interpretation to be given for the negative-energy solutions of the Dirac equation. He articulates what is going on as involving a kind of 'blurring' of the boundary between

[23] This section is taken from Bueno, French, and Ladyman 2002: 505–6.

[24] cf. Ismael and van Fraassen (2003) who, in anti-realist fashion, take symmetries to be merely a means to the identification of 'superfluous' theoretical structure.

M and the surplus structure (Redhead 2003: 129), but introducing families of structures and relationships of partial homomorphism makes this picture more precise.

But of course, we can give other examples where surplus mathematical structure is introduced—to aid computation, for example—but is not physically interpreted. Obvious examples would be those that involve the use of complex numbers, such as in the complex currents and impedances used when considering alternating currents, where physical quantities are embedded in the mathematical structure of complex numbers (2003: 128). Another nice example (2003: 128) comes from S-matrix theory, where scattering amplitudes considered as real-valued functions of energy and momentum transfer were carried over into the complex plane and their behaviour there used to construct systems of equations describing their behaviour in the physical 'realm'. Here the surplus structure is definitely not interpreted in terms of something physical.[25]

The point of this little excursion is that gauge freedom arises when we have automorphisms of *M*' that reduce to the identity on *M*, so that the relevant transformations act non-trivially only on the surplus structure. Nevertheless, given the relationship between this surplus structure and *M*, and hence with *P*, such transformations bleed through into the physical structure. Redhead maintains that this is precisely the case with Yang–Mills theories (2003: 130–2). Here, as we have noted already, the imposition of local gauge invariance requires the concomitant introduction of a new field—the gauge field—and this Redhead sees as an example of the requirements imposed on surplus structure 'controlling' physical structure (2003: 131). So, the relevant aspects of the latter are the charges or currents which are mapped onto the mathematical structure *M* which in turn is a sub-structure of *M*'. The local transformations act in the surplus structure and correspond to identity transformations on *M* and thus, correspondingly, on *P*. Here too, the surplus structure is heuristically useful but that does not of itself mean that it should be given physical content.

This view of gauge symmetry as being tied to a certain 'descriptive freedom' in our theory, and thus to redundant or surplus quantities, should be seen against the backdrop of what has been called the 'received view' of symmetry due to Wigner (Martin 2003: 49–50). As is well known, Wigner took there to be a 'great similarity' between the relationship that holds between the laws of nature and the relevant events, on the one hand, and that which holds between the relevant symmetry principles and these laws, on the other (see Wigner 2003a: 24). Two immediate questions spring to mind when faced with this assertion: what is the ground of this similarity? And, what is its nature?

Wigner himself places symmetries, laws, and events in a hierarchy, with the symmetry principles, of course, at the top. He then insists that the laws could not

[25] Another, more contentious example, would be that of the certain mathematical devices invoked in the consideration of critical phenomena that some see as playing an explanatory role (Batterman 2010), whereas others see them as features of surplus structure only (see Bueno and French 2011).

exist without the symmetries (2003b: 370), where by this he means that the latter are 'almost necessary prerequisites' (2003b: 370) for discovering and cataloguing the former, which in turn should be regarded as simply correlations between events. Finally, then, if we knew everything there was to know about the events, we would have no use for the laws (2003a: 24) and likewise, if we knew the laws completely and in all respects, the associated symmetries would provide no new information. Thus, if a final theory of everything were to be accepted, the symmetry principles would lose their place in the hierarchy, coming to be regarded as, at best, useful tools for deriving consequences from the final theory, in line with the 'Wigner programme', as sketched in Chapter 4 (Wigner 2003b: 370).

These claims mesh with two broad positions, one on the relationship between laws and symmetries and the other on the nature of laws. With regard to the first, symmetries are seen as prerequisites for, and hence imposing constraints on, laws. The alternative is to regard symmetries as by-products of laws (Lange 2009a), perhaps in the sense of being the manifestation of certain (higher-level) properties of the laws. I shall return to these positions in Chapter 10. With regard to the nature of laws, the philosophical divide is between Humean and non-Humean views, where the former hold law statements to be mere summaries of the regularities that involve, or the correlations between, the relevant events, so that laws are not metaphysically substantive additions to what we take there to be 'in' the world; and the latter holds the contrary line, that laws are something over and above the set of events and that, some forms of this kind of view insist, the laws 'govern' these events. Again, I shall return to these views in Chapters 9 and 10 but here I just want to note that Wigner held a combination of the 'symmetries-as-prerequisites' view from the first set of positions, with a Humean account of laws, from the second.[26]

The question now arises: how could a symmetry be a prerequisite for a summary of a correlation? One could envisage a view in which symmetry principles are metaphysically substantive additions to the ontology of the world, but laws are not, and the former condition the world in such a way that events are correlated appropriately (so, this offers one of the alternative combinations just mentioned as possible). However, this is not what Wigner himself seems to think as he regards symmetry principles as being akin to laws in the sense of summarizing the 'subtle properties' of the latter, so that if we know these laws, fully and completely, knowing these properties conveys no further information (2003a: 24–5). Thinking of symmetries in this way might lead to the symmetries-as-by-products view, but that would add a second line of tension, given what Wigner explicitly says about prerequisites.

The resolution is to be found in his distinction between 'geometrical' and 'dynamical' symmetries, where

[26] 'Wigner's theory of theories... takes observables, specifically probability functions, as fundamental. Laws are in effect nothing but convenient ways of encompassing the various probability distributions for observable outcomes' (Martin 2003: 51).

the former concern the invariance of all the laws of nature under geometric transformations tied to regularities of the underlying spacetime, while the latter concern the form invariance (i.e. covariance) of the laws governing particular interactions under groups of transformations not tied to spacetime. (Martin 2003: 50; see also Wigner 2003a: 25–7; 2003b: 368–9)

The former constitute the Poincaré group, whereas the latter include gauge invariance and as we shall see, this distinction, tied to Wigner's broadly Humean stance, creates the obstacle to understanding gauge symmetry from a structuralist perspective. It is the former, however, that he took to be genuine physical invariances, since they relate directly to the relevant physical events subsumed under the associated laws by virtue of representing certain features of the underlying space-time. So, for example, the invariance of all laws under spatial translations represents the fact that correlations among events depend only on the relative distances between these events and not on their absolute position. Dynamical symmetries, on the other hand, are specific to the relevant theory and do not, then, apply to all laws. On this view, electrodynamic gauge invariance, for example, concerns the specific laws of electromagnetism only (2003: 51). These symmetries are then taken not to relate directly to, or be underpinned by, the underlying events.[27]

Thus when it comes to geometrical symmetries, as expressed in 'principles of invariance', these do, in a sense, condition the way the world is, but not as metaphysically substantive additions; rather, they are summaries of certain features of space-time that make possible the constant correlations between, and regularities involving, events that are themselves summarized in law statements. As Wigner put it, '[i]f the correlations between events changed from day to day, and would be different for different points of space, it would be impossible to discover them' (2003b: 370). In this sense, then, *these* symmetries are prerequisites for laws. Dynamical invariances, on the other hand, cannot be, at least not in this sense. Although they may reflect certain features or properties of their associated laws, they do not, strictly speaking, impose requirements upon them[28] (however, this is not to say that such invariances may not be heuristically useful, as in the case of gauge invariance).

Taking this distinction on board, we obtain a slightly more nuanced account of Wigner's position, according to which geometrical symmetries impose requirements on laws, qua summaries of correlations between events, and they do so by virtue of themselves summarizing certain spatio-temporal features of these correlations, whereas dynamical symmetries merely reflect certain (second-order) properties or features of the associated laws, and hence might be regarded as mere by-products of them.

[27] Underlying this distinction is the further one between active and passive transformations, where the former relate to physical observers, but the latter are mere changes of description (Wigner 2003a: 26–7). A great deal has been written about this distinction already and I'm not going to add to this literature here.

[28] Thus it is not quite right, on this view, to say, as Martin does, that '[b]oth of these types of symmetries posit/embody a certain structure to some set of physical laws in placing restrictions on their possible forms' (2003: 50).

But this is not the only combination of views possible, as I have already indicated. First of all, let us briefly return to our first question, and ask on what basis Wigner could assert that there is a 'great similarity' between the relationship between laws and events and that which holds between laws and symmetries? One answer might be that epistemically we gain knowledge of laws and symmetries in broadly the same way. So, putting it crudely, just as we gain knowledge of laws through observing correlations between events and the associated regularities, so we gain knowledge of symmetries by 'observing' certain features or properties of, or pertaining to, all such correlations—as in the case of geometrical invariances—or only certain such correlations, as relating to specific interactions—as in the case of the dynamical invariances.[29] Indeed, in the former case, the 'observation' may be so basic or obvious that it is not made explicit until theoretical pressure forces it out into the open, as it were, but nevertheless these invariances are all 'products of experience', rather than a priori truths (Wigner 2003b: 368).

Now even if one were to accept the picture drawn here, one could still insist that the fact that we discover laws and symmetries in a similar manner does not imply that metaphysically they are the same, or similar.[30] Unless there is some alternative answer that metaphysically ties the symmetry–law relationship to the law–events one, the path is open to offer a variety of combinations in answer to our second question. Indeed, as we have seen, Wigner himself does just this. And returning to the suggestion made in that discussion, we could conceivably maintain a broadly Humean view of laws but a non-Humean view of symmetries. How could this be so? Well, one obvious way would be to argue that when it comes to what Wigner has identified as the geometrical symmetries, in so far as these represent features of space-time they, by virtue of that fact alone, go beyond the events and hence can be regarded as metaphysically substantive features of the world. Of course, a relationist would be uncomfortable with such a claim but one does not have to return to 'old style' substantivalism to make good on it—one could articulate it in terms of either 'sophisticated' substantivalism (see Pooley 2006) or, better in my books, some form of space-time structuralism (see Ladyman 2002 or French 2001). The latter would remove any hint of metaphysical 'cheating' here by insisting that taking symmetries as features of space-time is not to take them as features of some thing or further object, over and above those that presumably compose events (on the object-oriented stance) but as features of structure, and in so far as these are not 'regularities' in a straightforward sense, this is more akin to a non-Humean conception.

Or, one could insist that laws are metaphysically substantive and govern the relevant events but that symmetries—of whatever kind—are not, but are mere by-products of these laws, and thus add nothing to the furniture of the world. Of course,

[29] And however we think we gain knowledge of laws and symmetries we might take the relevant epistemology to be such that we can encompass both within some suitable account of truth.

[30] Thus, we might re-impose some form of discovery–justification distinction at this level.

one would have to drop the claim that symmetries, in whatever form, act as conditions on or prerequisites for the laws, and one would again have to say something about the geometrical symmetries and space-time. However, such a combination would presumably fit with the claim that the central features of space-time arise from the relevant dynamics, as encapsulated in the laws (Brown 2005).[31] Of course, one might feel that it is more natural to insist on the similarity of the relationships and either understand both laws and symmetries in Humean terms or take them both as non-Humean. In effect I shall be doing the latter, from a structuralist perspective, although not on grounds of naturalness and I shall also consider structuralist forms of the former combination.

Enough already. With that background in place, let us return to the issue of the apparent redundancy associated with gauge invariance and the problem of giving this symmetry physical content.

As far as Wigner is concerned, this is a dynamical symmetry and not physical, or at least, not in the way that geometrical symmetries are. Thus he writes that this invariance is 'artificial' and 'similar to that which we could obtain by introducing into our equations the location of a ghost' (Wigner 2003a: 26; Martin 2003: 51). Gauge freedom is merely a freedom in our description and of no physical consequence. Wigner's ghost is Redhead's surplus structure.

Now, Redhead identifies various ways of dealing with this surplus structure inherent in gauge theories (2003: 137–8). One would be to insist that we should just follow the practice of physics, which is to allow non-gauge invariant quantities to enter the theory via the surplus structure and continue to develop the theory by adding more surplus structure (and so we see the introduction of 'ghost' fields and the like). As we have seen, gauge invariance has been and can be expected to continue to be, heuristically very successful, so perhaps we should just acknowledge that success and not worry about seeking a physical counterpart to the formal principles that are introduced. However, this cannot be satisfactory to the realist. At best it would mean regarding the success of gauge field theories as simply a 'miracle'; at worst, as Redhead notes, it suggests some kind of Platonist–Pythagorean view of the role of mathematics in physics, from the perspective of which the relationship between mathematical and physical quantities and hence between mathematics and physics in general remains a mystery (2003: 138).

An obvious alternative, then, would be to (re-)formulate the theory in gauge invariant terms. Indeed, this is presupposed by the standard approach that characterizes gauge symmetry in terms of the covariance of the fundamental equations of motion for specific interactions and thus as tied to the presence of redundancy. It is not surprising then that on this view gauge transformations are seen as physically impotent, since 'any potential physical significance of the characteristic gauge

[31] One can also adopt this view of gauge invariance as nothing but a by-product of the specific dynamical field under consideration (see Martin 2003: 55).

symmetry has been washed away from the start' (Martin 2003: 52). Furthermore, there are costs involved. Thus, in the electromagnetic case, the gauge potential, or *A* field, would be replaced with the magnetic field *B* (= curl*A*). However, this leads to problematic forms of non-locality, as evidenced in the case of the Aharonov-Bohm effect, whereby charged particles are apparently affected by an electromagnetic field in regions where the field is zero (for detailed discussion see Healey 2007 and Smeenck 2009).

What is the structuralist to make of all this? How can she negotiate a way between the redundancy and profundity of gauge invariance?

Let us recall the point that the basis of the dismissal of gauge invariance as physically impotent lies in Wigner's (broadly) Humean account:[32] 'The associated transformations change nothing physical since they correspond to the identity transformation on observables' (Martin 2003: 51). In effect, what counts as 'physical' is tightly tied to what counts as observable. There are then two ways one can respond so as to allow gauge invariance to have some measure of physical significance. The first is to remain within this broadly Humean framework, but expand one's conception of the geometrical; the second is simply to drop this account and move to a different framework entirely.

With regard to the first response, one can regard gauge transformations as automorphisms of a kind of enlarged geometrical space via appeal to the mathematics of fibre bundles (see Martin 2003: 50 n. 65; for introductions to fibre bundles see Lyre 2004; Nounou 2003: 179ff). Of course, making such an appeal suggests that this bundle structure should be regarded as part of one's ontology. Certainly, the structuralist might welcome such an expansion and indeed, this is precisely the move that Lyre makes (2004). In particular, it is not just that the structuralist approach to gauge theories meshes with the group-theoretic representation of particles in modern physics. It is also that the historical development and application of group-theoretic structure suggests a structuralist response to the Pessimistic Meta-Induction that goes beyond the emphasis of ESR on the relevant equations (cf. again Saunders 1993). So, we recall Worrall's structuralist emphasis on the way in which Fresnel's equations are incorporated within and therefore, in a sense, drop out of, Maxwell's and—although Worrall does not pursue this—also from those of quantum electrodynamics. From the group-theoretic perspective, we begin with the group of transformations that encapsulate the gauge freedom inherent in Maxwell's theory and then note that when the latter is embedded into the wider framework of Dirac–Maxwell gauge theory and thence in quantum electrodynamics we move from that original gauge group to U(1), or, specifically, the Lie algebra associated with the latter (Lyre 2004: 22–3). From there one moves straightforwardly to the structure of the Standard

[32] Indeed, Wigner expressed his dislike for regarding gauge invariance as a symmetry principle (Martin 2003: 51).

Model and further, as Lyre notes, to developments leading to a viable theory of quantum gravity.

Of course the advocate of ESR could (and should) also draw on these developments, extending her epistemic purview to encompass these group-theoretic structures. Nevertheless, the problem of hidden natures remains: from the ESR perspective one would have to insist that the elements of the relevant group correspond, in some sense, to the objects that remain forever beyond our reach. As we saw in Chapter 4, Eddington argued that the group elements could not be separated from the relevant transformations and this 'package' view of group-theoretic structures can be extended to the 'moderate' form of OSR. From the more radical perspective one can perform something akin to the Poincaré Manoeuvre again (French 1997): at the level of the mathematics, we introduce the elements in order to be able to *define* and articulate the appropriate transformations, but once we have the latter, we 'read off' our ontology from these, and the relevant interrelationships, and discard the elements as mere heuristic devices or crutches that allowed us to 'get' to the group structure, which is where all the ontological action is.

Lyre also notes another form of underdetermination that arises by virtue of three possible interpretations of gauge theories, in terms of field strengths, potentials, and holonomies (Lyre 2004; see also Healey 2007). His conclusion is that if these interpretations are regarded from an object-oriented perspective then we have a strong form of underdetermination, with no way of choosing between these alternatives *on the basis of criteria appropriate to that perspective.*[33] Adopting a structuralist stance, however, the realist can take the fundamental structure of the relevant symmetry group, such as U(1) (and the associated fibre bundle), as a presentation of how the world is, and thus sidestep the underdetermination.

The second response to the apparent impotence of gauge invariance is to step outside the Wignerian framework and insist that the global–local distinction does not map onto that between the physically significant and the impotent, nor, crucially, should the latter be cashed out in terms of certain observables.[34] As I have already indicated, a realist would be motivated to take the success (including predictive success) of gauge field theories as indicative that they are representing (in part at least) how the world is (on pain of regarding this success as a miracle). Gauge invariance can be regarded as one of the elements responsible for that success and the fact that it cannot be given the same kind of 'geometrical' interpretation as the invariances captured by the Poincaré group speaks only to the failings of our (non-structuralist) realist imagination.

[33] Lyre insists that we can choose between them on the basis of structuralistically acceptable criteria and opts for the holonomy interpretation.

[34] There are well-known problems with regarding gauge invariance as observable. As Brading and Brown explain, it has only indirect empirical significance as a feature of both matter fields and gauge fields taken jointly and thus as a property of the relevant laws (Brading and Brown 2004).

Returning to Redhead's concerns then, we can perhaps understand what is going on here as similar to Redhead's own example of the negative-energy holes in the Dirac equation. We begin with certain mathematical transformations defined in what is, in effect and at this stage, the surplus structure. We then come to realize that the associated requirements 'control' (to use Redhead's own term) or impose certain restrictions on the embedded physical structure. The issue then is how to understand that imposition from a realist point of view, and from an object-oriented perspective it seems we are faced with a range of choices, as manifested in the form of another underdetermination. The structuralist resolution of this quandary is to urge that one's ontological commitment should be placed with the relevant structure, so that, in effect, what was originally taken to be surplus comes to acquire physical significance. Thus, one should take the relevant symmetry group—U(1), for example—and the associated fibre bundle as a presentation of (part of) the structure of the world. This means taking (statements of) these symmetries and the associated laws as more than mere summaries of the relevant (higher-order) properties and regularities respectively and as metaphysically substantive constituents of the world.

Returning to the issue of the significance of 'dynamical' structures which must supplement the group-theoretic representation of putative objects, consider the electron, for example, where the relevant structure is captured by either the Hamiltonian or Lagrangian formulation of electron theory, with the evidence for this structure given via the well-known 'historically stable properties' of the electron (Bain and Norton 2001). This dynamical structure, however, is not strictly group structure, since it is encoded not just in the invariants of the relevant groups, but also in the spaces that carry the representations of these groups. Thus, to give another example, the dynamics of the Yang–Mills theories touched upon earlier in this chapter can be encoded not just in the relevant invariants (twistors) but in the geometric structures defined over the projective carrying space (Bain preprint). Hence, the structuralist needs to incorporate the relevant dynamical structure into her account and thereby flesh out her understanding of the 'world structure' as multi-featured (French 2006).[35] In effect this is to acknowledge the breadth and complexity of the relevant structures, something that Falkenberg, for example, has also recently highlighted (Falkenberg 2007).

Let me now consider the third concern regarding this way of understanding OSR which presses on the relationship between symmetry and objectivity.

[35] Setting aside Curiel's arguments, if we were to accept the underdetermination between Hamiltonian and Lagrangian formulations then we could sidestep both this and the underdetermination over particle identity by adopting an appropriately complex ontology that includes both the group-theoretically characterized structure underlying the particles-as-individuals and particles-as-non-individuals packages, and the common symplectic structure underlying the Hamiltonian and Lagrangian formulations.

6.4 Concern 3: In Defence of Invariantism

The relationship between symmetries, with their associated invariants, and objectivity has long been noted. Weyl, for example, famously grounded objectivity on invariance with respect to the relevant group of automorphisms, understood in the context of the so-called 'Weyl programme' (Weyl 1952, rep. 2003: 23; cited in Ladyman 1998).[36] Based on this, Castellani has presented an 'objectivity condition' for the physical description of the world, namely invariance with respect to the space-time symmetry group (1993; 1998). The issue now for the object-oriented realist is how to move from objectivity to objects (Castellani 1993: 108). One option is to do so via Wigner's association of an 'elementary system' with an irreducible representation of the Poincaré symmetry group (Castellani 1993: 108), such that the set of states of the system constitutes a representation space for the irreducible representation (as articulated in the context of *his* programme). For quantum systems, the appropriate representation space will be the Hilbert space, of course. The labels of the irreducible representations are thus associated with values of the invariant properties characterizing the systems, as we have already noted.

Now, of course, this association does not immediately yield *objects*, at least not in the sense that the object-oriented realist understands them. First of all, if we identify Wigner's 'elementary systems' with elementary particles, then what we have is the group-theoretic construction of *particles*. But particles do not have to be conceived as objects, understood metaphysically. Indeed, as already noted, Wigner's 'association' can be taken as the basis for the structuralist reconceptualization of intrinsic properties such as mass and charge, whereby they effectively 'drop out' of the group-theoretic construction (as the labels of the relevant representations). It is then a further step to go from this to the claim that what is needed are objects, and that is a step that the advocate of OSR will insist we do not need to and should not take.

As Castellani notes, what this group-theoretic construction yields are classes or *kinds* of particles, not distinct objects (Castellani 1993: 109; 1998: 183–4). As she puts it:

> The invariant properties which are ascribed to a 'particle-object' on the basis of group-theoretical considerations - as, for example, definite properties of mass and spin are ascribed to a (quantum) particle which is associated with an irreducible representation of the Poincaré group - are necessary for determining that given particle (an electron couldn't be an *electron*

[36] Again, I am taking a 'Viking' Approach to such pronouncements. To do Weyl's statement historical justice, it should be understood in the context of his insistence that the only access to objective reality is via symbolic construction, a view that, as Bell and Korté note, brings him close to Cassirer's position (Bell and Korté 2011). Such symbolic constructions together with the relevant coordinate system provide the formal scaffolding on the basis of which we model the objective world. It is through invariance that the 'residue of ego involvement' represented by coordinate systems (in terms of which points—which have no individuality—can be defined via an ego-based ostensive act of pointing to the 'here-now') is rendered harmless.

without given properties of mass and spin), but they are not sufficient for distinguishing it from other similar particles. In addition to these 'necessary' properties (sometimes called 'essential' properties), one does need further specifications in order to constitute a particle as an individual object. (1993: 109)

However, we do not need to step outside of group theory to obtain this further specification. She argues that we can use the notion of an 'imprimitivity system' in this regard (as originally introduced by Mackey 1978).[37] The basic idea is to use the notion of a 'system of imprimitivity' associated with a symmetry group in order to determine 'individuating' observable quantities such as position and momentum and thus move from kinds to individual objects by supplementing the group-theoretic account.

Putting things somewhat crudely, we obtain an imprimitivity system in the following way: we associate with a system, in addition to the group G, a configuration space S (strictly a Borel space) on which G acts. A projection valued measure is then defined on S (where a projection valued measure is a mapping from a Borel subset of S to the relevant projection operator) and if the projection valued measure satisfies a certain identity ($U^{-1}{}_x P_E U_x = P^{-1}{}_{Ex}$; where P_E is a projection operator and U is a unitary representation) then the projection valued measure constitutes a 'system of imprimitivity' for U based on S. The importance of the system of imprimitivity associated with U is that it determines the structure of U as an induced representation (Mackey 1978: 71; Varadarajan 1985: ch. 9). In particular, if S is transitive and L is a unitary representation of a closed sub-group of G, then the equivalence class of L is uniquely determined by the pair U,P, where P is a system of imprimitivity for U and the commuting algebra for L is isomorphic to the subalgebra consisting of all bounded linear operators that commute with all P_E (Mackey 1978: 71–2). This amounts to a statement of the 'imprimitivity theorem' which has a number of important applications.

The virtues of imprimitivity have been extolled by Varadarajan, who writes that,

The approach through systems of imprimitivity enables one to view in a unified context many apparently separate parts of quantum mechanics—such as the commutation rules, the equivalence of wave and matrix mechanics, the correspondence principle, and so on. The same treatment leads moreover in a natural fashion to the notion of spin. (Varadarajan 1985: viii)

In particular, if S denotes physical space (3-dimensional, Euclidean, affine), and G is now the Euclidean group of all rigid motions of space, then the position of a particle, regarded as an 'S valued observable', can be described by a projection valued measure defined on S. The relevant projection operator is then the self-adjoint operator corresponding to the real-valued observable which has the value 1 when the particle is 'in' Borel subset/at a given position and 0 when it is not. If we impose the

[37] Although it is implicit in Wigner's 1939 work, and has been notably applied to the definition of physical particles by Piron (1976).

requirement that the description of the system be covariant with respect to G, then the projection operator must satisfy the identity which renders the projection valued measure a system of imprimitivity. Introducing momentum observables and applying certain group-theoretic results, one can then obtain the usual commutation relations, not by analogy with the Poisson brackets of classical mechanics but as a consequence of Euclidean invariance (Mackey 1978: ch. 18; Varadarajan 1985: ch. 11).[38]

Furthermore, one can show that every irreducible representation of the commutation rules is equivalent to the Schrödinger representation. The apparently special choices in the latter for representing position and momentum observables are in fact the most general ones possible subject to the commutation rules, if we assume irreducibility. On this basis, it is claimed, we can prove the isomorphism of Schrödinger wave mechanics (based on the Schrödinger representation) and matrix mechanics (based on the commutation rules only) (Varadarajan 1985: 151). And the results just keep on coming: if the relevant configuration space is affine, we get the Born interpretation of $|\psi|^2$ and an 'illustration' of complementarity in the sense that one can show that no single state exists in which both position and momentum can be localized sharply (Varadarajan 1985: 154–5).

As far as the current discussion is concerned, the important point arising from all this is that, 'All we need to discuss physical events are position observables and a dynamic group' (Mackey 1978: 195). In particular, through the imposition of a condition of covariance for observables, imprimitivity allows us to accommodate, in group-theoretic terms, the spatio-temporal location of particles (Piron 1976: 93–5). According to Castellani, this restores the notion of an object and thus we get the group-theoretic characterization (or for her, *constitution*) of not only kinds but individual objects (Castellani 1998: 190).

Now there is a sense in which this is not what the structuralist wants![39] But of course there are ways in which she can accommodate the central insight of Mackey's comment without being committed to objects in any robust or metaphysically substantive sense. Thus we might understand imprimitivity as giving a group-theoretic grasp on the position of a 'particle'[40] but insist that this does not yield objecthood.[41] In other words, we can buy into the whole group-theoretic analysis/reduction of 'objects' but simply resist the exportation of position, say, beyond the

[38] There is the possibility of further underdetermination here: 'Given any quantum system with a complex Hilbert space defining the logic, we may obtain another whose logic is defined by a real Hilbert space by simply composing the given one with a new independent system whose logic is the set of all subspaces of a real two-dimensional Hilbert space' (Mackey 1978: 197). According to Mackey, the ambiguity can be analysed and 'to some extent removed' by the application of group-theoretic notions.

[39] Thanks to Anjan Chakravartty for pressing me on this, early on in the discussions leading to this section.

[40] Perhaps understood as one of Bell's 'beables'.

[41] Beables don't give objects.

temporally limited domain of the immediately observable and into the realm of quantum objects as a whole.

Thus we can understand position as yielding, not individuality per se, but only a kind of 'pseudo-individuality' (as already indicated), or what Toraldo di Francia refers to as 'mock individuality' in the sense that one can pretend the particles are individual objects at the point of measurement, as it were, but only temporarily (Toraldo di Francia 1985; Dalla Chiara and Toraldo di Francia 1993). It is significant that this notion is articulated in the context of what can be taken as a form of structuralism,[42] according to which particles are regarded as 'nomological' objects in the sense that 'physical objects are today *knots of properties*, prescribed by physical laws' (1978: 63).[43] It is in this context that Dalla Chiara and Toraldo di Francia develop their view of quantum particles as 'anonymous' in the sense that proper names cannot be attached to them, although here too there is a tension between this and the underlying structuralism (for further discussion see French and Krause 2006: 221–5). However, the important point is that pseudo-individuality allows us to refer to 'objects', without compromising our structuralism:

> This is why an engineer, when discussing a drawing, can *temporarily* make an exception to the anonymity principle and say: 'Electron a, issued from point S, will hit the screen at P, while electron b, issued from T, will land at Q'. (1985: 209; Dalla Chiara and Toraldo di Francia 1993: 266)[44]

Indeed, one can tie this to the Poincaré Manoeuvre and take this idea of pseudo-individuality as allowing us to introduce a notion of pseudo-'object' as a descriptive convenience grounded in macroscopic position measurements, on the basis of which we can employ group theory, via the export of this notion into the quantum realm and the identification of the elements of the group with such 'objects', but which can subsequently be discarded once we have a grip on the relevant structures as described group-theoretically, leaving the latter as the focus of our ontology. Both this device and indeed Wigner's association discussed previously can be understood as ways in which we can maintain a form of eliminativism with regard to objects while still being able to talk about or refer to those features of the world that we standardly (but erroneously) associate with such objects. I shall present some further devices along these lines—although taken from metaphysics—in the next chapter.

This also allows us to respond to Suppes' 'obvious and practically important' point that, granted the important role of invariants in physics,

[42] Thus he refers to the process of 'objectuation' by which the mind 'decomposes' the world into objects (1978: 58; see also Toraldo di Francia 1981: 220). Crucially, 'objectuation is strictly connected with, or consists of, the mind's ability to distinguish *this* and *other*' (1978: 58).

[43] The similarities with both my view and the bundle conception of objects are obvious.

[44] We recall Eddington's point that we find it difficult to release our grip on this notion of objecthood (he expressed this point in terms of the entities retaining a 'legend of individuality'), in large part because of our experimental practices and, in particular, the role of position measurements.

> it is simpler and more convenient to make and record measurements relative to the fixed framework of the laboratory, rather than record them in a classical or Lorentzian invariant fashion. (2000: 1576)

This is presented as a conflict between invariance and efficient computation but it also bears on the metaphysical motivation for structural realism: in effect, such measurements yield a form of 'pseudo'-objecthood, precisely because of the lack of invariance, which cannot be imported into the quantum domain, as it were, on pain of running into the underdetermination problem. And as Suppes also notes, it would be a mistake to infer from the fact that scientists choose a convenient laboratory coordinate system in which to pin down the relevant pseudo-objects, that they are neglecting invariance.

The point then is that the advocate of OSR can adopt Weyl's characterization of objectivity without having to take the extra step from that notion to that of 'object', understood in a metaphysically robust sense. This offers a 'de-anthropomorphized'—or metaphysically non-substantive—conception of a physical object, just as the principle of general covariance did for Cassirer (Ryckman 1999). Thus,

> it is no longer the existence of particular entities, definite permanencies propagating in space and time, that form 'the ultimate stratum of objectivity' but rather 'the invariance of relations between magnitudes'. (Ryckman 1999: 606, citing Cassirer 1957: 467)

This association of objectivity with invariance is further reinforced by claims that the latter *explains* three crucial features that render a fact objective, namely (Nozick 2003; see also Earman 2004):

1. It is accessible from different perspectives.
2. There can be intersubjective agreement about it.
3. It holds independently of people's beliefs, desires, observations, measurements.

However, Debs and Redhead have raised a series of criticisms against this association (2007; for critical discussion see Nounou et al. 2010; and van Fraassen 2009). First, there is the problem of sorting out what is significant (cf. van Fraassen 2006): symmetries come in various shapes and forms and it is difficult, if not impossible, to know beforehand which will be heuristically fruitful or not. This seems an obvious point but it hardly impacts on the kind of objectivity claim just articulated. The core of the criticism is that no account has been given either for why some symmetries are physical, others mathematical, some dynamical, others accidental, etc., or for why some are fruitful and others not. Indeed, it is claimed, 'history suggests' that no such account will be forthcoming and the significance of certain symmetries must be taken as a 'brute fact'.

However, if you're a realist, then this significance, understood appropriately broadly, is 'explained', again, *by the way the world is*. This would be the ultimate 'brute fact'! If significance is meant as something akin to heuristic fruitfulness, then retrospectively we give the same answer—gauge invariance has turned out to be so fruitful because that's the way the world is structured—and prospectively, we can only say 'that's why it's called heuristics' since we can't know ahead of time which will work, and we can't give an algorithm for scientific discovery. As for 'explaining' the differences again we will have to appeal to the sort of 'line drawing' given in response to Roberts' concerns. And again, when it comes to distinguishing a physical symmetry, as represented by a mathematical group that is applied, from a non-physical one, as represented by a group that is not applied, we simply have to refer to the structure of the world. Ultimately we have to stop somewhere in our explanatory endeavour and if the question is why one group represents the world and not another, the realist's answer will be that that is the way the world is. If this seems unsatisfactory, I think it seems so for reasons that have nothing to do with the role of invariants in establishing objectivity.

Secondly, there is the problem of choosing 'The Definitive Group' (Debs and Redhead 2007). Here the worry is that different aspects of the physical world are associated with different symmetries but identifying those that are universal is difficult. As an example consider the contrast between the hydrogen atom with relativistic space-time, where we have two models structured by very different symmetry groups (2007).[45] In the latter case, a kind of fruitful heuristic leapfrogging occurred, but not in the former. Now, this might be expected given the very different physical systems concerned, and of course sometimes structures and symmetries are exportable from one domain to another very different one (consider, for, example the renormalization group in the context of the development of quantum field theory, where the relevant representation was imported from condensed matter physics; see Fisher 1999). It might well be that a set of symmetries applicable to one system turns out to be applicable to another very different kind of system. As Debs and Redhead acknowledge, ultimately this is determined on a case-by-case basis and it is empirical success that plays a fundamental role in this determination, but of course, no one but the sociologists of science expected it to be determined in any other way!

Thus if one is a convergent or non-pluralist realist, one will insist that ultimately we will arrive at the set of fundamental 'universal physical symmetries'. The worry now is how to pin down this set, when it seems that all we have to go on is their heuristic fertility. But, of course, we don't just have that, we also have empirical success and although perhaps a complicated story will need to be told about how that flows up from the phenomena to the symmetry principles, that is surely not unusual in the philosophy of science. So the answer to their question, 'If these symmetries are so selected due to their heuristic effectiveness, then why add to this the notion that they are associated with objectivity?', is that they are *not* so selected and the 'adding to' here simply reflects the difference between heuristics and justification.

[45] Guay and Hepburn (2009) suggest that the former is better represented via groupoids and relate these to an extension of the concept of symmetry in terms of equivalence classes.

Thirdly, 'invariantism' can be understood as tied to the search for a unified theory but, it is claimed, objectivity should be something that is independent of such a goal (Debs and Redhead 2007: 71). However, one could presumably still be an 'invariantist' and a Dupré-style pluralist or a Cartwrightian dappler (Cartwright 1999)—each domain or 'patch' would have its own set of symmetries in terms of which objectivity would be given. Certainly, one could still retain subject-independence within this framework and a Cartwrightian would surely object to the claim that nothing could be more subject-independent than a Grand Unified Theory (GUT). Similarly if one were an ontological non-reductionist, one would insist that each 'level' could have its own symmetries—if such could be made sense of.

Taking the standard convergent realist line, it is still not clear why a problem arises. The worry seems to be that we could not have 'full' objectivity until the GUT is known, and 'partial' objectivity is unacceptable. But one could adopt a broadly fallibilist stance that allows us to accept that at least some of what we currently take to be objective may turn out not to be—so parity goes out of the window, to be replaced in some sense by Charge conjugation-Parity transformation-Time reversal (CPT)—but incorporating specific partiality, so we have good grounds for believing that at least certain features represented by current theory count as objective.

Debs and Redhead insist that according to the invariantist approach classical physics must fail in its objectivity because of the relevant lack of invariance—to which the appropriate response is surely 'yes, yes it does!' Again, even though the models of classical physics no longer count as objective, we can still say they're pragmatically useful, approximately accurate within the appropriate limits, etc., or even that they are partially or pragmatically true (da Costa and French 2003). And we can still make objective claims that are provisional if we adopt the appropriate fallibilist stance(s). From such a stance, complete objectivity would indeed be an ideal, to be reached once we have the GUT, but an understanding of objectivity need not offer more than this to be useful.

Debs and Redhead dismiss (complete) invariantism as a 'tantalizing illusion' and insist that it must be regarded as conventional or contextual (cf. van Fraassen 2006). Thus they argue that the 'objective identities' of objects could be construed as objective features of some model but these are clearly not invariant. Hence, the objectivity of such identities must be conventional. However, as an eliminativist I see no grounds for regarding such identities as objective to begin with; rather, objectivity is appropriately grounded in the invariants that group theory presents. But of course, talk of eliminativism alarms some people; after all, how can it be that the appearances are illusory, where these include not just familiar everyday objects like tables and people, but also the 'objects' of science, such as genes, molecules, and even elementary particles. In the next chapter I hope to allay the fears on this score by indicating ways in which we can retain talk of such objects while maintaining a non-object-oriented, structuralist ontology.

7

The Elimination of Objects

7.1 Introduction

My aim in this chapter is to defend an eliminativist attitude towards objects by setting out some of the metaphysical devices that we can draw upon, in the spirit of the 'Viking Approach', in the articulation of OSR. I shall begin by considering the relationship between 'everyday' objects and their constituent physical entities, and use this as a springboard to examine the relationship between structures and objects, from the perspective of OSR. In doing so I hope also to indicate, more generally, how metaphysics and the philosophy of science can be brought into a more productive relationship, following the discussion in the previous chapter.

7.2 Dependence and Elimination: Tables and Particles

Consider, as an exemplar of an 'everyday' object, the table at which I am sat. Two obvious questions arise: What is the relationship between everyday objects like tables and the entities posited by physics? And: What is the relationship between those entities and the structure posited by OSR? One answer to the first of these questions would be to say that the table is somehow *dependent* upon the relevant assembly of physical entities (whether these are taken to be particles, fields, strings, or whatever). However, as Correia notes, in his useful survey (2008), the term 'dependence', as deployed in metaphysics, covers a whole family of properties and relations (see also Lowe 2005; Rosen 2010). Broadly speaking, it may be taken to denote some form of 'non-self-sufficiency':

> A dependent object... is an object whose ontological profile, e.g. its existence or its being the object that it is, is somehow derivative upon facts of certain sorts – be they facts about other particular objects or not. (Correia 2008: 1013)

This sense of being derivative can be captured via the alignment of dependence with entailment, as expressed by what Rosen calls the 'Entailment Principle' (Rosen 2010: 118): if x is dependent on y, then y entails x. One can then distinguish three forms: existential, essential, and explanatory dependence (Rosen 2010: 118; Lowe 2005). Existential dependence obtains when the existence of the object requires that a condition of a certain sort be met; essential dependence obtains where the object would not be the object that it is had a condition of a certain sort not been met (Correia 2008: 1014); and explanatory dependence obtains where the object stands in a certain kind of explanatory relationship with other objects.

Taking existential dependence first, its denial captures the following intuition: object *a* could have existed even if object *b* did not and if this is the case, we can say that *a* is ontologically independent of *b*. Thus my table could have existed even if the chair on which I am sitting did not, and in this sense is independent of it. However, my table could not have existed if its constituent particles/fields/strings/whatever did not, and in this sense is existentially dependent upon them (Correia 2008: 1015). One can read the sense of dependence here in terms of 'rigid necessitation', so that the table rigidly necessitates its specific constituent particles. Sortal considerations enter with 'generic necessitation', in the sense that my table generically necessitates the existence of fermions. Likewise, redness generically (but not rigidly) necessitates red things and a methane molecule generically necessitates carbon and hydrogen atoms.

Similar considerations apply to essential dependence, so one can distinguish 'rigid essential involvement', such that, for some relation, *a* is essentially related by that relation to *b*, and 'rigid essential necessitation', whereby *a* is essentially such that it exists only if *b* does (2008: 1017), together with their generic counterparts. Finally, 'explanatory dependence' holds in forms such as 'if *a* exists, then this is in virtue of the existence of *b*' and 'if *a* exists, then this is in virtue of some feature of *b*' (Correia 2008: 1020).

Now, not all of the notions of dependence currently in play possess the appropriate feature of derivative-ness, or fundamentality. So, *a* rigidly necessitating *b* does not imply that the existence of *a* is derivative upon or less fundamental than that of *b*, for rigid necessitation is not asymmetric (Correia 2008: 1023). Thus, take Socrates and his life, for example: Socrates' life depends on the existence of Socrates and vice versa, yet Socrates and his life are not identical since they each possess properties (weighing so many kilograms, being so many years long) that the other does not (Lowe 2005). Moving to the essentialist notion or that of explanatory dependence may help, because if the obtaining of *b* is essential to *a*, then the identity of *a* may be said to be derivative upon *b*. Thus, we might capture the asymmetry involved here by asserting that *a* is dependent upon *b*, iff the identity of *a* is dependent on the identity of *b* (Lowe 2005). Likewise, if the existence of *a* is objectively explained by *b*, then *a* is less fundamental than *b* (Correia 2008: 1023).

Of course, an obvious issue with explanatory dependence as it stands is how one should understand 'in virtue of'. One option would be to take it as primitive, with the relevant derivativeness built in (see Rosen 2010: 113). Alternatively, one might reasonably suggest that 'in virtue of' acts as a kind of umbrella phrase, to be cashed out or explicated in specific terms depending on the context. In this case, the inherent derivativeness would be dependent on the specific nature of the explanation, which, in the cases I am interested in, would draw on the relevant physics. Thus if the solidity of my table is explained by the way in which electrons occupy the relevant atomic states, which in turn is explained by the Pauli Exclusion Principle, or, more fundamentally, the anti-symmetry of the relevant wave-functions and the role of Permutation Invariance, then that solidity can be said to be less fundamental than, or derivative upon, those features associated with symmetry. Or, shifting from explanatory to essential dependence, Permutation Invariance would be essential to what the table is; or, again, thinking of existential dependence, we would say that the table could not exist without Permutation Invariance.[1] Of course, the latter is not an object, so what we have is necessitation in terms of a kind of symmetry, which the advocate of OSR understands as a feature of the structure of the world.

A possible worry here is that the kind of dependence that 'in virtue of' signifies effectively evacuates all there is to *a* in favour of the relevant features of *b*. If all there is to *a* holds in virtue of, and hence is explained in terms of, features of *b*, then what is left that has any independent existence? And if there is no feature left over, then we have no grounds not to eliminate *a* from our fundamental ontology.[2] Thus, the dependence of the solidity of my table on the existence and properties of electrons together with Permutation Invariance motivates the elimination of tables from our fundamental ontology.

However, this may be too quick. One might insist that the explanation of *a* by *b* simply implies that *a* is less fundamental than *b* (Correia 2008: 1023), not that all there is about *a* can be restated in terms of *b*. One might, for example, flesh out this insistence by describing *a* as 'merely factual' and *b* as 'fundamentally real' (Fine 2001) and then argue that being 'merely factual' does not signify elimination in favour of the 'fundamentally real'. Of course, labelling where *a* and *b* sit in some metaphysical hierarchy does not obviate the original concern. So, in the case of explanatory dependence, if all the facts about *a* hold in virtue of and are explained by facts about *b*, then we can certainly mount a case that *a* is at best derivative upon *b*, or may even be eliminable in favour of *b*. A similar conclusion can be pushed from the claim that *a* essentially rigidly necessitates *b* so that the identity of *a* is dependent upon *b*. Not surprisingly perhaps, these conclusions have been resisted and in what follows I shall consider two examples of this resistance—one historical, one current—in order to indicate how one might respond to them in a way that is relevant to our overall discussion.

Thus, the alternative is to answer 'nothing' to the question of what is left that has any independent existence and understand the relationship between *a* and *b* in

[1] It also could not exist without electrons, but from the perspective of OSR these will be conceptualized in structural terms (that is, in terms of the relevant laws and symmetries, such as embodied in Permutation Invariance).

[2] Note the elimination is with respect to our *fundamental* ontology—to suggest that tables, chairs, people, particles, whatever should be eliminated from that ontology is not to suggest that we may not speak of such things, or pragmatically negotiate our way around them, or whatever.

'eliminativist' terms.[3] Now eliminativism seems to make people nervous, perhaps because it has been taken to imply that our claims about the appearances must be regarded as simply false and thus that we are all guilty of entertaining and asserting falsehoods.[4] However, there is an alternative: we can reject tables, people, everyday objects in general as elements of our fundamental ontology, whilst continuing to assert truths about them. I shall indicate two ways in which we can do this shortly. Before we get there, however, let's limber up, as it were, with a consideration of how one might be an eliminativist with regard to everyday objects, such as tables.

7.3 Eddington's Two Tables and the Elimination of Everyday Objects

Now, we have been here before, of course, with the (in)famous case of Eddington and his 'two tables'. In the introduction to his popular exposition of the structuralist understanding of modern physics, based on his Gifford lectures (1928), he compares the 'commonplace' table which has extension, is coloured, and 'above all' is substantial, with the 'scientific' table, which is mostly empty and is not substantial at all (1928: xi–xiii). It is the latter that is 'really there', whereas the former is an illusion (1928: 323). Presented thus, we seem to have a nice example of scientific eliminativism. This is certainly how Stebbing views it in her dismissal of Eddington's claims as 'preposterous nonsense' (1937: 54). Her core objection is that the object of scientific description is not the 'table', as this term is used in common discourse, and thus there cannot be two tables, with one granted ontological priority over the other. Furthermore, the 'scientific' cannot duplicate, and consequently replace, the everyday, since the properties of the latter, such as colour, cannot be duplicated via entities that do not possess such properties.

Now, in evaluating Eddington's claim it is important to pull together and consider arguments from across his works, both scientific and popular, in order to produce a (more) rational reconstruction of his position. Two features then become clear. The first is that like many who have sought a radical ontological reconceptualization, Eddington struggles to find a language that is not corrupted by the very ontology he is trying to replace.[5] This ontology that he is trying to get away from is one of things

[3] Wolff (2012) argues against the position set out in this chapter on the basis of the assertion that all ontological dependence relations are non-reductive. Rather, she insists, reduction must involve supervenience. Here, I think, there is just a basic disagreement between us. As McKenzie (2014) notes, the point of supervenience claims is to 'liberate priority attributions from specific claims regarding the nature of the relata'—and thus the mental can be said to supervene upon, but not be eliminated in favour of, the physical, for example. Given the combination of priority plus reconceptualization/elimination that OSR appeals to, dependence would seem to be the preferred option and indeed, she deploys Fine's analysis in this regard to try to make sense of and undermine eliminativist OSR.

[4] This would amount to a form of 'error theory'.

[5] The cost of constructing such a language is evident in the difficulty one encounters in trying to understand his final work which attempted to construct a form of quantum gravity (1946).

and, in particular, substances. This brings us to the second feature, which is Eddington's structuralism, something that Stebbing fails to grasp, as covered in Chapter 4.[6] The crucial feature of 'everyday' objects that Eddington wants to eliminate from our ontology is their substantiality and, as with other structuralists of the time, such as Cassirer, his structuralism can be characterized in those terms. How one expressed that elimination was a central problem for Eddington but it can be understood as an appropriately contextualized version of the issue we are facing here, namely how to characterize and represent the relationship between 'everyday' objects and the underlying structures that physics presents to us.

Stebbing's attack has been taken up again more recently by Thomasson (2007) who defends an ontology of ordinary objects against eliminativist arguments. She explicitly addresses the impact of science on such an ontology, identifying two forms of this impact (2007: ch. 7): according to one, associated with Eddington, science and the 'everyday' are in conflict; according to the other, associated with Sellars, they are merely rivals. With regard to the first, there can only be conflict if the two sides are talking about the same thing.[7] However, here again, sortal considerations enter the picture as Thomasson argues that reference to things is fixed via some categorical framework. Hence, she maintains that,

> scientific theories... do not use sortals such as 'table', and if science and common sense are using sortals of different categories, the 'things' picked out by the two descriptions cannot be identical. (Thomasson 2007: 142)

One might try to present the conflict in terms of some neutral sense of 'thing' but 'thing' in that sense would not then be a sortal term and could not be used to establish reference. Or one could appeal to a common notion of 'physical object' or 'occupant of a spatio-temporal region', but, she argues, the first finds no place within physics itself, and the second is hardly common in everyday descriptions. Hence there is no conflict between science and ordinary discourse: both have their distinct ontologies.

[6] Relatedly, she completely misses what Eddington took to be the fundamental implications of the new quantum mechanics with regard to the individuality of particles. Perhaps this is because she relied on her colleague, William Wilson, for her understanding of quantum physics (1937: xiii). Wilson is perhaps most well known for his work on the quantum conditions of the 'old' quantum theory and Stebbing clearly drew heavily on his paper 'The Origin and Nature of Wave Mechanics' (1937), which makes no mention of the kinds of implications that Eddington and Cassirer (and indeed the likes of Born, Heisenberg, and Schrödinger) were concerned with. These are relevant precisely because in so far as they were understood in terms of the non-individuality of the particles they were taken to rule out the possibility of such particles being ontologically characterized as objects.

[7] This is where Thomasson differs from Stebbing, who focuses on predicates, such as 'solid', and argues that unless we understand what this means, we cannot understand what the denial of solidity means, and we can only understand it if we can 'truly say' that an everyday object such as a plank is solid. Of course, one does not need to rely on Eddington's rhetoric to advance a form of eliminativism in this case, as we shall see.

With regard to the Sellarsian view of a rivalry between the 'scientific image' and the 'manifest image', in which the former has primacy over the latter, Thomasson again argues that any account of what there is presupposes a certain sortal framework. Such accounts can only offer a complete description in terms of that framework in the sense of covering all the things in those categories. However, the scientific and manifest images presuppose different sortal frameworks and hence cannot be complete in any way that renders them rivals (2007: 148). Consequently, acceptance of the scientific image does not require rejection of the ontology of the manifest.

Eddington's position is also undermined, she claims, not least because on a structuralist interpretation, there is a 'lack of conflict between the merely structural properties physics imputes to the world and the qualitative content involved in ordinary world descriptions' (2007: 139). Now, the distinction between structure and content is one that has arisen repeatedly in discussions over structural realism, as we saw in the earlier chapters, but it evaporates as far as OSR is concerned, since all relevant content is taken to be cashed out in structural terms. In so far as the 'qualitative content' that Thomasson refers to goes beyond this, it becomes part of the more general issue having to do with the relationship between the scientific and the 'everyday'.

Here a number of concerns arise, not the least being that Thomasson's account creates a vastly inflationary ontology. Let me be clear: it is not that Thomasson is claiming that ordinary objects are somehow derivative; rather, they count as metaphysically robust elements of our ontology, just as elementary particles are. As a result her metaphysics is entirely detached from the relevant physics, since the latter incorporates an assortment of physical relations that hold between, for example, protons, neutrons, and electrons, atoms and molecules, molecules and polymers, and so on. One option is to explore the possibility of meshing the metaphysics with the physics by constructing metaphysical relations that effectively track the physical ones; another, as we shall see, is to radically reconfigure the relevant ontology so as to remove the necessity for positing certain such relations. Either way, we keep the metaphysics and physics in touch with each other, as it were, rather than cleaving them entirely apart as Thomasson does.

The issue then is whether the establishment of such a relationship effectively guts the ontology of the 'manifest' framework by reducing it to the scientific. Consider a general metaphysical characterization of such relationships in terms of 'grounding', say: *a* is said to be grounded in *b* in the sense that *a* holds in virtue of *b*, without it being the case that only *b* exists. Thus the 'fact' of there being a table in front of me (or Eddington) is grounded in facts about the relevant aggregate of quantum particles in the sense that the former fact holds in virtue of the latter (see North 2013: 26). Now, explanatory relations such as this crop up elsewhere of course, and offer a broader framework than, say, causal accounts, whilst not trivializing the relationships as deductive accounts do. However, as we saw in our brief discussion of dependence earlier, one worry here is that if we take this relation seriously, metaphysically speaking, then the kind of dependence that 'in virtue of' signifies effectively evacuates all there is to *a* in favour of the relevant features of *b*. Of course, one might point to standard examples, such as the explanation of the shadow cast by the flagpole in terms of its height, the angle of the sun, and some elementary geometry and insist that this does not imply that the shadow does not exist. However—leaving aside issues as to the nature of shadows—this just pushes the issue back a step or two: once I have given the best and most complete explanation available, articulated in terms of quantum field theory perhaps, then what is there to a shadow, as an object in its own right, that is not cashed out in terms of features that are more fundamental?

Talk of 'facts' here may actually obscure the issue: granted that the fact expressed in the claim 'there is a table in front of me' is a 'real', albeit non-fundamental fact (North 2013), this does not imply that the table itself should be taken as an element of our ontology. Consider the property that Stebbing focuses on in her critique of Eddington, namely solidity. As already noted, this holds in virtue of the relevant physics as expressed in the Exclusion Principle and, more fundamentally, the anti-symmetrization of the relevant aggregate wave-function. In this case one might then insist that the latter feature of quantum mechanics entirely explicates the solidity of everyday objects and in doing so eliminates the predicate from the scope of our fundamental ontology. Of course, as we shall see, one may still utter truths about tables, how solid they are, and so on and these truths may be regarded as further facts beyond those that are fundamental, but one can still have all this and deny that the entities exist. I shall return to this point shortly.[8]

Eliminativism about ordinary objects may seem a radical position to adopt[9] but it is one that meshes with our understanding of contemporary physics, according to which there is only a limited number of certain fundamental kinds of elementary particles and four fundamental forces—everything else is dependent on these. I aim to take this picture seriously, in the sense of indicating, in at least a preliminary way, how an appropriate metaphysics might be constructed on this basis.[10]

Now one reason this seems such a radical line to take is that we appear to have good grounds for claiming that 'Tables exist' and a dilemma is generated: according to eliminativism, tables don't exist and yet the statement 'Tables exist' appears to be true! Indeed, the fact expressed by such a statement might well be taken to be 'Moorean' in the sense that we have better knowledge of it than the premises of any argument that seeks to deny it. In that sense, it trumps any attempt at eliminativism.

[8] There is also the concern that Thomasson appears to have introduced a form of sortal relativism into this context. This has obvious problematic implications for realism, something that Schaffer takes up in his critical review (2009).

[9] Actually it may not seem such a radical position to some: many metaphysicians adopt a deflationary ontology, including nihilists of course. Nevertheless, the reaction I get whenever I mention it (much less argue for it!) is surprising for its intensity.

[10] There are of course important issues here as to what we mean by 'fundamental'; see McKenzie 2011 and 2014.

However, adopting such a line here would not simply undermine scepticism as in Moore's 'here is a hand' case, but would undermine the kind of reductive analysis that physics appears to push us toward.

Let me now briefly sketch different metaphysical manoeuvres we can deploy to help resolve this dilemma.

7.4 Metaphysical Manoeuvres

7.4.1 Manoeuvre 1: Revise our semantics

We could adopt a form of error theory, according to which the sentence 'Tables exist' is understood to be simply false but it is allowed that we can still pragmatically use such sentences. Such approaches can be found in the philosophy of mathematics and ethics (see Miller 2010): one can reject the claim that the relevant objects exist, or one can admit that they exist but deny that they instantiate the relevant properties. Thus, in the philosophy of mathematics one can find forms of fictionalism that deny that mathematical objects exist and according to which the statements of mathematics are strictly false. Nevertheless mathematics serves a pragmatic purpose in helping derive relevant conclusions, and the relevant statements can be taken as 'true-within-the-derivational-context' or more broadly, within the 'story' of mathematics, just as statements about Sherlock Holmes, for example, are true within the stories of Arthur Conan Doyle. Likewise, one could insist that ordinary objects do not exist, that all our statements about them are strictly false, but that nevertheless beliefs about such objects serve a pragmatic purpose and the relevant statements can be regarded as 'true-within-the-narrative-we-construct-for-our-everyday-lives'.

Alternatively, one could adopt something like the error-theoretic account one finds in ethics: there, it is not denied that people exist (at least not typically) but the error-theorist insists they do not have the moral qualities usually attributed to them and hence the declarative statements one finds in ethics are strictly false. Now the argument for such a view depends on the claim that there are no objectively prescriptive qualities (see Miller 2010 for a nice summary) and the qualities attributed to everyday objects certainly do not seem to be prescriptive. Furthermore, adapting something like this for everyday objects would lead to the conclusion that there are tables, but they do not possess the properties they are usually taken to have, such as solidity, for example. One could certainly maintain that solidity can be reduced to the anti-symmetry of the collective wave-function, as indicated previously, and thus that in so far as it is regarded as more than that, nothing is solid (contra Stebbing and Thomasson), but then the table, as an object, would possess neither the properties it is usually said to have, nor those the latter are reduced to, since these are only attributable to quantum particles and their aggregates.

7.4.2 *Manoeuvre 2: Revise our notion of existence, truth, and/or ontology*

Here are some alternative ways we could account for the appearances—that is, our apparent experience of tables—and maintain the truth of the relevant sentences: introduce some notion of derivative existence; deploy a form of truth as indirect correspondence; introduce truthmakers.

7.4.2.1 MANOEUVRE 2A: DERIVATIVE EXISTENCE

So, we could maintain that the sentence 'Tables exist' is true but take the sense of 'exist' here to be derivative. This is not, perhaps, a well-trodden metaphysical path to take, given our standard understanding of existence. A notion of derivative existence that is more than just a way of speaking does not seem to feature prominently in the metaphysicians' toolbox, and for good reason perhaps, since it would require modifications to the standard syntax and semantics associated with the existential quantifier.

However, Eddington can be thought of as adopting something like this kind of view in his application of structuralism to the concept of existence itself (see, again, French 2003a and Chapter 4). Thus he rejected 'any metaphysical concept of "real existence"' (1939: 162) and introduced in its place a 'structural concept' of existence (1946: 266). This followed from his analysis of claims such as 'Tables exist' as half-finished sentences, requiring completion in structuralist terms.[11] Hence, atoms and electrons, for example, 'exist', in this derivative sense, since they are analysed as aspects of structure.

The question then is, what about the structure of the world itself, does that exist? To say that this exists would result in another half-finished sentence by Eddington's lights, for what further structure could the physical structure be a part of? Eddington maintained that this question never actually arises within his epistemology: having described the nature of physical knowledge, understood itself as a description of the physical universe, nothing further is added to our knowledge of it if one were to say 'and the physical universe exists'. He then went on to consider the structure of existence itself, characterized as having only two values and thus represented in terms of idempotent symbols (French 2003a: 249–50). Interestingly, this takes him towards the occupation number interpretation of quantum field theory, couched in terms of a group-theoretic analysis from which particles effectively emerge. Returning to the issue of the two tables, Eddington was explicit that it was by analysing existence in this way that one could respond to the concerns of philosophers such as Stebbing: 'Tables exist', on this view, must be understood as a half-finished sentence, to be completed by incorporating structure. The full sentence will then be 'Tables exist as

[11] Stebbing's critique was published before this later work of Eddington and hence she makes no mention of it.

features of a certain structure' and in this sense their existence can be understood as derivative.[12]

7.4.2.2 MANOEUVRE 2B: TWEAK TRUTH

On Eddington's view, statements such as 'Tables exist' cannot be taken as either true or false, since they are incomplete. Taking such statements to be non-truth-apt might be seen as forcing too radical a revision of our standard semantics, so an alternative would be to continue to take them to be true, but explicate truth in something other than the standard correspondence sense. Horgan and Potrc canvass just such a view in their defence of what they call 'austere' realism, which also eliminates 'everyday' objects, but on the grounds that they are vague and since ontological vagueness is impossible, they must be removed from our ontology qua objects (Horgan and Potrc 2008; see French 2011b).[13] What is important for my purposes here is Horgan and Potrc's use of contextual semantics:

> Numerous statements and thought-contents involving posits of common sense and science are true, even though the correct ontology does not include these posits.... Truth for such statements and thought-contents is indirect correspondence. (Horgan and Potrc 2008: 3)

Note that they accept that tables, for example, are not to be included in our 'correct ontology' but we can continue to utter statements about them and regard these statements as true, but with truth understood not in terms of correspondence along the usual Tarskian lines, but in terms of indirect correspondence. This is understood as semantic correctness under contextually operative semantic standards (2008: 370), in terms of which the relevant statement is made true not by some truthmaker but 'by the world as a corporate body' (2008: 3). Thus the claim 'There are tables' is true, in the 'indirect correspondence' sense, under the contextually operative standards governing 'ordinary' usage. However, these are not the standards appropriate for the context of 'serious ontological enquiry'. If we designate in italics those posits which feature in this enquiry, then 'There are tables' is true but there are no *tables*. In particular, 'There are tables' is true, under the contextually operative standards governing common usage and 'There are no *tables*' is true, under the much rarer semantic standards that apply to 'direct correspondence', where this involves the standard Tarskian account of truth. The typical reaction

[12] We can also usefully apply this analysis to the quasi-particles of condensed matter physics, which arise from the collective effect of a macroscopic aggregate with an atomic lattice structure, such as a crystal (for a useful analysis, see Falkenberg 2007, esp. pp. 243–6). Both the dynamical properties of quasi-particles and their independence arise from certain approximation procedures applied to the excitations of the relevant collective (Falkenberg 2007: 240). Without the collective, the quasi-particles would not exist; hence Falkenberg refers to them as 'fake entities'.

[13] It is not clear that this argument can be extended to the objects of scientific ontology, since, at least as far as quantum objects are concerned, these cannot properly be regarded as 'vague' (rather than indeterminate); see Darby 2010.

of many to the elimination of objects can then be dismissed as a competence-based performance error (2008: 122).

Within this semantic framework Horgan and Potrc survey and dismiss various potentially viable austere ontologies (2008: ch. 7) and conclude that there can be only one concrete object—the 'blobject'—about which statements are true in the standard correspondence sense. This obviously yields a radically minimalist ontology in one sense, although in order to capture the observable features of the world, the blobject must manifest considerable spatio-temporal structural complexity and local variability. I shall briefly return to this later.

Furthermore, although this is an interesting way of resolving our dilemma, it raises an obvious worry about the context dependence of this notion of truth, namely that it leads to a form of relativism with regard to the content of the relevant statements (Korman 2008). Thus, suppose Julie is talking in our 'everyday' context and Kate in that of 'serious ontological enquiry'. Each utters the sentence 'tables exist'. According to Horgan and Potrc, Julie said something true (but in the indirect sense) and Kate something false (in the direct sense). If the content of the sentence is invariant across context (2008: section 3.5), then the truth and falsity of that content must vary with context, and relativism appears to result. However, the examples that Horgan and Potrc consider—that cover both diachronic and synchronic meaning change—all involve differences governed by the relevant standards, whether those of direct or indirect correspondence. In the case of Julie and Kate, we have different standards brought into play (we recall that on this view truth is just semantic correctness, under operative semantic standards), rather than simply different contexts, and hence the possibility of relativism is denied. Instead what we have is precisely what Horgan and Potrc are seeking to capture, namely the elimination of tables, as objects of serious ontological enquiry, whilst maintaining the truth (in the indirect sense) of our everyday statements about tables. That is not relativism. Nevertheless, one might still feel uneasy about tampering with truth in this way, so let us consider a further option that retains truth as we know and love it but introduces truthmakers.

7.4.2.3 MANOEUVRE 2C: TRY TRUTHMAKERS

The final option we shall consider retains both our standard understanding of existence and the standard interpretation of truth in terms of direct correspondence but urges us to reconsider what it is that makes statements such as 'Tables exist' true.

According to the Quinean view of ontological commitment, with its famous slogan 'to be is to be the value of a variable', we should be committed to those things that lie within the domain of the quantifiers if the relevant sentences of the theory are to be held as true. However, this not only requires an appropriate regimentation of the theory concerned such that the relevant variables are made manifest, but the mode of regimentation may itself bear on this issue of ontological commitment. The debate over whether a form of 'thin' individuality can be ascribed to quantum particles—touched on previously—and a weak form of the Principle of Identity of Indiscernibles sustained, depends, in part, on not only differences as to the formal framework chosen for the regimentation but also whether such regimentation is a prerequisite for such commitment to begin with (see French and Krause 2006: ch. 4). Furthermore, the metaphysician may find that the Quinean criterion operates on too high a level to address the ontological questions she has in focus. Thus, this approach is of no help in helping resolve the debate between those who think that every collection of things composes something, and those who hold that none do (Cameron 2008: 4). And this is because the relevant variables in our regimented theory will pick out 'things' at the level of tables, dogs, and electrons, rather than composite parts; that is, it applies at too high a metaphysical level. Of course, some might well insist that it is at precisely this level that our ontological commitments should lie and that thinking of the Quinean commitment in this way reveals what is problematic about such metaphysical debates—namely that, in these Quinean terms, they are ontologically empty. I'm going to leave that issue to one side because my concern here is just to lay out some of the manoeuvres developed by the metaphysicians that the structuralist might find useful.

So, according to the alternative 'truthmaker theory', the ontological commitments of a theory are not whatever is referred to by the variables of an appropriately regimented theory, but are just those things that have to exist in order to make the relevant sentences of the theory true. On the standard understanding of this account, the truthmaker for the claim '*x* exists' is always *x* (see, for example, Armstrong 2004), and thus in the case of 'Tables exist', we must be committed to the existence of tables. However, one can modify this approach in order to shift ontological commitment elsewhere:

> I think one of the benefits of truthmaker theory is to allow that <*x* exists> might be made true by something other than *x*, and hence that '*a* exists' might be true according to some theory without *a* being an ontological commitment of that theory. (Cameron 2008: 4)

When it comes to the relationship between complex objects and their constituents, this has mainly focused on the issue of whether we need to take as *true* those sentences that refer to the former, with the attendant commitment to such objects. However, the worry here is that,

> serious ontological questions are being decided by linguistic facts; whether we are committed to complex objects is being decided by whether or not sentences concerning them can be paraphrased away into plural quantification over simples. What's wrong, in my opinion, is the Quinean idea that we have to resist the literal truth of 'there are tables' if we want to avoid ontological commitment to tables. (Cameron 2008: 5)

Thus the idea here is to retain truth (à la Tarski) for such sentences but avoid an inflationary ontology by taking the constituent objects themselves to make it true that there is a sum, or composite, of those objects. What makes the sentence 'Tables exist' true are whatever we take the fundamental constituent objects of tables to be: molecules, atoms, elementary particles, table parts, whatever. Metaphysicians employ a generic term to cover those objects that are fundamental in the sense that they themselves have no proper parts—they call them 'simples', which is perhaps unfortunate because in some cases these fundamental elements of our ontology will not be simple, at least not physically. However, bearing that point in mind, I shall use the term here.

Note first, that it is clearly no contradiction on the Cameronian view of truth-makers, even adopting a disquotational view of truth, to maintain that 'Tables exist' but deny any ontological commitment to tables (2008: 6).[14] What we are committed to when we utter such a sentence is whatever it is that makes it true, and on this view that would be the relevant metaphysical simples. Secondly, although this approach may appear to mesh with the idea of derivative existence, the suggestion that tables exist in such a sense is just a way of talking, for what really exist, and all that really exist, are the relevant metaphysical simples (2008: 7).

So, we can accept that 'Tables exist' is true but refrain from any ontological commitment to tables, because 'Tables exist' is made true by the relevant 'simples' (arranged table-wise, one might say, although the notion of 'arrangement' here will have to be fleshed out using the relevant physics,[15] in particular the Pauli Exclusion Principle—or, better, the anti-symmetrization of fermionic wave-functions[16]). This line on our dilemma retains the literal (and non-contextual) truth of sentences and captures the thought that what we should really be focusing on, in setting out our fundamental ontology, are not tables, chairs, and so forth, but the fundamental entities of which they are composed.

Now there are well-known worries about metaphysical simples—whether they must be understood as point-like, for example, or can be extended (see Callender 2011). More significant for this discussion is the concern over whether they must be broadly spatio-temporal, in the sense of being localizable in space-time. Insisting that

[14] Returning to the broader issue that has to do with how we read off our ontology from our theories, we recall that the Quinean insists that our ontological commitments are revealed by what the relevant sentences quantify over. Cameron's approach rejects this: our ontological commitments lie with whatever must be included in our ontology to ground the truth of the relevant sentences. The former requires the theory to be presented in an appropriately regimented form; the latter requires a clear view of what 'grounding the truth' consists in such that it is clear what should be included among our commitments. In cases like that of tables, the relevant physics helps us to get a grip on this grounding but when it comes to physics itself, we may find that grip slipping.

[15] In order to rule out sums of tables, for example, Cameron himself suggests that what makes the appropriate claims true are the relevant simples together with certain (non-mereological) relations holding between them, such as spatio-temporal relations (2008: 14). There seems to be no in-principle objection to extending this to other kinds of relations, such as are embodied in the Permutation Invariance. Of course, from the structuralist perspective it would not be quite correct to think of these relations as holding between simples understood as ontologically distinct from those relations; rather, it is the structure itself that would constitute the simple.

[16] Here I am suggesting that physics can be deployed to help enhance some metaphysics, rather than the converse.

they must be raises obvious difficulties if the relevant simples are taken to be quantum particles (so, can a photon be a simple?) and brings into the picture something that is not prima facie a simple and may be subject to analysis itself, namely the spatio-temporal background (certainly the structuralist will want to give this a particular interpretation). But in this context at least I see no reason why we cannot release simples from such a (spatio-temporal) constraint and allow them to be the kind of 'building block' from which one constructs space-time, elementary particles, and so on. This should become clearer when we consider structuralist simples later in this chapter.[17]

7.5 Ontic Structural Realism and the Elimination of Particles (as Objects)

Having canvassed various manoeuvres that we might adopt when faced with our dilemma regarding tables, let us now consider a similar dilemma regarding particles: the ontic structural realist insists that all there is, is structure and the objects of physics are at best reconceptualized, or even eliminated altogether, depending on which variant is chosen.[18] This yields two forms of our dilemma: following the example of high-energy particle physicists we may wish to assert that 'particles

[17] There is a further concern that the kind of metaphysical nihilism associated with simples is undermined by the suggestion that science could reveal layer after layer of fundamental 'atoms'—from atoms to electrons and nuclei, from nuclei to protons and neutrons, to quarks and so on (see Wasserman 2009). Cameron himself shies away from denying the existence of tables. But even if one did, it is not clear how powerful the inference is from the relevant observation of the history of science to the conclusion that science will never reach a layer of entities whose lack of further proper parts would entitle them to be called 'simples'. If the latter are taken to be associated with some notion of fundamentality, then there is a better argument against this which draws on the bootstrap approach to elementary particles (see McKenzie 2014). However, this is entirely consistent with the structuralist line adopted here (McKenzie 2011 and 2012).

[18] Brading and Skiles note that allowing for these variants introduces a further form of underdetermination, in the sense that physics underdetermines the correct metaphysics of structure, in the sense of either an eliminativist or reductive conception (2012). Thus, they argue, the very argument that OSR relies upon can be used against it! That the insertion of metaphysics into our realism brings further underdetermination with it is a fair point, although as Chakravartty suggests (see our discussion in Chapter 3), it is perhaps inevitable. However, it is not just that OSR relies on any old underdetermination as motivator; recall: it arises as a response to the specific underdetermination regarding individuality which, its advocate insists, undermines the fundamental status of objects within realism. (Brading and Skiles, as I have noted, do not see the force of this form of underdetermination, because they seek to detach objecthood from individuality profiles, in a move that takes the realist closer to OSR.) Thus, one could argue that this further form of underdetermination has a different status and rather than undermining OSR, presents us with a choice: eliminativist or non-eliminativist. Certainly it is not clear that there's anything inconsistent in adopting different attitudes towards these two forms of underdetermination. Now in the case of the underdetermination regarding individuality in the quantum context, there is an element of commonality between the horns, namely the relevant structure. Here it's not clear that it makes much sense to talk of a similar commonality, given the different nature of the 'horns', but in this case my response is to advocate the eliminativist horn over the other. Again, I see nothing inconsistent in demanding commonality in the one case and accepting one of the horns in the other, given the differences between the two.

exist', yet according to the ontic structural realist, either there are no particles (as objects) at all, or at best they are metaphysically 'thin' with their identity cashed out in relational terms.[19] Here we seem to have something similar to the table example—from the structuralist perspective particles *as objects* do not exist but we still want somehow to accommodate talk of them. In particular, we want to accommodate statements such as 'Particles exist', or 'Particle x exists', while acknowledging that fundamentally or ultimately, they are merely aspects of structure and hence do not.[20] Again, it seems, we can deploy the metaphysical tools already used. Let us return to the notion of dependence.[21]

7.6 Priority and Dependence in OSR

I shall take as a core feature of OSR the claim that the putative 'objects' are dependent in some manner upon the relevant relations. We can express this as follows:

> Each fundamental physical object depends on the structure to which it belongs.[22]

There are then three obvious options in terms of which the notion of dependence can be articulated.

> Option 1: the identity of the putative objects/nodes is (symmetrically) dependent on that of the relations of the structure and vice versa.

With this option, neither 'objects' nor the relations are held to have ontological priority; both are interdependent on the other. In this case the following holds:

> x depends$_R$ for its existence upon y = $_{df}$. Necessarily, x exists only if y exists (see Lowe 2005).

As we have noted already, an example of such interdependence can be found in Eddington's structuralism and in recent years has been espoused by various people as a form of Moderate Structural Realism (MSR) (Esfeld 2003; Pooley 2006; Rickles 2006; Esfeld and Lam 2008 and 2010; Floridi 2008). Here the putative objects, as fundamental relata, are conceptually necessary and hence cannot be eliminated, but nevertheless all there is to these objects are the relations that they bear. In other words, their (putative) intrinsic properties and identity are given entirely by these relations and thus by the structure. Now, MSR must assume numerical diversity as a primitive in order to account for certain features of physics and one might wonder if

[19] The particle notion is problematic in the context of QFT, as is well known (see Fraser 2008; French 2012a).

[20] Thus Cao criticized OSR for eliminating particles and thus rendering physicists' talk false (Cao 2003); as was pointed out, it is not particles-as-elements-of-the-scientific-lexicon that are eliminated but particles-as-metaphysical-objects (French and Ladyman 2003).

[21] The following is drawn from French 2010.

[22] Where the use of 'it' here should not be taken as referring to objects as elements of our fundamental ontology.

this is tantamount to reintroducing some form of primitive identity. There is also the worry that if, according to MSR, all there is to objects are the relations in which they stand, then there is nothing to objects at all, and the position collapses into eliminativist OSR (French 2010a; Chakravartty 2012).[23]

Let me elaborate: we recall the Russellian point that the obtaining of a relation requires the prior grounding of the identity of the relata: in order to appeal to such relations, one has had to already individuate the entities which are so related and the numerical diversity of these entities has been presupposed by the relation which hence cannot account for it. If this is how the central claim of MSR regarding the conceptual necessity of such relata is cashed out, then not only must this view face the demand that the Russellian insistence itself needs a non-question-begging defence but a tension also arises with the further claim that the very identity of the relata is given via the relations. Indeed, it seems difficult to maintain a symmetric interrelationship in this case but once we acknowledge the relevant asymmetry, we move to the second option:

> Option 2: the identity of the putative objects/nodes is (asymmetrically) dependent on that of the relations of the structure.

Here the relevant sense of dependence can be captured thus:

> Fundamental physical objects depend for their existence on the relations of the structure = (necessarily) the identity of such objects is dependent on the identity of these relations. (Lowe 2005)

Thus, for example, it has been argued that the identity of space-time points is appropriately given by the relations that hold between them, yielding a form of 'contextual' identity that supports a 'thin' sense of objecthood (Stachel 2002; Ladyman 2007).

Is this sense 'thin enough' for OSR (cf. Chakravartty 2012; Wolff 2012)? Of course, Option 2 is still incompatible with the 'thick' conception of individuals in opposition to which OSR was originally proposed. But there might still remain the worry that even granted such a sense of dependence, there might be more to the object than is given by the relations of the structure. What is needed, it might be said, is some justification for the claim that the identity of the object depends on the structure, *and nothing else* (Wolff 2012). Here the onus issue arises again: the non-structuralist asks for just such a justification; the structuralist asks what else *could* fix this identity? Indeed, as far as the structuralist is concerned, this demand assumes precisely that which she denies, namely that there is anything 'beyond' the structure that pins down the identity of objects. Certainly, it is hard to see what that could be. The 'thin' notion

[23] McKenzie deploys Fine's analysis of dependence to articulate a reciprocal relationship between putative objects and structures that also supports a form of Moderate OSR (McKenzie 2014); again, I fear this may collapse into the eliminativist version.

of objecthood itself is understood as thoroughly structuralist in so far as objects are not assumed to be individuated independently of the nexus of relations in which they stand; rather their identities are taken to be dependent on those of the relevant relations alone in accordance with the characterization given here. The onus is on the non-structuralist to indicate what else could serve to nail the identity down.

Nevertheless, the concern remains that such a 'thin' notion may amount to no notion at all.[24] As in the case of MSR, if one must conceive of quantum particles and space-time points as bare relation bearers with nothing to them, as it were, over and above the relevant relations, one starts to lose one's grip on what this 'thin' notion is, and how these views are really different from the supposedly more 'radical', eliminativist form of OSR. In particular these alternative forms posit objects as relata on conceptual grounds only, to serve as bare relation bearers, but all their properties are cashed out in relational terms, so the question arises, what precisely is it that is doing the bearing? One can posit whatever you like on conceptual grounds but for it to have any worth in this context, it needs a physical correlate and there is no physical correlate to this aspect of the putative objects. In other words, 'thin' objects appear to be merely conceptual objects only (cf. Chakravartty 2012).[25]

Furthermore, these moderate or contextual forms of OSR cannot recover the relevant facts about how many such objects there are (Jantzen 2011). More specifically, in the absence of identity relations, no set of relational facts is sufficient to fix the cardinality of the collection of objects implied by those facts.[26] Here the problem

[24] Chakravartty usefully explores the space of possible positions between a 'thick' conception of objects (such as that underpinned by a notion of substance, for example) and eliminativism and concludes that there is simply no room for a viable 'thin' conception (2012). As he notes, his conclusions do not impact on eliminativist OSR, but that still faces the problem of explicating how we can have concrete relational structures with no relata.

[25] Wolff has a different concern (2012): if we adopt this option and take the structure to be the relevant quantum state then the notion of particle becomes state-dependent, so that, for example, talk of the different possible states an electron must be in has to be understood as talk about different possible electrons. And even if that is acceptable, the *kind* that the particle falls under is not state-dependent in this way and hence whether Option 2 yields a notion of object 'thin' enough for the structuralist depends on whether she is happy with a non-structural conception of kind-hood. But of course, that weak discernibility only holds for fermions (if we discount the attempted extension to bosons on the grounds that it introduces peculiar operators) and only then when the fermions are in the 'right' kinds of states (such as singlet states) is just further grist to the anti-object-oriented mill! Furthermore, the notion of structure the advocate of OSR has in mind is broader and certainly encompasses kinds, since these are given, or better perhaps, *presented*, group-theoretically and hence structurally. Thus, if one were to favour a 'thin' notion of objecthood in this context, I see little to worry about in Wolff's concern.

[26] With first-order languages, this is because without the identity relation, it is possible to add any number of indistinguishable objects to the universe of the relevant model. This can be blocked by adding identity as a primitive binary relation but that goes against the spirit of OSR (Jantzen 2011: 441). In second-order languages we can define a binary relation coextensive with identity but only on the assumption that the relevant models are all 'full', in the sense that they contain every possible relation that can be defined extensively on the domain. However, while this might seem reasonable in the mathematical case, it is clearly not when it comes to models representing the physical world, since it would not only lead to a massively bloated ontology but would undermine the whole structuralist project, as we would know a priori that the world must be the most complex structure possible (2011: 441).

has to do with the way in which the notion of cardinality is dependent on identity (Jantzen 2011: 441–2; French and Krause 2006: ch. 7). Shifting to an alternative notion of cardinality is of no help, if we take '[t]he clarity and foundational role of the classical notion of cardinality throughout metaphysics, mathematics, and the sciences [to] outweigh the metaphysical gains that may follow from replacing it' (Jantzen 2011: 443). Of course, that last point is hardly likely to impress the advocate of OSR, of whatever form, since the kinds of concerns arising from modern physics—particularly quantum mechanics—that motivate her ontological stance also motivate these alternative notions.[27]

These concerns do not apply to eliminativist OSR, since this rejects a premise of the argument to the effect that 'any successful ontology of objects must be capable of expressing the claim that a determinate number of objects exists in the universe or in some portion of the universe' (Jantzen 2011: 439). But the advocate of eliminativist OSR does not think there is any determinate number of *objects* in the universe or any portion thereof.[28] Now, this is not to say that I don't think one can make statements about the number of *particles* in the universe, or some portion of it. One can certainly conceive of such particles in object-oriented terms, apply a set-theoretic formalism and come up with a cardinal number but, I insist, this conception should not be regarded as fundamental. At the fundamental level there are no objects, only structures, and ultimately it is in these terms that particles should be understood. And in those terms, the notion of cardinality will not be applicable.[29] Indeed, I take Jantzen's concerns to apply to both the moderate structural realist and the advocate of 'thin' or contextual identity. The way to avoid them, of course, is to reject the underlying object-oriented presupposition to begin with, and adopt a position according to which there are no objects at all, whether thick or thin.

Option 3: the very constitution (or 'essence') of the putative objects is dependent on the relations of the structure.

Essentialism has not typically been viewed all that favourably in the context of modern physics[30] but if we take it in the comparatively innocuous sense in which it is understood in mathematical structuralism, then we can characterize the relevant sense of dependence as follows:

[27] One such is that of quasi-set theory (French and Krause 2006) in which a form of quasi-cardinality can be defined (Domenech and Holik 2007). This has been rejected on the grounds that it implicitly assumes an identity relation and hence we return to an ontology of objects with primitive identity (Jantzen 2011; but see Arenhart 2012). Whether this objection has any force is beside the point since I take quasi-set theory to buttress only one horn of the metaphysical underdetermination that motivates OSR (namely the package of non-individual objects) and not as the appropriate framework for OSR itself.

[28] Jantzen lumps me in with those he criticizes (2011: 434), which is unfortunate but at least his argument gives me another stick to beat the moderate with!

[29] Which is all to the good, since we know that it is problematic in the quantum-field-theoretic context anyway.

[30] But see McKenzie 2014 who, as I have said, uses Fine's essentialist account of dependence to make sense of OSR's claims and push it towards the moderate form.

x depends$_E$ for its existence upon *y* = $_{df}$. It is part of the essence of *x* that *x* exists only if *y* exists.

Our putative objects only exist if the relevant structure exists and the dependence is such that there is nothing to them—intrinsic properties, identity, constitution, whatever—that is not cashed out, metaphysically speaking, in terms of this structure. This yields eliminativist OSR: there are no objects, thick or thin, and no identity, contextual or otherwise.[31]

Here we face a form of our earlier dilemma: even if we adopt eliminativism, we may still want to talk about objects and utter true sentences that apparently feature them. Now, in the spirit of the 'Viking Approach' to metaphysics, there are various strategies or approaches we can appropriate, as already indicated in the case of tables.

Thus the Eddingtonian approach would allow us to continue to assert that 'Particles exist...' (expressed in the 'practical language of elementary particle dynamics') but insist that we must understand this in the structural sense of existence; that is, the sentence must be understood as incomplete, with its completion articulating the claim that particles only exist as aspects of structure.

Or we could understand 'Particles exist' as (contextually) true in the indirect correspondence sense but false in the context of 'serious ontological enquiry'; that is, there are no *particles (as objects)*, just structure or aspects thereof.

Or we could take 'Particles exist' to be (literally) true but maintain that what makes the sentence true are not particles as objects; that is, the truthmakers are structures or aspects thereof (arranged, to put it one way, 'particle-like').

In this last case (which has the advantages of retaining our standard understanding of truth), the relevant metaphysical simples obviously cannot be particles-as-objects, or their metaphysical correlates. One could follow Quine (1976) in his assertion that physical objects have metaphysically withered away under the glare of quantum mechanics, leaving only space-time points. The latter would then be our 'simples'. However, this depends on a particular understanding of quantum mechanics as requiring particles (qua objects) to be non-individuals, a requirement that, ironically, the application of Quine's own criterion of ontological commitment in support of a 'thin' notion of object shows can be resisted. Given that this latter notion is itself a structuralist one, whether one builds one's structural realism on this directly or takes it as comprising one horn of the metaphysical underdetermination that has also been taken to power OSR, one might be inclined to understand the 'simples' themselves in structuralist terms.

[31] Wolff argues that Option 3 is ruled out on the grounds that ontological dependence relations are non-reductive (2012). Rejecting Option 1 as 'strange', that leaves 2, which, she maintains, leaves open the question as to just how structuralist the position obtained via this option really is. However, while I agree that Option 2 is problematic, as indicated earlier, I obviously don't share her opinion on 3, since, as also indicated earlier, I would insist that if *x* is dependent on *y in the right sort of way*, then *x* can be eliminated in favour of *y* and what we have in the case of physical particles is just the right sort of dependence that can sustain eliminativism.

Two further broad options then present themselves: one can take the relevant 'features' of structures as acting as the appropriate 'simples' or truthmakers. These features will obviously not be the kind of thing that metaphysicians have in mind, where they typically think of this notion in broadly 'atomic' terms. Here they will include symmetry principles and fundamental laws and the truthmaking relation will be reversed of course, in so far as it is not objects and properties that make true law statements and the like, on this view, but rather the laws, and symmetries, that ground the properties and behaviour of the putative objects. This is actually an important feature of my account to which I shall return in Chapter 10. But nor will these simples be spatio-temporal, unless one views the physical structure with all its features as sitting in or contained by space-time. It has long been part of the structuralist programme to incorporate space-time within this ontology (Auyang 1995; French and Ladyman 2011; Muller 2011), and the structure of the world has been taken to include space-time structure, although the details of that inclusion are waiting on a viable theory of quantum gravity (Rickles and French 2006).

Alternatively, one might want to say that there is only one 'simple', namely the structure of the world in all its glory, considered as a single entity. This invites obvious comparisons with 'blobjectivism'. The problem now is that faced by all forms of monism: how to account for the apparently manifest complexity and variety of 'the appearances'. As Horgan and Potrc note, one cannot say that physical magnitudes, in all their huge variety, are instantiated by *parts* of the blob, since strictly speaking, it has no parts. Instead, they refer to 'manners of instantiation', in the sense that the blob itself instantiates in a certain manner (and, in particular, in a spatio-temporally local manner) the relevant properties and relations (2008: 169). However, there is the obvious concern that this metaphysical move is merely parasitic upon (and therefore adds nothing to) the account offered by physics with regard to the relationship between the physical correlate of the blob[32] and the relevant physical magnitudes. More acutely, perhaps, the notion of a 'manner of instantiation' remains obscure (Schaffer 2012).

If the idea of structure, of features of structure, functioning as metaphysical simples is less than compelling, then there are further options that one might consider, including the following.[33]

7.7 Bringing Back the Bundle

Thus one might try to stick with truth, standardly understood, resist truthmakers, and offer some form of metaphysical account in terms of which we 'recover' the relevant features we are interested in, in this case, particles, from our base ontology,

[32] Healey has suggested this might be the quantum field.

[33] Here I am particularly grateful to L.A. Paul for discussion both via email and at the Leiden conference where aspects of this chapter were first presented.

in this case, structures, or features thereof. There are various routes one might take, but here I shall consider three that have particular relevance in the structuralist context.

As noted previously, the early structuralists, such as Cassirer and Eddington, expressed their ontological commitments in terms of opposition to what they saw as the generally accepted substantivalist views of the day. This naturally leads to comparisons with another well-known anti-substantival ontology, namely the so-called 'bundle' view of objects, according to which the latter are nothing more than bundles of properties (French 2001). Indeed, Chakravartty's 'semi-realism' (2007) incorporates just such a view. Specific forms of the bundle theory will then vary according to the account of the nature of properties, their instantiation, and so forth. Chakravartty prefers a dispositionalist account (further details will be presented in Chapter 9); others opt for trope-theoretic formulations (I shall return to this shortly; see Morganti 2009). Whatever form one adopts, some modification will be required when importing it into the quantum context. Standardly the Principle of Identity of Indiscernibles has been allied to the bundle view as a kind of metaphysical guarantor of the discernibility of these object-bundles in the absence of substance, which rules out qualitative duplicates, but that Principle faces well-known problems here (see French and Krause 2006: ch. 4). Saunders' revival of the Principle in Quinean form may offer a way forward and the consequent inclusion of relations into the bundle, although taking this view away from the original Leibnizian vision brings it closer to a structuralist conception, which in turn meshes with Chakravartty's approach, for example.[34]

The question now is, can this 'bundle' view of objects be allied with an appropriate metaphysics that is consonant, at least, with a structuralist base ontology?

Here I shall outline three options: trope theory, network instance theory, and 'mereological bundle theory' (MBT).[35]

The basic idea behind the first is that a 'trope' is a particular instance of a property, such as Springsteen's awesomeness, and the proclaimed advantage is that, with both particulars and properties constructed out of, or reduced to, bundles of tropes, we get a parsimonious one-category ontology (see Bacon 1997).[36] As with most such accounts, trope theory needs some principle to tie the bundle together. In this case, the Identity of Indiscernibles, as standardly formulated, would be inappropriate (since tropes are particulars and not universally instantiable), so, typically, some

[34] Nevertheless, other considerations that support the structuralist conception may undermine the bundle view. McKenzie has pointed out that the role of symmetry in elementary particle physics yields an ontological picture that is significantly different from the bundle view since the relevant symmetry relationships specify both the kinds of particles and the compositional relationships that hold between these kinds.

[35] In all three cases we have particulars, of a kind, without objects and by setting these options out, I hope to satisfy Nola's request for some metaphysical 'bush-clearing' (2012).

[36] Tropes also do useful service in acting as truthmakers for non-existential propositions about particulars.

relation of 'compresence' or 'togetherness' is invoked. So, a putative object would be a bundle of tropes related via compresence.

However, compresence clashes with physics. Thus, it has been argued that it is neither necessary nor sufficient:

> it is not necessary because a trope bundle may be widely distributed, as in particle pair formation where paired tropes constituting electromagnetic polarisation or spin may be vastly separated yet mutually dependent. It is not sufficient because more than one trope bundle can be compresent as when two or more electrons occupy the same shell of an atom. (Simons 2000: 148)[37]

In other words, compresence cannot do the job of bundling because of quantum non-locality and indistinguishability!

Alternatively, a primitive 'foundation' relation has been introduced:

> An electron must have a certain mass, charge and spin, and in addition is variably endowed with a position relative to other things and with a velocity and acceleration in particular directions at any time. When individual tropes require other individual tropes we say they are *rigidly dependent* or *founded* on these. When founding is mutual then a group of tropes must either all exist or none do. The mass, charge and spin of an electron must coexist, they require each other and form a bundle. A bundle consisting of all the tropes mutually founding one another directly or indirectly we may call a *nucleus*. (Simons 2000: 148; see also Simons 1994)[38]

Here the core issue is accounting for the fact that certain properties—mass, charge, spin, etc.—appear together in our physics, a fact that Chakravartty attributes to their 'sociability' (2007). I shall return to the latter notion in Chapter 9 but on this issue OSR can come to the aid of the trope theorist by replacing, or supplementing, the notion of 'foundation' with a group-theoretically informed structuralist account of this 'sociability'. Moving in the opposite direction, and thinking again of tools the structuralist can take down off the metaphysicians' shelf, trope theory may offer a sympathetic framework for a structuralist understanding of properties.

Of course, compresence may not disappear from the picture entirely, as even in the context of modern physics we do retain (putative) objects that appear well localized.[39] If the trope bundle theory is sufficiently 'flexible', then perhaps it can cover both the 'pseudo-objects' that manifest via scintillation screen flashes and the like and structures in general: a pseudo-individual is a bundle of compresent tropes, whereas a structure, or 'kind-structure', in the aforementioned sense, is a bundle of tropes which are not compresent. Trope theory may also be congenial to structuralism in so far as some trope theorists emphasize and defend the irreducible nature of

[37] For further criticism see Mertz 1996: 27–8.

[38] Tropes may also require other tropes as members of a kind and in such cases, instead of 'founding', we have 'generic dependence', with the tropes generically required forming a 'halo'.

[39] At least post-measurement.

relations (see, for example, Mertz 1996). And, of course, we again don't have substance in the picture nor do we have the Identity of Indiscernibles.[40]

A related alternative that has itself drawn on aspects of OSR in support is 'network instance realism' (Mertz 1996). This rejects the 'tyranny of the monadic' and takes monadic predicates to be the limiting case of *n*-adic predicates. The latter are not repeatable—here we see a similarity with trope theory—that is, they are individuated to specific *n*-tuples of properties. Ontic predicates are not to be conceived as 'in', that is, as internal constituents of, their subjects, although the predicates' characterizing intensions are 'in' them as constituents; i.e. an intension can be a non-predicable constituent of each of multiple predicates, but the subsuming predicates are neither universal nor in their subjects. An ontic predicate on this view is a simple entity with a dual nature: one aspect corresponds to a combinatorial state to or among one or more subjects; the other aspect is a content or intension ('sense') that delimits the predicate as to kind and, when the predicate is polyadic, the number and order of the unified subjects.

The basic ontological units are then individuated relation (including property) instances, each of which is a simple entity having the abstractable dual aspects of outwardly directed and unrepeatable predicability that is correlative with a repeatable content or intension. These instances are *necessary for* and *sufficient as* both ontology's 'primary substances' and as the 'cause *sine qua non*' of all plural wholes, including, needless to say, structures. In particular, Mertz explicitly addresses the issue of the relationship between relations and relata and insists that the existential dependence of relational instances on their relata results not from some defect of *being* (*ens*), but rather derives from their positive status as ontically productive and unifying principles—recalling again Eddington's view, the relations unify the relata in an ontologically significant whole (structure). Objects, qua natural entities, do not exist per se, but rather as, or as Mertz writes, abstracted or constructed intensional nodes in, sub-structures in the all-encompassing physical structure of the world.

Thus what we have is a metaphysical picture in which structures and relations are taken to be ontologically primary and 'objects' constituted out of these as intensional nodes in the network of relational instances. And of course, when we identify the relevant relations, at the level of physical research, we assume that there are 'underlying' relata but this should not mislead us in taking the latter as ontologically primary. This clearly offers a framework that is congenial to OSR.

Finally, let me consider 'Mereological Bundle Theory' (MBT). The key move here is to regard 'our knee-jerk way of thinking about the things physicists describe as "objects" or "particles" as little material-like hunks of stuff [as] fundamentally mistaken' (Paul 2010 and preprint: 35–6). According to this account, the world is not built from the bottom up, 'spatio-temporal hunk by spatio-temporal hunk', as it

[40] For further work on trope theory in the context of current physics, see Morganti (2009).

were, but rather should be conceived of in terms of a one-category ontology in which the only category is that of properties, with 'objects' understood as bundles of these. Instead of invoking primitive and hence rather mysterious relations of 'compresence' or 'foundation' to tie the bundle together, MBT understands bundling in terms of fusion where this captures the idea that it involves the *creation* of objects.[41] Everyday objects and those that can be spatio-temporally located in general are effectively created by fusing the relevant properties with spatio-temporal location, where the latter is also understood in property terms, rather than as a 'sui generis entity' (see Paul preprint). The relationship between property fusion and spatio-temporal fusion is crucial for understanding how putative objects can be composed of property parts and also smaller spatio-temporal parts (preprint). In particular, property parts are no different in kind from spatio-temporal parts—the former are not to be understood as abstract, with the latter as concrete; rather properties, or at least some of them, and in particular those that are everyday objects, are concrete (I shall return to this issue in the next chapter). This also sheds light on the nature of fusion: it does not somehow produce concrete entities out of abstract ones but rather just creates the one (object) from many (properties). All fusions, on this account, are fundamentally qualitative fusion.

What about the individuation of objects and, in particular, the role of the Principle of Identity of Indiscernibles, which, as I've noted, is problematic in the context of modern physics? One option is for the bundle theorist to simply deny that the identity and individuality of objects has to do with qualitative properties, even if the object is nothing but a bundle of such properties. Thus she could insist that identity facts do not supervene on any qualitative properties but simply on the object x itself (Paul preprint). Of course, this amounts to a form of primitive individuation but it does at least avoid a lot of 'ontologically heavy machinery'.[42] It is worth noting that the motivation here is to accommodate the kinds of symmetries that the structuralist sets such store by:

> the primary ontological choice one must make, given the seeming possibilities of various sorts of qualitative symmetries, is not between ontologies but between accommodating the possibility of these symmetries or not. Only if one chooses to accommodate the possibilities, must one then choose between ontologies: between a universe with primitive grounded differences and a multiplicity of categories, or a universe with primitive ungrounded differences and a single category. (preprint: 28)

However, a well-known problem now arises, namely the possibility of multiple, qualitatively indiscernible particles existing in the same state (forthcoming; see also

[41] The creation of bizarre or generally unwanted objects can be avoided via appropriate restrictions.

[42] In this regard MBT would run up against the concerns expressed by Dasgupta (2009). Even without the heavy machinery, appealing to primitive individuation introduces extra danglers into the picture, but if PII is deployed, unacceptable constraints are placed on the sorts of general facts that can hold since PII rules out certain situations as impossible.

French and Krause 2006). One option is to extend Saunders' approach to bosons (Muller and Seevinck 2009). Alternatively, one could argue that such states do not have the quantitative structure their name implies: what we have is a property instance of 'two-boson-ness', where the latter is an example of what Armstrong called 'fundamentally intensive properties',[43] in the sense that they lack structure and cannot be reduced to co-instantiations or co-occurrences of multiple instances of unit properties such as 'being a boson' (Paul preprint: 30–6). Thus, the bundle view can accommodate the possibility of multiple, qualitatively indiscernible particles by accepting structureless intensive properties and in effect denying that we have two, or more, objects in such states—a move that also meshes with QFT (preprint: 33–4).[44]

There is a cost of course: that of introducing many intensive properties, with a consequent inflation of our property-based ontology. Of course, the alternative objects-as-distinct-from-properties ontology is likewise vast in terms of the number of items it entertains but at least it presents fewer *kinds*: the kind 'boson', under which fall numerous objects, as opposed to numerous 'kinds' of property, such as two-boson-ness, three-boson-ness, and so on. Furthermore, the denial of internal structure does not sit well with the experimental 'facts': we can manipulate such states and obtain what appear to be single particles from them. Of course, between observing the flash on the scintillation screen and asserting the existence of a single particle a number of inferential steps must be laid down, but something needs to be said about how the property instance of 'two-boson-ness', say, can yield an instance of 'one-boson-ness' (perhaps one could say that an operation of 'de-fusion' is involved).

Still, the structuralist would be sympathetic to the anti-substantivalist stance that lies behind this form of bundle theory, particularly in so far as it offers a one-category ontology in which the distinction between objects (qua bearers of properties) and properties themselves evaporates. Indeed, if the latter include, as they should, relations and non-monadic properties in general, then the distinction between bundle theory and a structuralist ontology may reduce to cigarette paper thinness, as already noted. Furthermore, as with trope theory's 'foundation relation', the co-occurrence of certain properties lends itself to a structuralist understanding. So properties, it is claimed, differ from objects in that the former may be co-dependent in ways that the latter are not (Paul preprint). This has

[43] A well-known example of an intensive property would be 'being sweet'.

[44] A standard way of understanding fields in this context is in terms of field quantities instantiated at, or smeared over, space-time regions (for a discussion of possible ontologies for QFT see French and Krause 2006: ch. 9). Typically the latter are given some form of substantivalist interpretation, with the former taken to be properties-as-universals possessed by or instantiated in this substance. Taking the field to be a bundle of qualitative and spatio-temporal properties is an interesting step and bears comparison with Auyang's structuralist view of physical structure and space-time structure as emerging together as aspects of the world-structure, a view that is also similar to Eddington's (Auyang 1995).

been taken to block the reconceptualization of the latter in terms of the former. But of course, co-occurrence does not imply ontological co-dependence:

> It just means that there are certain facts about the universe that result in certain connections: for example, that anything with mass also has extension. (Paul preprint: 15)

Adopting a structuralist perspective offers a more robust response: the supposed ontological independence of objects is problematic to begin with. Cashing out this independence in terms of the grounds for identity and individuality leads to the metaphysical underdetermination in the quantum context that OSR aims to overcome. Dropping this presumption of independence (derived ultimately from reflections on everyday objects as bits and pieces of matter banging about in the container of space-time) then removes the source of the worry. Furthermore, the suggestion that the connections should be understood via the role of laws (Paul preprint) can be bolstered by a structuralist understanding of this relationship.

As we shall see, I shall suggest that we should reverse the current understanding of the relationship between (intrinsic) properties and laws by taking the latter to have ontological priority as features of the structure of the world, with the former as derivative, or dependent. On this view, the 'connections' are precisely those that the structuralist will want to highlight as physically significant (such as that between spin and particle kind as given by the relevant statistics, for example), together with the relevant symmetry principles.[45] Again, the properties that characterize both the kinds and their interrelationships are connected to these symmetries in such a way that the meaning of a physical quantity such as spin can be understood as deriving from its representation in terms of the eigenvalues of the generators of the relevant group algebras and the (second-order) properties of these quantities is given by the associated structure.

In this context we might then bring together blobjectivism and the bundle theory under the structuralist umbrella. A 'global' bundling of the relevant polyadic properties understood in group-theoretic terms will yield the blob as structure of the world, with a 'local' bundling of the relevant properties giving us the putative 'objects'. Of course, there still remains the issue of accounting for the complexity of the appearances, but here we can supplement the metaphysics of 'manners of instantiation', or fusion, with physics-informed OSR. Again, we can move in two directions: we can supplement and reinforce the metaphysics with the relevant physics; and we can use the former as tools to help us understand the latter, in the context of OSR. My principal aim has been to illustrate the range of moves, views, and strategies that are available and, in particular, to indicate some of the metaphysical options that the advocate of OSR can take down off the shelf, as it were.

[45] Thus Kerry MacKenzie's concern about the bundle theory in this context noted in note 34 may be alleviated by modifying bundle theory in this way.

7.8 Conclusion

All of these moves come with some cost. However, at the very least they can be used to assuage some of the concerns associated with the kind of revisionary ontology that structural realism presents. In particular, we can still say things about everyday objects while maintaining that only elementary particles exist, either by adopting the division between truth as indirect- and direct-correspondence, or by some form of truthmaker theory with simples. Proceeding down a metaphysical level, we can still say things about elementary particles while maintaining that there are no objects, only structures.

At this point, one might well feel that we have proceeded too far down, into what Magnus calls 'deep' realism (2012). However, I believe that deploying such metaphysical moves is absolutely crucial if we are to develop forms of realism that are appropriate for current physics. As I noted in Chapter 3, 'physics-lite' metaphysics runs the risk of floating free from any contact with modern science (Ladyman, Ross, et al. 2009: 7), but on the other hand, metaphysics-lite realism runs the risk of incomprehension. Certainly it is not enough to pose a revisionary ontology, without articulating that ontology in metaphysical terms. And one of the things I want to emphasize is that, however one views the current state of metaphysical research, it lays out for us an array of tools and manoeuvres that we can deploy in the service of that articulation.

Less obviously, perhaps, the humility that has to be adopted towards many features of today's metaphysical views allows them to be insulated from physics (cf. Ladyman, Ross, et al. 2007: 22). Consider the question whether the metaphysicians' simples are individuals or not. Quantum physics can't answer that, because of the underdetermination touched on previously. The correct response, as I have argued here, is to reduce the level of humility that has to be adopted, in order to bring these metaphysical views into closer accordance with the relevant physics. The central example here is that of the notion of 'object': removing that from our pantheon resolves the metaphysical underdetermination and moves our metaphysics closer towards modern physics. But to make sense of an object-less ontology, we need to draw on the kinds of moves I've sketched here. Talking of objects and properties or compresence and foundation in the absence of a consideration of the relevant physics is just armchair metaphysics-mongering; but simply pointing to the physics leaves us with just a set of equations, at worst, or at best, a partial interpretation cashed out in crude metaphysical terms that sit uneasily with the physics itself. What I've tried to do here is indicate a possible 'third way' in which the physics motivates a certain kind of realism and we then draw on the range of options available to help make metaphysical sense of it. This is not the only way to proceed, but proceed we must if we are to construct a proper *philosophy* of physics.

There are further issues to explore in this articulation of OSR, of course, and in particular I need to explicate further the notion of structure in terms of the laws and symmetries of physics. Before I get to that point, I need to first clarify the manner in which this structure differs from mathematical structure, which will allow us to indulge in a useful compare-and-contrast exercise with structuralism in the philosophy of mathematics. Now, one way of articulating the difference is in terms of the notion of causality that physical structure might be supposed to exemplify. However, modern physics is notoriously inhospitable to such a notion and hence I will also need to say something about how the advocate of OSR views it. This will finally prepare the ground and take us to our account of the nature of laws and symmetries. But let us first discuss the difference between the structures the ontic structuralist is interested in and mathematical structure and in particular, the objection that the two are so blurred that the structuralist is condemned to be a Pythagorean!

8

Mathematics, 'Physical' Structure, and the Nature of Causation[1]

8.1 Introduction

It is often said that mathematics describes its domain only up to isomorphism, and this has been interpreted to mean that it only describes the structure of that domain. With the mathematization of science it is natural to extend this thesis to scientific knowledge and then the latter too comes to be conceived of as structural knowledge. Of course, in both cases the same old question arises: whether this limitation of knowledge to structure is simply epistemological or reflects the fact that there is nothing more to be known. The supposed philosophical incoherence or ungroundedness of the latter position has been the most fundamental objection raised against both mathematical and scientific structuralism throughout the histories of these tendencies. Here again, many have supposed that even if mathematics describes only the structure of the natural numbers, the latter must nonetheless have intrinsic natures in order to be said to have structure. Thus, many philosophers reject the idea of 'pure' structure as incoherent, where structure is understood in the sense of a domain of objects lacking any non-relational or non-structural properties whatsoever.

Underlying this claim of incoherence is the presupposition that to be an object is to be intrinsically so. From the perspective of OSR this is question-begging and based on little more than metaphysical prejudice. However, there is another problem to be faced, namely that if intrinsic natures are taken out of the picture and a 'purely' (however that is understood) structural description advocated, then it may become hard to discern any difference between the physical world and the mathematical world. Indeed, given the mathematization of science, and physics in particular, the structural description of the physical world may appear to be entirely mathematical, as we have seen in the case of the group-theoretical 'presentation' of fundamental structure in the quantum domain. In this case the concern arises that from the structuralist perspective, the physical collapses into the mathematical. Let's call this

[1] A fair-sized chunk of this chapter is taken from an early draft of French and Ladyman (2011) and I am grateful to James Ladyman for agreeing to let me use material that we eventually decided was a bit of a digression from the overall theme of that final paper.

(with typical wit and originality) 'the collapse problem'. The argument, put briefly, is that if only the structure of mathematical theories is relevant to ontology in mathematics, and only structural aspects of the mathematical formalism of physical theories are relevant to ontology in physics, then there is nothing to distinguish physical and mathematical structure. Hence, the concern runs, the structural realist must conclude that the world is a mathematical structure.

How should we respond to this problem? One option is to bite the bullet and accept the conclusion. Thus, Tegmark (2006) explicitly embraces a Pythagorean form of OSR in arguing for what he calls the *Mathematical Universe Hypothesis* (MUH), namely that our physical world is an abstract mathematical structure. Beginning with the standard realist claim regarding a mind-independent external 'reality', he argues that for any description of this reality to be complete—in the sense of a 'Theory of Everything'—it must be well defined not just for us humans (presuming that's who's reading this book) but for non-human sentient entities as well. Hence, the description must be accessible in a form that is devoid of contextual 'human baggage'. This 'baggage' manifests itself via the terms that both provide the interpretation of the equations of these theories and connect the theoretical structures of the theory to the empirical sub-structures and, ultimately, observations. Thus, eliminating such baggage in order to arrive at what Tegmark calls a 'complete description' will yield a description that is entirely mathematical. Since this mathematical structure is a Theory of Everything it will be isomorphic to external reality. However, Tegmark insists, two structures that are isomorphic are identical; hence, external reality is a mathematical structure.

Note that to insist that what distinguishes the physical from the mathematical is the relevant *interpretation* of the latter is to beg the question here. Of course, it is not enough for Tegmark to simply reduce the 'baggage allowance' when it comes to this interpretation—he must also show that one can in effect obtain the empirical sub-structures and associated observations in purely mathematical terms. And indeed, Tegmark attempts to demonstrate how 'familiar physical notions and interpretations' emerge as implicit properties of the structure itself. Here, the role of symmetries is crucial and it follows from the MUH that any symmetries in the mathematical structure correspond to physical symmetries.[2] Tegmark then proceeds from both top-down and bottom-up directions, in the hope that, meeting in the middle, as it were, one can connect up observation with high-level symmetry. Thus he recalls the familiar points about the role of group representations and, again, the

[2] Since symmetries correspond to automorphisms of the structure, diffeomorphisms—such as lie at the heart of the so-called 'hole argument' in General Relativity—and gauge symmetry in general, do not count as physical symmetries for Tegmark but merely correspond to redundant notation that can then be dismissed. I have tried to argue against this dismissal in Chapter 7. Relatedly, and with regard to surplus structure, Tegmark is not arguing that the (supposedly) physical world manifests *all* mathematical structures; only that it is *a* mathematical structure. As we'll see, the totality of all mathematical structures corresponds to a 'multiverse' of worlds.

Wignerian identification of elementary particles with irreducible representations of the Poincaré group (see McKenzie 2011). In his terms, this shows how 'baggage' such as mass and spin emerges from the mathematics and he further notes the point that 'symmetries imply dynamics' in the sense that the latter can be identified with the transformation corresponding to time translation (one of the Poincaré symmetries), which in turn is dictated by the irreducible representation (McKenzie 2011). Using the 'empirical observation' that we can view intersubjective quantities we call angles, distances, and durations, these can all be related or reduced to properties of the mathematical structure. Furthermore, the invariance of these laws under the associated group is not to be regarded as a 'starting assumption' but rather a consequence of the MUH. As to why the structure has those symmetries, this amounts to asking why our world has this structure and not some other and perhaps the best that we can do in responding is to appeal to some form of the anthropic principle.[3]

However, even if we grant much of what Tegmark asserts, we might still resist the conclusion that everything about this, the actual, world, can be obtained from the relevant symmetries, since there are features of this world that are assigned to the relevant initial conditions and thus cannot be obtained from those symmetries and associated laws. (We shall return to this point in Chapter 10.) However, Tegmark has a response: the MUH leaves no room for initial conditions, since by definition it is a complete description of the world. Furthermore, he insists, history shows that what count as 'initial conditions' have been steadily pushed back, spatially and temporally, so that they can now be regarded as simply telling us which structure we happen to inhabit. If one accepts the claim that all mathematical structures exist, and that each corresponds to a world (in some sense) then one obtains an ultimate multiverse of such worlds, with the initial conditions reduced to a kind of 'multiversal telephone number'.[4] Of course, to say this is ontologically inflationary would be an understatement.

Nevertheless, one might resist the claim that all the meaning of terms like 'spin' and 'mass' can be extracted from the mathematics and that this 'baggage' can be ditched. And in doing so, one does not have to dismiss the crucial role played by the relevant mathematics in grounding this meaning. Consider, for example, Morrison's point regarding the role of experimental practices in establishing the meaning of spin (Morrison 2007). From this perspective, the non-mathematical baggage cannot be jettisoned. Of course, Tegmark can simply respond by insisting that although these practices played an important role in confirming and helping us get a grip on the

[3] See note 2; why the world is *this* mathematical structure and not *that* is not a question that can be answered in structural terms!

[4] Greene (2011) gives a taxonomy of different types of multiverse, including this one. One can perhaps see it as arising from a form of the 'Principle of Plenitude' (Cushing 1985), where 'physical possibility' is extended to the limits of mathematical possibility.

relevant theoretical features, and thus, through them, on the fundamental mathematical structure, once that structure is confirmed and we have a grip on it, we can show that all the meaning of spin can be cashed out in these terms. Furthermore, given his claim that these experimental practices can be derived from this structure,[5] to insist that they are in some sense non-mathematical in their foundations is to beg the question.

Relatedly, Jannes (2009) points out that one and the same mathematical description may cover two very different physical objects as in the case of a harmonic and anharmonic oscillator, which are mathematically equivalent when described in Hamiltonian terms, the difference depending on the coordinate system that is chosen. Thus the physical content of the system cannot be exhausted by its purely mathematical description. However, Tegmark would presumably reply that once one considers the relevant systems more fully, with further details included, the difference will be grounded in such details, themselves articulated, described, and ultimately conceived of in structural (and hence mathematical) terms. In other words, once one moves away from such toy examples considered in isolation, any appeal to non-mathematical physical content will be undermined by the description of such content via the relevant equations.[6]

A further response might be to turn the question-begging charge against Tegmark. It is an important step in his argument that since the mathematical structure is a Theory of Everything (ToE) it will be isomorphic to external reality; but two structures that are isomorphic are identical and hence, external reality is a mathematical structure. Now, one might use 'isomorphic' loosely, or as a *façon de parler*, in saying, for example, that a theory or physical model—indeed, perhaps one built out of wire and tin like Crick and Watson's—is isomorphic to some aspect of the world, or some system.[7] But for Tegmark's argument to work and for the identity claim to follow, he needs to use 'isomorphism' in the strict sense in which isomorphisms only hold between mathematical structures. However, that would be to assume precisely that which he aims to show, namely that reality is a mathematical structure.

[5] Something that Tegmark does not actually show but, in his own words, merely 'hints' at.

[6] Jannes' type of objection crops up again and again in discussions about structuralism and one can adopt Tegmark's response for these other occasions as well.

[7] A well-known criticism of the model-theoretic approach was that it assumes that isomorphisms hold between set-theoretic models and physical systems which, of course, is strictly nonsense. Pointing out that the relationship between any formal representation and the physical systems that it represents cannot be captured in terms of the former only (French and Ladyman 1999) led to the accusation that the structuralist who relies on such representational devices cannot give an appropriate account of the relationship between representations and the world in terms of those very representations. My response is that all current forms of realism must face this accusation, not just OSR.

Finally, and perhaps most profoundly, one might question the assumption that any interpretation of the equations, or, equivalently as far as Tegmark is concerned, any meaning assigned to the relevant terms amounts to '*human* baggage'. It is this that underpins his claim that a ToE that is well defined for non-human sentient beings (as it must be on his account) will be purely mathematical. But of course, one can argue that the meaning that the term 'spin', say, acquires though its connection with observable phenomena can be regarded as independent from the particular contingent circumstances of the relevant experimental practices that have to do with our human 'situatedness' (as organic, carbon-based life forms with two arms, one head (cue Zaphod Beeblebrox), etc., residing on 'the third stone from the sun' (cue Jimi Hendrix), and so on). And hence that sentient beings living under very different circumstances will assign 'spin' the same meaning.

Now of course further argument must be given, in particular to establish that last point, but presumably Tegmark would also want to rule out the possibility that the meaning obtained under such different circumstances would be different, since that would also undermine his Pythagorean realism. A more pressing concern is that by extracting those aspects of the meaning of 'spin' that are independent of the contingent circumstances surrounding our humanoid experimental practices, we are simply reinforcing the claim that this meaning is ultimately structurally grounded, and the structure is just mathematical. Thus we might say, crudely perhaps, that it is part of the meaning of 'spin' that particles that possess this property behave in a certain way when passing through a magnetic field perpendicular to their trajectory (such that particles with spin up are deflected one way, and particles with spin down, another), where we refrain from giving details as to the nature of the experimental arrangement, or at least those details that have to do with our human nature. But if both the particles and the magnetic field are conceptualized in structural terms, where these terms are presented via the relevant mathematics, then what about this acquisition of meaning is specifically physical rather than mathematical?[8]

This brings us back to the fundamental question: how do we distinguish physical structure from mathematical structure? And there is the further issue whether any such distinction can itself be understood in structural terms; if not, then it seems we must admit a non-structural element into OSR.[9]

[8] One might also relate Tegmark's project to the claim that the fundamental ontology of the world is digital. Floridi argues against such a claim, pressing the point that 'digital and 'analogue' are just two different 'modes of presentation' in the context of his 'informational' structural realism (2011; see also Bueno 2010).

[9] A similar complaint is made by Cao (2003); for a response see French and Ladyman (2003). There is an analogy here with the theory of universals and the problem of exemplification. Saunders (2003c) claims that there is no reason to think that ontic structural realists are committed to the idea that the structure of the world is mathematical but does not say much more. Ladyman, Ross, et al. (2007) assert that no account can be given of what makes the world-structure physical and not mathematical.

8.2 Distinguishing Mathematical from Physical Structure: First Go Round

The mathematical might trivially be distinguished from the physical in that there is more of it; there is more mathematics than we know what to (physically) do with, which is what Redhead expressed with his notion of 'surplus structure'. However, as we have already noted, and as we shall see, some of this surplus structure will be taken to correspond to physical possibilities, so that certain arrays of mathematical structures, in group theory for example, will be taken to encode the relevant modality associated with physical theories. And as we have also just noted, more radically, according to Tegmark, all such surplus structure corresponds to physical structure in the extended sense that all such structures correspond to 'worlds' in the multiverse.

A second response—almost as trivial—would be to insist that physical structure is *interpreted* structure. That's going to cut no ice with the likes of Tegmark, however, for the reasons already given. Relatedly, we might simply draw on a primitive distinction between instantiated and non-instantiated structure and align physical structure with the former and mathematical structure with the latter. The worry now is that given this distinction, OSR seems to fall on the wrong side. Thus Morganti (2011) accuses the ontic structural realist of fatally conflating certain general, abstract properties with the relevant concrete property instances. So, talk of invariants across group transformations, in terms of which objects and properties are identified within OSR, sits at a level where object and property tokens are simply not to be found. Consider yet again the distinction between bosons and fermions. Although this distinction can be articulated via the relevant group representations, Morganti insists that it cannot ground the 'actual' properties of an 'actual' boson or fermion, any more than the 'actual causal features' of actual coloured material objects can be reduced to the general features shared by abstract concepts such as 'greenness', 'redness', and the like.

However, the latter comparison seems misconceived and Morganti's worry overall looks suspiciously question-begging. First of all, it is clearly not the case that the advocate of OSR is arguing that the distinction between bosons and fermions given in terms of Permutation Invariance can ground *all* the 'actual' properties of these particles. It can't ground mass or charge, for example; but what it can ground is the kind distinction—the 'bosonness' or 'fermionness', if you like—and also—given the Spin-Statistics Theorem—the integral or half-integral character of their spin. And how it does so is via the relevant structural relationship as revealed by an appropriate analysis of the theory of quantum statistics. Now Morganti objects to this because although such an analysis might deflate the general/particular distinction, it leaves the gap between the abstract and the concrete. But that is precisely what I am trying to close here, of course, and as we saw in Chapter 7, the structuralist can appropriate notions of properties-as-concrete from either trope theory or 'mereological bundle theory'. Furthermore, I am suspicious of talk of 'actual' properties of 'actual' particles when the notion of 'actual' remains unarticulated; as we shall see, in response to similar criticisms from Psillos, my contention is that the structure of the world is indeed 'actual'.

This worry about question-begging might seem to attach to the very notion of instantiation itself, if this is taken to imply a commitment to objects, understood in the thick sense. But we do not need such thick objects to possess these properties in order for them to be instantiated (Paul preprint). Any lingering concern might be assuaged by reading 'instantiation' as 'making manifest' and indeed this understanding may help with our more general concern: if the distinction between instantiated and uninstantiated is read as that between manifested and unmanifested, then in so far as the distinction between physical and mathematical structure depends on this deeper distinction, it can be understood as simply a reflection of the distinction between manifested and unmanifested structure. Manifested structure can further be read as the structure of this, the actual, world and unmanifested structure can be understood as both surplus (à la Redhead) and also as encoding a range of further possibilities. I will return to this understanding in Chapter 10 but consider again Permutation Invariance: the manifested structure is represented by the bosonic and fermionic representations and the unmanifested by all the rest, including paraparticle representations. These can be seen as surplus, and as corresponding to a range of possibilities, some of which were of course entertained in the mid 1960s and hence can be regarded as 'close' (under some suitable metric) to the actual world.

Although useful, this distinction between manifested and unmanifested still does not fully ground the physical–mathematical distinction and allow us to respond to Tegmark. Perhaps then the broader distinction between the abstract and concrete should be brought into play here. Unfortunately, as Rosen has argued, establishing a firm ground for this distinction is also problematic (Rosen 2001).

Thus, one way of grounding it is to appeal to some process of abstraction, so that we begin with concrete entities and obtain, somehow, via this process, abstract entities by (of course) 'abstracting away' certain features of the concrete. However, the nature of this process is either unclear, or involves problematic features, having to do with the particular philosophy of mind assumed in talk of 'obtaining' abstract entities (Rosen 2001). Furthermore, depending on what we take to be the concrete entities we start with, the likes of Tegmark are going to insist that this characterization is either question-begging or fundamentally skewed in leaving elementary particles, say, on the wrong side of the divide. Certainly, it is not at all clear that abstraction in the sense suggested here plays any role in scientific theorizing—a point that I shall return to shortly.

Shifting our attention in the other direction, we might focus on what makes the physical *concrete*. So, we might insist, crudely, that physical structure is concrete in that it can be related—via partial isomorphisms in the partial structures framework, say—to the (physical) 'phenomena'. This is how 'physical content' enters our theories and allows them to be (at least partially) interpreted (again we recall Morrison's point about experimental practices and spin). But of course, this content must itself be understood as fundamentally non-mathematical. One way of securing this would be to argue that there are mind-independent modal relations between phenomena (both possible and actual), where these relations are not supervenient on the properties of unobservable objects and the external relations between them; rather this structure is ontologically basic (French and Ladyman 2003; Ladyman, Ross, et al. 2007). This in itself renders structural realism distinct not only from standard realism but also from constructive empiricism.

However, this option is not open if one is an eliminativist about phenomena, in so far as the phenomena has to do with, or is composed of, 'everyday' objects, such as tables, for example. From such a perspective, there is nothing to such objects that cannot be cashed out in structural terms, and so there is nothing intrinsically concrete about the phenomena and our problem returns. Indeed, the very structuralist moves that are appealed to in order to demonstrate how even phenomena involving positions, etc. (caveat the remarks on localizability to follow shortly) can be brought within the structuralist pale can be taken as gutting such phenomena of their intrinsic concreteness (if one is an eliminativist, which is a big 'if' for some folk). Furthermore, if we acknowledge the structuralist *bone fides* of imprimitivity systems (see Chapter 6) then those features that are typically associated with the concreteness of phenomena—namely, position and momentum—can also be brought within our framework, and the problem of establishing the distinction returns. Indeed, the manner in which they are captured gives further succour to Tegmark and his ilk!

Such appeals to imprimitivity also bear on the two further obvious ways of securing this distinction, which involve the requirements that abstract objects be non-spatial or causally inefficacious, or both (Rosen 2001). However, even if we set aside these appeals (with position observables representing the spatial nature of the concrete, and momentum its causal efficacy), cashing out these requirements raises problems that bite particularly hard in the current context. So, consider one way of doing this: abstract entities do not exist in space-time the way that concrete entities do. Now a lot depends on how we understand the idea of existing 'in' space-time. If this is taken to mean that an entity has a determinate spatio-temporal location throughout its existence, then, as Rosen indicates (2001), quantum entities might be seen as providing counterexamples. And even if one is prepared to take a stand on how we should understand the Uncertainty Principle, and indeed, quantum mechanics in general, such that quantum particles can be said to always have determinate positions, significant and well-known problems with localizability arise once one moves to quantum field theory (for an overview see Kuhlmann 2006). And when it comes to the world-structure, obvious issues arise with regard to the relationship between this structure and space-time. Certainly if the latter is also regarded as fundamentally structural and, furthermore, as intimately bound up with the putatively physical structure we are concerned with here, then articulating the concreteness of this physical structure in terms of its relationship with spatio-temporal structure is not going to be straightforward.

As for causal (in)efficacy, we shall consider this in a lot more detail shortly but again as Rosen notes, the crucial issue is to characterize the distinctive way in which concrete entities 'participate in the causal order', and, as we shall see, achieving this in the physical context is also deeply problematic.

Before we do, there is the further issue of whether the distinction between structure and non-structure can itself be articulated in purely structural terms.

8.3 Structure–Non-Structure from a Structuralist Perspective

This apparent problem for OSR is clearly stated by van Fraassen:

> It must imply: what has looked like the structure of something with unknown qualitative features is actually all there is to nature. But with this, the contrast between structure and what is not structure has disappeared. Thus, from the point of view of one who adopts this position, any difference between it and 'ordinary' scientific realism also disappears. It should, once adopted, not be called structuralism at all! For if there is no non-structure, there is no structure either. But for those who do not adopt the view, it remains startling: from an external or prior point of view, it seems to tell us that nature needs to be entirely re-conceived... (2006: 292–3)

Note the iterative nature of this point: we begin with a 'something' that is structured and that appears to have unknown qualitative features and we (that is, the structuralists) remove the latter, leaving only the structure. But by doing so, van Fraassen claims, we remove the basis of the distinction between structure and non-structure and hence OSR collapses into 'standard' scientific realism.[10]

However, we must be careful with the multiple senses of 'remove' here! In the first sense, with regard to the removal of the unknown qualitative features, we are talking about an ontological removal—something (objects with individuality profiles) that was presumed to be in our metaphysical pantheon, is now argued not to be. But in the second, when van Fraassen argues that the basis for the distinction between OSR and standard realism has been removed, we are talking about a conceptual sense. This second sense does not follow from the first. Indeed, the contrast between structure and what is not structure can still be articulated even after OSR has been accepted: one can adopt an iterative framework, for example, such that one can say

[10] So, on the one hand, the accusation is that OSR collapses into mathematical structuralism; on the other, we are told that it collapses into standard realism. At this point, I can't help but recall the accusations made against the Campaign for Nuclear Disarmament in the 1980s, that according to the right-wing press in the UK it was funded by 'Kremlin gold', while according to Soviet propaganda we were all American stooges. The conclusion drawn was that we had to be doing something right!

that with the first iteration leading to OSR, one notes the distinction between the structural and unknown qualitative features of the something (physical system, whatever...) one is analysing. This allows the relevant contrast to be drawn: OSR argues for the ontological priority of the structural over the unknown qualitative; ESR and standard realism deny this and commit to both. With the second iteration we remove the unknown and qualitative from our metaphysical pantheon: ontologically it does not exist, but that does not mean it cannot be invoked in order to articulate the distinction.

If one likes (which van Fraassen would not), one could go modal and make the distinction in terms of what would exist were ESR to be the correct stance (which it is not). Either way, van Fraassen's conclusion that if there is no non-structure, there is no structure either, *understood as a reductio of the argument for OSR*, is an ontological conclusion that does not follow from the supposed failure to draw the relevant contrast. But there is a sense in which there is something to van Fraassen's claim: OSR offers a kind of monistic ontology at the fundamental level, in that even though it asserts that there are different kinds of structures, there is only one category of 'thing' and hence the ontological distinction between the structural and non-structural has disappeared. Perhaps all forms of monism face this sort of issue. Certainly if one cleaves to the view that to describe something as structured is to presuppose something that is not structure, one is going to have problems getting a grip on the claim that all that there is, is structure. One form of relief is to adopt the iterative approach just sketched. Another is to accept that the world is as it is and the best way to describe that is in structural terms, where these may still leave something to be desired (even if it is not always clear what).

Secondly, the fact—if it is such—that the distinction between manifested and unmanifested structure cannot itself be drawn in structural terms does not as it stands undermine structuralism. It is no part of OSR or of other members of the structuralist tendency in general that all terms, concepts, features, elements, or whatever have to be *defined in* or *reduced to* structuralist terms. The core feature of OSR, we recall, concerns the structuralist reduction of and, according to one form, elimination of objects and such a feature and its associated claims is certainly compatible with further non-structural features and their associated claims. Thus, one could be a structuralist about objects but a non-structuralist, even a quidditist, about properties, arguing for a form of 'bundle theory' which includes relations and n-adic properties in general but takes their identity to be given not by the role they play in the relevant laws and symmetries but by some quiddity. There is more to say here but the point to be emphasized at this stage is that adopting a structural analysis of physical objects does not imply extending that analysis to all metaphysical features (although there may be a certain metaphysical 'harmony' in doing so). Still less does it imply extending such an analysis to such features as the distinction between manifested and unmanifested (for further discussion of this issue see French and Saatsi 2006).

Let us return to the issue of trying to draw a line between mathematical and physical structure.

8.4 Back to the Problem of Collapse

Interestingly, given the motivations for OSR, the problem of collapse has also been motivated by considerations drawn directly from the foundations of physics. In particular, the apparent implication that quantum objects must be regarded as non-individuals in some sense has often led to comparisons with mathematical objects. We recall that Cassirer described electrons as 'points of intersection' of certain relations and thereby drew an explicit comparison with geometrical objects; a comparison that goes back to Poincaré and the influence of the *Erlangen* programme. Of course, this comparison was effected, in large part, by the rejection of physical substance, which formed such a fundamental component of early 20th-century structuralism. With substance out of the picture, and an emphasis on the structural aspects of theories, it is natural to compare physical objects with mathematical ones. Thus the great physicist Heitler, who did so much to provide the underpinning of the reduction of chemistry to physics, argued that with the 'loss' of individuality, quantum objects had become more akin to mathematical objects, an argument that was also later echoed by Resnik (1997).

One might think that this comparison is undermined by the claim which supports the aforementioned metaphysical underdetermination, namely that quantum particles can after all be regarded as individuals, even if only in a 'thin' and contextual sense as indicated in Chapter 2, whereas mathematical objects—regarded perhaps as mere positions in a structure—cannot. However, Leitgeb and Ladyman (2008) argue that even completely indiscernible mathematical objects may be regarded as individuals in the 'thin' sense. Permuting structurally similar objects in a mathematical structure results in exactly the same structure. Hence, if primitive identity facts are posited in mathematics, they must respect a form of Permutation Invariance as applied to mathematical structures such as edgeless graphs (2008). It has been suggested that positing a kind of primitive identity that allows for this, by virtue of being contextual rather than intrinsic, makes for a consistent form of mathematical structuralism (Ladyman 2007).

The question as to whether the individuality of putative objects in mathematics and in physics is significantly different is an open one. Certainly, primitive contextual individuality can be defended in the mathematical context whereas in that of physics it may be argued that individuality must be grounded in qualitative relations that give rise to a form of discernibility that respects the symmetries of the theory. However, even if it turns out that the same notion of putative individual object can be rendered appropriate for both mathematics and physics, this in itself does not break down the distinction between mathematical and physical structure. I will consider in some detail the characterization of physical structure as causal shortly, but before I do, it is worth taking a brief look at the extant varieties of mathematical structuralism, to see where the similarities and differences might lie. Reck and Price (2000) have helpfully reviewed recent discussions and classified the resulting positions in terms that help facilitate such a comparison.

8.5 Mathematical Structuralism, its Motivations, and its Methodology

Thus, the following 'intuitive theses' can be taken as sitting at the core of mathematical structuralism: (1) that mathematics is primarily concerned with 'the investigation of structures'; (2) that this involves an 'abstraction from the nature of individual objects'; or even, (3) that mathematical objects 'have no more to them than can be expressed in terms of the basic relations of the structure' (Reck and Price 2000: 341–2). As we shall see, whereas (1) and (3) are analogous to claims made by the proponents of OSR, abstraction does not play a crucial role in the latter, or at least not in the way it does for mathematical structuralism. It is certainly not the case that we begin with physical objects and then 'abstract' from their 'natures' to arrive at the structures that scientists investigate. I shall return to this shortly.

Nevertheless, the 'structuralist methodology' which Reck and Price identify as motivating mathematical structuralism does bear some resemblance to core features of structural realism; namely, what is typically focused on in practice are the structural features of mathematical entities and the 'intrinsic nature' of these entities is taken to be of 'no real concern' (Reck and Price 2000: 345). However, as they note, one then has to ask the question: 'How should we understand such a structuralist methodology in terms of its philosophical implications?' (Reck and Price 2000: 346). And the methodology itself will be neutral with regard to the different answers, in that a range of epistemological, semantic, and metaphysical positions are consistent with such a methodology. So one option they identify would be to adopt a minimalist, deflationary view which asserts that all there is, is the mathematical formalism, understood as a set of empty signs (Reck and Price 2000: 347). A 'thicker' line would be some form of 'relative structuralism', according to which reference to mathematical objects is relative to the choice of model, but the truth of mathematical statements is non-relative because all such models are isomorphic[11] (Reck and Price 2000: 348–54).

This form of structuralism meshes nicely with the general eliminativist tendency that motivates, in part, a structuralist philosophy of mathematics, which has to do with a claim of no privilege: Thus consider the conjunction of the Dedekind–Peano axioms formulated in second-order logic on which arithmetic is founded. These are satisfied by a range of equivalent set-theoretical models, each of which is capable of

[11] When it comes to categorical theories, at least.

playing the role of the natural numbers. Given that, none of these models should be taken as privileged in this regard. Here we can draw a nice comparison with the underdetermination that motivates OSR (cf. Benacerraf 1965: 284–5).

So, we might say that if elementary particles are (metaphysical) objects, then they must be objects with a particular 'individuality profile', to use Brading's phrase. But if an electron, say, is really an object with one such profile rather than another—an individual rather than a non-individual—then it must be possible to give some cogent reason for thinking so, where by 'cogent' we mean some reason grounded in the relevant physics.[12] However, no such reason can be given. Likewise, Parsons (1990) has noted that there is more than one identification that can be made of numbers with 'logical objects' and that there are no principled grounds on which to choose one over the other. Now, relative to the purposes of doing physics, one individuality profile will do as well as the other; relative to the purposes of defending a metaphysically informed form of realism, however, such fundamental ontological ambiguity is not tolerable and we should shift to the view that there are no electrons- or, more generally, elementary particles-, as-objects.

Analogously, the (relative) mathematical structuralist's response to the 'no privilege' claim is that if none of the relative choices of model is preferable to any other, then we should conclude that there are no natural numbers (Reck and Price 2000: 354). Of course, according to the relative structuralist, the number '3', for example, refers to the base element of some chosen model and as Reck and Price go on to note, one can move to a 'universalist' form of structuralism by insisting that '3', say, refers to *all* base elements of the relevant models. Again this involves a process of 'abstracting away' from the peculiarities of particular models, and mathematical statements are now understood as making assertions about all (relevant) objects, functions, predicates, and so forth. One can identify a further important eliminativist element here in that 'the assumption of a special, unique system of objects, to be identified as "the natural numbers", is avoided or "erased"' (Reck and Price 2000: 358), but '3' is effectively quantified out in the relevant expression and treated as a variable, rather than an ambiguously referring term.[13]

In particular, one of the more well-known structuralist views, known as 'pattern structuralism' is a form of universal structuralism, since it focuses on the patterns instantiated or exemplified by different relational systems (Reck and Price 2000: 363–4). According to one variant, such patterns are *composed of* 'positions' or 'nodes' whose identity is given entirely by their role in the structure; as Resnik puts it 'they have no identity or distinguishing features outside a structure' (Resnik 1997: 201). According to another, associated with Shapiro, the structures and the relevant

[12] And of course, I maintain that the position that holds that it is simply an unknowable truth which profile the electron has, is likewise 'hardly tenable' from a realist perspective.

[13] One might draw an analogy here to the move behind the Ramsey sentence representation by which theoretical terms are replaced by existentially quantified variables, as discussed in Chapter 5.

positions exist *over and above* the patterns that instantiate them.[14] In both cases, thesis (3) of mathematical structuralism is satisfied, and in so far as this form of structurally given identity can be regarded as contextual and relational, one can draw a useful comparison with the 'thin' view of objects advocated by non-eliminativist OSR. Of course, one might insist that entities which cannot be individuated as objects independently of the role they play in a structure, should not be regarded as individual objects, or even as 'acceptable entities' at all. Again we see here the shadow of Russell, who insisted that if something, a number, say, is to be anything at all, it must be *intrinsically* something. And again, it smacks of the pernicious influence of object-oriented metaphysics: the whole point is to argue that these 'nodes'—be they numbers or electrons—are not, indeed, individual objects (at least not in any 'thick' sense), but just that, mere nodes in a structure.[15]

There is a great deal more to say but all I want to do here is establish a context in which to consider the criticism that OSR shares certain theses with mathematical structuralism and that these similarities undermine the former. Put bluntly my response will be that such criticism fails to spot certain crucial distinctions between the two that block the importation of concerns from mathematical structuralism into the domain of its physical analogue. We will then return to the problem of collapse.

8.6 Crossing the Bridge from Mathematical Structuralism to Physical Structuralism: Abstraction and Properties

So, a number of critics have tried to import into the debate over OSR the central thesis of Shapiro's view, namely that a number is simply a place in the number structure and that that structure exists independently of any exemplifying concrete system (see, for example, Busch 2003; also Wolff 2012). The analogue of this latter independence claim is then seized upon as generating problems for the ontic structuralist, with regard to the apparently abstract nature of this structure, the relationship between the structure and the exemplifying system, and the role of objects in characterizing the last. However, it is unclear why OSR must share this thesis. In particular, while it might seem plausible in the mathematical context, where it underpins and shapes the debate over the applicability of mathematics, its analogue with regard to physical structure would certainly need further argumentation. And indeed, in so far as the putative 'exemplifying concrete system' is the world, I shall

[14] Of course, this is not an eliminativist view in so far as, first of all, a new kind of abstract object is postulated as existing prior to and independently of instantiation; and secondly, there is a special, unique system of objects, to be identified as 'the natural numbers', namely the natural number pattern (Reck and Price 2000: 366).

[15] As far as the pattern structuralist is concerned, such positions in patterns are objects in a weak sense in that they can be referred to with singular terms, can be said to have relations to other such patterns, but do not possess any 'intrinsic' nature over and above being such a position in the pattern.

deny it. This is not to say that appropriate sub-systems can't be identified that might be said to 'exemplify' features of the relevant physical structure, in so far as one can identify and isolate sub-systems in physics in general.[16] But it is not the case—at least not within the form of OSR that I advocate—that the structure I am concerned with exists independently of the physical world. I shall return to this issue shortly.

This comparison with mathematical structuralism has generated not only a skewed understanding of OSR but also the raising of concerns which can be dismissed as inapplicable. So, for example, it has been argued that the very idea of structure presupposes some elements that 'make up' that structure in some sense, just as numbers make up the relevant mathematical structure (Busch 2003). The underlying thought here is that it is unclear how we are to understand what a structure is if it does not involve reference to objects. However, to repeat the point made previously, while it is of course correct that *any* mathematical formulation assumes some domain of quantification and while it may be the case that in order to *represent* structures, via set theory, or other parts of mathematics, or whatever, we have to invoke certain elements, one should resist the implication that is usually made from *description* to *ontology*. As I have already indicated, one can adopt the 'thin' notion of object (which might seem particularly appropriate in the mathematical context), or perform the Poincaré Manoeuvre, or introduce an alternative formal framework, such as category theory, but in whatever case, one can avoid commitment to an object-oriented domain of quantification. In particular, that any mathematical formulation assumes some domain of elements should not be taken to imply any metaphysical thesis as to the nature of those elements and, in particular, should not be taken as precluding a structural understanding of them.

The concern has also been raised that since mathematical structuralism lacks proper criteria for the individuation of numbers, OSR must similarly lack criteria for the individuation of objects, and hence should be rejected as a result. But this concern is surely misplaced: either any such criteria are rejected from the outset, according to the eliminativist view I advocate, or the criteria are themselves understood in structural terms, according to the form of OSR that incorporates a contextual notion of individuality.

This concern can be related to the Dummettian criticism that to insist that numbers have only structural properties is to 'fall prey to mysticism', in some sense. Now one way of understanding this claim is that on Shapiro's view, the structure is taken to be somehow 'free-standing' but thus is mysterious. However, it is unclear how this applies to the kind of structure considered in OSR. A number of commentators have taken it to be mysterious what the latter's ontological status

[16] This is of course a major issue in both statistical mechanics, where Reichenbach's approach to understanding the statistical mechanical analogue to the Second Law of Thermodynamics requires the (at least) pragmatic isolation of sub-systems, and in quantum mechanics, where decoherence allows one to identify (at least) temporary sub-systems such that one can meaningfully talk of measurement outcomes.

could be but in one sense, this is not mysterious at all—the structure we are interested in is the physical world![17] In that sense, then, it is free-standing, but does not stand free from physical systems, in the sense that underpins the concern.

Another concern has to do with the core idea associated with mathematical structuralism that it doesn't matter what kinds of objects—beer mugs, tables, whatever—instantiate the relevant mathematical structure, if any. But since it clearly does so matter when we consider physical structures—or so the argument goes—the advocate of OSR cannot simply adopt the strategies of mathematical structuralism[18] (Wolff 2012); indeed, she may not even be entitled to claim that her project is truly structuralist! But of course, I agree that the advocate of OSR should not take over the strategies of mathematical structuralists wholesale; indeed, my point in this section is to emphasize the dangers of doing so. Nevertheless, there is a commonality residing in the core idea that the entities we are both concerned with have no identity or distinguishing features beyond those conveyed by the structure. Indeed, to insist that it should matter to the physical structuralist what kinds of objects instantiate her structure runs the danger of begging the question again: of course it matters what kind of *particle* we are considering—boson or fermion, say—since that will involve a different aspect of the structure of the world, but to think in terms of *objects* instantiating that structure is again to bring into the metaphysical picture precisely that which OSR denies.

Similar concerns of a skewed comparison arise with regard to the role of abstraction in mathematical structuralism. Thus, following Shapiro, a system can be defined as a set of objects among which certain relations hold; a 'pattern' or 'structure' is then the 'abstract form' of a system taking into account certain idealizations.[19] Carrying this definition back over the bridge into the consideration of OSR, it might then be concluded that the latter is committed to an abstract notion of structure and again the issue of distinguishing physical from mathematical structure arises. Note, however, that in effect the 'mathematical structure' will be arrived at via an abstraction from physical patterns or structure. Now, of course, abstraction and idealization play a significant role in the construction of scientific models. However, that the *representation* of physical structure involves such abstraction does not imply that the structure itself should be regarded as abstract, in the way that a mathematical structure might. Of course, this still leaves the issue of how to understand the *presentation* of the relevant structure via group theory, for example, which is the core topic of this chapter.

[17] One suggestion is that it is the supposed lack of causal efficacy of such free-standing structures that lies at the heart of the mystery and I shall return to this shortly.

[18] This is part of Wolff's argument that objects cannot be reduced to structures, whether via supervenience or dependence, as discussed in Chapter 7.

[19] One can of course simply block this particular comparison by denying that mathematical structures are abstract in this sense.

This view of structure as arising from abstraction then feeds into the further concern that the ontic structural realist's supposed indifference to the instantiation of 'first-order' (intrinsic) properties such as mass and charge, implies that she will not take such instantiation seriously (Busch 2003). The worry here is that 'second-order' structural properties are relations among first-order ones which in turn are possessed by objects, and if the latter are abandoned, then the whole structural edifice is put in jeopardy. We have already indicated our dissatisfaction with this view and again there is potential question-begging here. Obviously a structuralist will have to say something about intrinsic properties and also about instantiation, but one can understand the latter (or the notion of 'manifestation') without taking it to require the existence of objects.

Now this view, of structure as arising from abstraction, and as 'free-standing' in the sense of existing independently of the systems that exemplify it, can of course simply be denied. One could insist, alternatively, that systems are ontologically prior to structures and talk of 'the' structure is understood either as talk about *any* system structured in a certain way or talk about *all* systems structured that way. However, the point has then been pressed that on either understanding of 'structure', objects are required, contrary to the underlying metaphysics of OSR (Psillos 2006; cf. also Wolff 2012).[20] This seems obvious for the second alternative, since the systems that are taken to be ontologically prior to the structures are composed of objects and relations. It is also claimed to be true for the conception of structure as abstract, since this was developed precisely to secure the existence of mathematical objects. There is the further point that this form of structuralism takes mathematical objects—such as numbers, for instance—to have only those properties conferred upon them by the structure: there is nothing extra to a number over and above its relationships within the number structure. This, it is suggested, does not appear to be generally true of physical objects: height, for example, may be the only relevant property when it comes to classifying people according to how tall they are, but they clearly possess other properties about which the structuralist should remain silent (Psillos 2006). So the fact that we focus on certain properties for certain purposes should not lead us to suppose that these are the only properties the objects possess, and hence, it is argued, the structuralist should remain agnostic on the issue of whether the properties that feature in the structure are the only ones there are. Thus, whatever account of structure the structuralist chooses, she must admit objects and also and at least the possibility of properties not captured by her structure.

However, the comparison with mathematical structuralism is again misleading. First of all, as I've already said, the role that abstraction plays in the construction and

[20] For further discussion and criticism of Psillos' claims, particularly with regard to mathematical structuralism, see Landry 2012. In particular, she insists that he is wrong to assume that one must *begin with* a 'domain of elements' or 'ontology of individuals' that *fixes* the relevant interpretation. On this, at least, Landry and I agree.

development of models and theories in science should not be taken to imply that the structures which these theories and models are representing should be taken as abstract. Thus, this role does not support the claim that the 'structure of the world' exists independently of any exemplifying concrete system. This is not to say that there might not be reasons for holding such a claim and in the spirit of maintaining a 'Big Tent' approach, I'm prepared to entertain the possibility that some form of 'structure as abstract' version of OSR is viable. Nevertheless it brings serious problems in its wake, most notably to do with the lack of causal efficacy of this kind of structure, as we shall shortly see.

Likewise, I'm prepared to admit that one could develop a 'structured system' form of OSR, according to which concrete systems composed of physical objects and their properties are taken to be ontically prior to the relevant structure. If the latter has epistemic priority, or indeed, monopoly, then such a form shades over into ESR. But as we have noted, this comes at a metaphysical price as its object-oriented agnosticism raises the level of metaphysical humility. Dropping the epistemic monopoly of structure and allowing access to the objects of the system reduces the level of humility but now we no longer have a form of ESR but something akin to the standard form of object-oriented realism (albeit with a structuralist emphasis).

The view that I wish to maintain and develop does not sit comfortably within either of these alternatives dragged over from mathematical structuralism.[21] Putting things in broad terms, the 'quantum structure', say, does not exist independently of any exemplifying concrete system, it *is* the concrete system. But this is not to accept that the *system*, as such, and as typically conceived of as composed of objects and relations, is ontically prior to the structure. Indeed, the central claim of OSR is that what appears to be a system of objects and relations should be reconceptualized as a relational structure; that is, it is the structure that is (ultimately) ontically prior and also concrete. Hence, the conception of structure as abstract is rejected also. What might be seen as 'havering' between the two alternatives here is actually an attempt to articulate a notion of (physical) structure that does not fit into the categories imported from the philosophy of mathematics.[22]

This then allows me to respond to the criticism that the kind of structure focused on by advocates of OSR is abstract (group-theoretic) structure but what is needed for realist purposes is instantiated, 'empirical' structure (Slowik 2012: 52–3). Taking the structure to be non-specific, general, and *ontological* (Slowik 2012: 53) would be to accept a contradiction in terms, but 'liberal' ESR (discussed in Chapter 2) avoids this

[21] In this context it's worth noting that, within mathematical structuralism, Landry has argued that both alternatives are problematic and has argued in favour of a category-theoretic version of mathematical structuralism that understands a category as a *schema* for what we say about structured systems (Landry 2011 and 2012). Here we might recall the brief discussion of category-theoretic forms of structural realism in Chapter 5.

[22] cf., again, Wolff (2012) who notes that if one were to take the 'structured system' option, as far as OSR is concerned all that would remain of the system would be the relevant relations.

by insisting on strict neutrality with regard to the underlying ontology (taking the structure to be *epistemological* and grounded in a specific, non-general ontology; Slowik 2012: 53).

Now, as I have emphasized, the structure I am concerned with is indeed physical and 'manifested', in the sense of being instantiated in the absence of (thick) objects. Thus it is not 'non-specific' or 'general' and hence there is no contradiction in terms. This motivation for liberal ESR thus evaporates. Furthermore, the notion of a specific, general but epistemically accessible structure being grounded in a specific, non-general but *inaccessible* ontology remains obscure. Obvious questions arise, such as how can the grounding relation relate the accessible with the inaccessible? Nevertheless, the issue remains as to how this emphasis on physical and manifested structure meshes with the role of group theory and I shall be returning to this in subsequent chapters. Certainly one can say that the abstract vs concrete distinction manifests again in the debate about whether one should invest group-theoretic structure or the associated group representations with ontological significance. Again, I take both the representations and the group structure to be elements of the 'fundamental structural base' in terms of which elementary particles can be conceptualized. The representations represent (!) what is actual, while the group in general encodes the relevant possibilities (see Chapter 10) and in this sense, again, the aforementioned distinctions do not quite cover what I have in mind.

Before we move on from this issue of the way in which the comparison with mathematical structuralism may skew our understanding of OSR, we need to consider a dilemma that has been raised in precisely this context (Psillos 2006): if the ontic structural realist adopts the view of structure as abstract, she is unable to capture the causal features of the world which many realists would consider central; if she goes for the alternative version, on which it is systems that are structured, she has to admit something non-structural into her structuralism, namely the systems.

Here's how the dilemma unfolds: 'natural' physical structures of the kinds the structural realist would be interested in capture the natural—that is, causal-nomological—relations among the objects of a system. But abstract structures are incapable of doing this—they have no causal unity and play no causal role, because they are abstract. Thus we must adopt the alternative. But then, importing the basic structuralist postulate of mathematical structuralism, what is it that privileges a particular structure as *the* structure of the system concerned? The choices all seem to involve a non-structural element: one could argue that the 'right' structure is the one that gets the causal-nomological relations right. Alternatively, one could insist that the 'right' structure is the one that appropriately 'saves' the phenomena, following the suggestion that what renders a structure physical rather than mathematical is that it can be related to the phenomena which effectively provide the physical content (French and Ladyman 2003). But again, this appears to admit a non-structural element (Psillos 2006).

Now, we shall come to the issue of accommodating causal roles shortly but let me ask the question: would the admission of a causal aspect or role introduce a non-structural element—of a kind that would undermine OSR? One could surely allow that n-adic properties in general, and the kinds of relations in particular that we are concerned with, have—in some sense—a causal aspect and although that causal aspect may not itself be structural this does not introduce the kind of non-structural element that is problematic. Indeed, it is not clear what it would be for the causal aspect of a relation, as distinct from the relation itself, to be structural, or non-structural to begin with. Consider charge, for example, understood from the structuralist perspective and the claim that it has causal efficacy: in so far as this efficacy can be considered over and above the nature of charge as a property and its nomic role, etc., it is not clear that it makes any sense to wonder whether this causal efficacy is structural or not. Similarly, to agree that there must be some epistemic principle on the basis of which we can privilege a particular structure as *the* structure of a certain domain is not to admit that a relevant aspect of one's supposedly structuralist ontology is non-structural. Suppose we were to argue that we should select the 'right' structure on the basis of Bayesian confirmation theory—does that introduce a non-structural element sufficient to undermine OSR? Surely not, not least because it simply would not make sense to say that the probabilities of Bayesian confirmation theory are in any relevant sense structural. The point is, again, that not all non-structural features of one's philosophy of science or scientifically informed metaphysics are even candidates for the ontology posited by OSR and thus cannot stand as counterexamples.

The alternative of taking the 'phenomena' as grounding the necessary privileging is a bit trickier, however. Of course it all depends on what one means by 'phenomena' in this context. There is the sense in which the likes of van Fraassen use it, as shorthand for events, processes, etc., composed of or involving observable objects and their properties. If the ontic structural realist has a reductionist view so that such objects and properties are themselves understood as composed of, or involving, unobservable objects and properties, then, as indicated in the introduction to this chapter, the 'phenomena' would also be regarded as reducible to structure. Likewise for the alternative view famously developed by Bogen and Woodward (1988), according to which 'phenomena' are constant and stable and repeatable across experimental contexts and, significantly, not observable. But of course, one might also relate 'phenomena' in an older sense to 'sense data' and historically, of course, there was considerable discussion about the nature of sense data and some of this touched upon structuralism. Thus, putting things crudely, one could insist on a broadly atomistic view of phenomena in this sense, which would suggest they should be regarded as non-structural. Some structuralists, such as Eddington, for example, bit the bullet and argued that even sense data were inherently relational.[23]

[23] Others drew on the findings of Gestalt psychology to reject the atomistic view of sense data (adopted by Russell, for example, as the basis for his 'upward' path) and adopted a broadly structural view of phenomena in this sense. As mentioned in Chapter 4, note 57, Cassirer then explored the possibility of adopting a group-theoretic approach to Gestalt psychology itself (Cassirer 1944; Cei and French 2009).

Given the history of the debate over sense data, the ontic structural realist might resist being dragged down this road, but even if she does venture along it and eschews the Eddingtonian picture, it is not clear that the introduction of sense data, for example, undermines her view of quantum objects. At worst, this offers another form of structuralism that incorporates some non-structural aspect at this level. And certainly it is not clear that one should be concerned about introducing such impurities into the structural vision when phenomena are described as 'physical'—the term is introduced to help establish the difference between mathematical structuralism and the kind I am interested in but as I've already said, it would be bizarre to insist that the distinction between mathematical and physical structure should itself be given in structural terms lest the ontic view be fatally undermined. And of course, that the structure has physical content only appears to introduce a non-structural impurity if one holds the position to begin with that physical structures must be formal, that is, mathematical.

My claim, then, is that drawing a comparison with structuralism in mathematics can confuse matters when it comes to OSR. Of course for Tegmark there would be no such confusion, since for him 'physical' structure is mathematical! Let me now finally turn to the possibility of drawing the distinction between the two by appealing to causal efficacy. Here again it has been claimed that such a notion requires an object-oriented ontology.

8.7 Causation without a Seat

A number of commentators have argued that individual objects are central to productive conceptions of causation and hence to any genuine explanation of change (see Busch 2003, Psillos 2006, and Chakravartty 2003b). Objects, it is alleged, underpin or even provide the 'active principle' of change and causation. I shall grant the existence of such an active principle to begin with and shall survey some ways in which the structuralist can accommodate it (see French 2006; also Ladyman 1998 and 2002 and French and Ladyman 2003) before briefly reviewing some well-known arguments against its articulation in the context of modern physics.

How should causation be understood within a structuralist framework in general terms? Once again Russell has been turned to as offering a prima facie promising answer in the form of a 'structure-persistence' account, according to which events, themselves understood as complex structures, form causal chains, or 'causal lines', as Russell called them, and the members of such chains are taken to be similar in structure (Russell 1948: ch. 6; Psillos 2006). Here we have structural persistence without qualitative persistence. However, as Russell himself noted, this appears to fail as one can easily give examples of apparently causal changes that do not seem to involve the persistence of structure, such as the explosion of a bomb (Psillos 2006). As he says, '[w]hen a charge of dynamite explodes, all the structures change, except the atoms; when an atomic bomb explodes, even the atoms change' (Russell 1948: 416).

Two features of this account need to be emphasized: first, Russell's considerations of persistence and identity through time in this analysis explicitly assume the rejection of substance and are generally amenable to a structuralist understanding. Secondly, one might wonder what he is doing discussing causation at all, given that he famously argued that the notion, as understood by philosophers, had no place in the 'advanced sciences', as we shall see. Here, however, he is discussing what he called the 'primitive concept' of causation, which he took to retain some validity within appropriate limitations and to thus be useful for approximate generalizations and pre-scientific inductions. Indeed—and this is significant for my brief consideration here—although this later work of his included a more up-to-date (if, not surprisingly, positivistically inclined) account of quantum mechanics than *The Analysis of Matter*, he explicitly acknowledged that the examples deployed to help reinforce what he meant by causal lines, etc., were from what he called the 'beginnings of science' (1948: 420), rather than the advanced sciences themselves. Thus as we shall see, in so far as OSR draws upon the latter, it can be argued that it offers no comfortable home for this 'primitive concept'.

Furthermore, an obvious objection to this structure-persistence view is that, as presented, it approaches the issue in entirely the wrong way. OSR does not advocate analysing all macroscopic causal processes in a structuralist fashion—we have to recall the motivation for this position in quantum physics. In essence, OSR piggy-backs on the reduction of such chains or 'lines' in terms of ultimately quantum processes and then insists that those have to be understood in structuralist terms. Imagine again two particles of the same charge approaching one another and being mutually repelled. OSR would take the currently accepted theoretical description of that process—whether in terms of field-theoretic interactions or the exchange of force particles or whatever—and would simply insist that rather than thinking of this description in terms of causally interacting physical objects, we think of it in terms of a system of relations some of which might be described as causal, where that notion is appropriately characterized in this context. Of course there is more to say here about the metaphysics behind such a picture and the manner in which structure has to be in some sense dynamical, but the idea is not to analyse events into series of similar structures, but rather to view the interactions between particles in structuralist terms.[24]

The further point has been raised that in analysing a process in terms of that whose persistence renders the process causal, the most 'natural' account should involve objects and properties; hence persistence cannot be purely structural (Psillos 2006). Of course, that such an account is, perhaps, the most familiar, particularly for philosophers, does not render it the most 'natural' in any significant sense.[25] One

[24] Persistence is of course problematic in the quantum context anyway, given the difficulties associated with the notion of a spatio-temporal trajectory.

[25] And of course it is precisely thinking in terms of objects which makes persistence so problematic in the quantum context.

might recall the old adage that to talk of change presupposes some*thing* undergoing change but the structuralist would precisely resist the move to some account of the 'thing' underlying change in terms of either substance or individual objects; rather what we have is the world-structure with a particular configuration, if you like, or family of relations at one time and a different configuration, or a different arrangement or family of relations, at either earlier or later times.

There is the further concern that any structuralist approach which views causation in terms of a relation between isomorphic systems cannot distinguish between cause and effect, precisely because of the isomorphism (Psillos 2006). Now of course specifying what it is about events that renders one the cause and another the effect is a major metaphysical issue that I cannot hope to address fully here. Still, an obvious response to begin with would be to say that the cause is that which precedes the effect in time. Again the objection has been raised that since 'being in space and time' are non-structural properties, this would be to admit something deeply non-structural into the OSR picture. But in what sense is being in space or time a non-structural property?

Setting aside quantum considerations for a moment, how should the structuralist analyse the statement that object *a* is 'at' position *x* at time *t*, or relativistically, 'in' space-time? One might approach this by suggesting that there is a relationship between *a*, point *x*, and time *t*, or the relevant space-time point. Now, taking *a* to be reconceptualized in structural terms, it is not at all clear that there is any block on this relationship also being understood in such terms. Of course, on a substantivalist view of space, time, and/or space-time, the relationship would be *with* something non-structural, but again it is not clear why that should be deemed to undermine OSR as formulated here. There seems to be nothing incoherent in holding a structuralist view of physical objects and a substantivalist view of space-time, particularly in its current 'sophisticated' guise (see Pooley 2006).[26] On the other hand, if holding a substantivalist view were deemed sufficient to render the apparent property 'being in space-time' non-structural, one could go for either the relationist or structuralist options. In the former case spatio-temporal relations are ultimately reduced to relations between material objects and if the latter are appropriately reconceptualized, again, it is hard to see what has entered the picture that is non-structural. Alternatively, if one felt the need to be a completely consistent structuralist about not only physical objects but space-time as well, then one could follow the lead of the earlier structuralists, such as Cassirer and Eddington, who understood General Relativity (GR) in such terms, or, more recently, Stein who argued that Space, or more generally, space-time, is '*an aspect of the structure of the world*' (1977: 397; his emphasis; see also Auyang 1995; Slowik 2012; Ladyman 2002; French and Krause

[26] Although one might have to perform some fancy footwork with respect to the relationship between mass-energy and space-time curvature in General Relativity.

2006: ch. 3).[27] Further consideration of structuralist approaches in the context of space-time theory can be found in Rickles et al. (2006). Certainly it would seem that there is no obstacle here to OSR based on spatio-temporal relationships.

This brings us on to the further concern as to how the structuralist would understand the *relata* of causal relations in general. Typically these are taken as either events or facts, but in either case these are, again typically, further decomposed into objects and their properties, or generally regarded as particulars. Hence, it is concluded, causal claims require objects and properties as truthmakers (Psillos 2006).

Now of course, this touches on a range of issues to do with the nature of events and facts, and, in particular, arguments to the effect that only 'immanent' objects—those that are situated in space-time—can causally interact. Hence if facts are true propositions, as usually understood, they cannot perform such a role. The standard response is to propose a substitute for facts and the obvious choice would be objects (ultimately, elementary particles and aggregates of them) that provide the immanent basis required. Again there is nothing here to trouble the ontic structuralist who could again adopt an iterative approach and acknowledge that, on first analysis, the relata of causal relations are objects—understood in the usual, non-reconceptualized sense (e.g. electrons, quarks, etc.)—but would then insist, of course, that such objects are to be further understood as mere nodes in the structure. It might be objected that this threatens to lead to circularity: causal relations hold between causal relata which are resolved into objects which are themselves nothing but 'nodes' of sets of causal relations. The worry dissipates if we think of levels of metaphysical analysis: we begin with sets of events or facts the relationships between which we analyse in terms of causal relations holding between putative objects, such as, reductively, electrons, protons, etc. (at this point we're still assuming the concept of causation makes sense at this level). From the perspective of OSR we then analyse those in terms of the relevant structure, including the laws and symmetries pertaining to the entities concerned, and in so far as this is relational, we do end up with supposedly causal relations *sans* causal relata. Ultimately, as far as the structuralist is concerned, there is nothing but the structure—that is, the set of relations—and there seems to be nothing to prevent these having causal powers. Certainly it is unclear why this should be more of a problem for the structuralist than the non-structuralist who, ultimately, must hold that objects, or more basically, bare substances must have such powers. I shall return to this shortly.

[27] Stein notes the historical antecedent for such a view in the well-known passage in one of Newton's unpublished papers where he writes that 'Space is an affection of a thing *qua* thing' (1962: 136; Stein 1977: 396). Stein interprets this as follows: '*the fundamental constitution of the world*—its "basic lawful structure"—*involves the structure of space*, as something to which whatever may exist must have its appropriate relation' (1077: 396; his emphasis). Put in current terminology, what this means is that 'Whatever exists of a physical nature...must be appropriately related to a space-time manifold with a fundamental tensor-field satisfying the Einstein equations' (1977: 397).

Thus the conclusion that causal claims require objects and properties as truth-makers seems unwarranted. Furthermore, if the decomposition of events and facts is into either everyday or the physicists' objects, then the structuralist is not denying that we can continue to refer to such objects as a *façon de parler*, as it were, only that these must themselves be regarded as reconceptualizable in structural terms. More importantly, even if one were to regard events, say, as decomposable into objects, the metaphysical underdetermination still applies, with the choice of whether to understand those objects as individuals or non-individuals with nothing in the physics to guide you (French and Krause 2006: ch. 9). Best then to eschew objects entirely; one can still insist that an 'event', such as a light coming on, is to be taken as metaphysically decomposable into objects—photons, say—and their properties—energy levels—but these objects are themselves to be understood as structural. And of course the standard view of events as particulars does not appear to present any special problems: an event on this view is to be identified, not via its spatio-temporal location, but rather in terms of its 'causal location' within the causal net of the world.[28]

Is it the case, however, that the structuralist framework can accommodate causal powers and more specifically, the inherent 'activity' of causality? Chakravartty, for example, argues that an important explanatory role served by objects is to provide a 'means of change' (2003). The idea here is that change is represented in science via dynamical equations—such as Newton's laws or Schrödinger's equation—and such equations in turn represent relations between properties and these, again in turn, are properties of objects. Thus objects are central to any explanation of change. In particular, without them we are left with 'explanatory gaps' between subsequent states of affairs, as we have no account of the active principle that transforms one concrete set of relations, say, into another. Objects, then, have 'ontological clout'.

Now, one option is to adopt a structuralist form of the Humean 'regularity' view which rejects such active principles, and simply accept an analysis of events in terms of brute successions of structures (I shall come back to this). There would still be Russellian concerns regarding structure preservation to be faced but I've already indicated how one might do that. However, although being a realist about structures and an anti-realist about causality does not seem to be an incoherent combination, this is not the only alternative. Again, I'm not convinced that the claim that causality has some 'active' component *requires* a metaphysics of objects and properties.

Thus consider the question: where might this active principle be located? With the object or the properties? If the former then we obviously need to press a little further and ask for an account of objecthood which could accommodate such activity.

[28] Bartels' accommodation of this within the context of QFT could easily be adapted to the structuralist cause (Bartels 1999; see also French and Krause 2006: ch. 9), so that events are understood as instantiations of smeared (field) properties at space-time points. We've already touched on the issue of accommodating space-time earlier and Auyang gives a nice structuralist account of QFT in which neither the field event structure nor the space-time structure is given ontological priority but both 'emerge together' as aspects of the world-structure (Auyang 1995; this is very similar to certain features of Eddington's view).

Obviously the idea of objects as ultimately bare substrata can't do the job; and equally obviously the view of objects as bundles of either properties or tropes forces us to look closer at the latter as the source of this activity. There are then further options: either each property that is causally active has some causal principle particular to that property, or kind of property, that is involved in the conferral of causal power on the possessors of that property; or there is some generic causal activity which together with the other features of properties confers such powers.

This again raises important issues, to do with the relationship between properties and their causal powers or capacities, whether different properties can have or bestow the same power, whether the same property can bestow different powers on its instances, and so forth (for a useful introduction see Swoyer 2000). Indeed, it has been suggested that to be a property is to *bestow* causal powers; that it is through such bestowal that properties are identified; that this bestowal involves some intimate relationship between the property and its powers and, of course, that, at least when it comes to the properties that we are concerned with here—that is, 'fundamental' properties like charge and spin and mass and so on—they are interrelated in ways that are appropriately described by physical laws and theories in general (see Chakravartty 2007). This view finds its most well-developed expression in the dispositionalist account of causation, laws, and properties that I shall consider in the next chapter. The point I want to emphasize here, in the spirit of maintaining the 'Big Tent' outlook, is that there appears to be no insurmountable obstacle to the structuralist herself appropriating such a view but insisting that the powers are held by the relations, say, rather than any underlying objects (see, for example, Esfeld 2009). Indeed, she can respond to Chakravartty's concerns by insisting that the explanatory buck stops at a point down the chain before we reach objects. That is, she can insist that the 'active principle' in question lies with the relations and properties themselves and it is these which carry the clout, thus effecting a kind of ontological economy. We shall return to consider the benefits of this kind of manoeuvre in Chapters 9 and 10.

This brings us back once more to the metaphysical analysis of properties themselves. Again crudely speaking, one might think of at least two alternatives: first that there is nothing to properties, metaphysically speaking, but their causal powers (this corresponds to the 'Dispositional Identity Theory' that I shall discuss in Chapter 9); or one might think of a property as composed, in some fashion, of various features, including some form of particular or generic causal activity and something else, some 'quiddity' perhaps, which makes the property the particular property that it is. At this point one might wonder if, in adopting this latter view, one is again committed to some non-structural feature. Psillos, for example, has pointed out that structuralists might be naturally sympathetic to causal structuralism which holds that all there is to a property is its causal powers and no more (Psillos 2006); that is, it denies quiddities (see Esfeld 2009). I think sympathy is the right attitude, but I don't think it would be inconsistent for someone to be a structuralist about physical objects, but accept that properties and relations have some quidditical aspect.

The worry about denying quiddities and adopting a form of causal structuralism instead is that it seems to entail a kind of 'hyperstructuralism', according to which the causal powers of a property are seen to be purely structural, with the result that we end up with 'nothing but a formal structure' (Psillos 2006). But again, one can imagine someone being a structuralist about objects and their properties but drawing a line at a structuralist account of the causal powers of those properties. Indeed, if one thinks of structure in relational terms, it is hard to see precisely how such an account could proceed—would the causal powers of properties and relations themselves be properties and relations? Even if that could be spelled out, there appears to be nothing incoherent about being a structuralist with regard to the objects and their properties but not about the latter's causal powers. Indeed, perhaps the most intuitively plausible form of structuralism is precisely one according to which objects and their properties are metaphysically dissolved into a 'multilayered' network of relations, where certain of these relations are 'causally empowered' and where this empowerment, for want of a better word, is inherent to the relation. Is that inherent empowerment non-structural? Yes, in the sense that it is not itself a structure or describable in structural terms (if it were so describable an obvious regress would threaten);[29] no, in the sense that it is precisely an aspect of the world-structure.

This whole discussion has granted the assumption that one can make sense of the notion of causation in the context of modern physics from which OSR draws its primary motivation. Now let us briefly consider whether that assumption can in fact be maintained.[30]

8.8 'Seats' and Structures without Causation

The debate over causation has been characterized as an 'amiable jumble' of approaches, stances, and accounts (Skyrms 1984), arising from the lack of agreement on what causation is, or how it should be characterized: 'Is it a matter of the instantiation of regularities or laws, or counterfactual dependence, or manipulability, or transfer of energy?' (Beebee et al. 2009: 1; see also Hall 2011 for a useful introduction to some of the issues). One reason for this lack of consensus is that here we have a notion that is originally derived from reflection on some subset of the 'pushes' and 'pulls' of everyday life[31] but that cannot survive extension across the diversity of

[29] So there is a sense in which Psillos' concern is a form of Roberts' (2011) as discussed in Chapter 6, in so far as the former is about the structural describability of causality and the latter about the structural describability of structure.

[30] The following draws heavily on my essay review of *The Oxford Handbook of Causation* (French 2011c) and does not proceed much beyond what can be found in Ladyman, Ross, et al. 2007.

[31] I once attended a talk given by a certain famous philosopher who attempted to demonstrate the 'reality' of causation by punching the convener of the seminar on the arm! No doubt the reader can think of alternative, less macho, demonstrations.

these external 'forces' (where the term is understood here in a generic, layperson's sense), much less its importation into the context of modern physics.[32]

As a result of this lack of consensus, some have concluded that no univocal analysis of causation is in fact possible (Beebee et al. 2009). Alternatively, one can identify three broad strategies: pluralism, of some form or another; the view that causation is an 'essentially contested' concept; and a broadly epistemic stance, which takes causation to be a feature of the representations we deploy. These are not intended to be exhaustive, nor exclusive, but they provide a useful framework for the discussion to follow. In particular, I am loath to tie OSR to a particular such account so in the spirit(s) of both the 'Big Tent' attitude and the Viking Approach, I shall indicate how it might mesh with various strategies.

So, causal pluralism holds that 'causation' denotes not a single kind of relation but a diversity, where this can be understood as manifested 'in the world, in some sense, or simply at the level of the concepts we employ' (Godfrey-Smith 2011: 327–8). 'Ontic causal pluralism', as I shall call it, holds that the phenomena that 'causation' picks out are 'in some deep sense, plural in character' (Godfrey-Smith 2011: 328).

Two ways of cashing out this pluralism are 'horizontally', in terms of distinct domains to which different accounts of causation, and indeed, perhaps different theories, apply; and 'vertically', as it were, in terms of different levels, to which theories from different sciences apply. The former is exemplified by Cartwright's 'dappled' realism or 'metaphysical nomological pluralism' (1999), according to which laws are not regarded as universal but apply to and thus delineate distinct domains. As an example, consider quantum and classical mechanics: on this view it is not the case that the former has superseded the latter but rather that they correspond to distinct features of the world that lie 'side-by-side', as it were (1999: 361).[33] There is an obvious tension that arises with regard to the delineation of domains here: take any entity, such as an electron, and the question arises, which domain does it belong to? In some contexts its behaviour can be described in classical terms, in others via quantum theory. Does this mean that it belongs to both domains? Answering in the affirmative raises the concern over consistency: does the electron possess both classical and quantum properties? Adopting a structuralist stance may help dissipate this tension. If we drop the underlying assumption of an object-oriented ontology and accept that within the classical domain the appropriate ontology is one of classical structures, and correspondingly, quantum structures for the quantum domain, with domain-specific particles understood as dependent on and thus conceptualized in terms of the appropriate structures, then we might say that we have distinct particles-as-nodes in the relevant structures in the different domains: in a sense, then, we

[32] Thus the causal singularist takes specific relations associated with terms such as 'scrape' and 'burn' and presumably 'punch' to be semantically prior to the general term 'cause'.

[33] Bokulich compares this to Heisenberg's account of 'closed' theories (Bokulich 2008: ch. 2) which can also be understood as a form of ontic pluralism.

would have classical and quantum electrons coexisting in the world (albeit, a dappled structural world). Indeed, with structures understood in terms of the relevant laws and symmetry principles, Cartwright's invocation of the relevant laws as delineating the associated domains meshes quite nicely with this suggestion.[34]

The second form of ontic pluralism is one that applies across different levels, in so far as these can be distinguished. Thus, it might be the case that 'causation' denotes one kind of relation when it comes to the phenomena described by physics—or indeed, no relation at all—and a different kind of relation when it comes to biological phenomena, or those covered by economics, or the social sciences, say. Such a view might draw succour from the failure of accounts of inter-theoretic reduction between levels, although typically ontological reduction is still maintained despite this failure (see, for example, Le Poidevin 2005). In that case, if genes and biological molecules are ultimately reducible, ontologically, to chemical entities and the associated bonds, and both of the latter are taken to be reducible to elementary particles and the arrangements of electrons imposed by the anti-symmetrization of the wave-function, respectively, then it is hard to see quite how distinct, level-dependent relations of causation are going to get their purchase.

Alternatively, one could deny this form of ontological reductionism. If *A* is taken to be ontologically reducible to *B* only in the case that the causal powers conferred by possession of *A*-properties are exhausted by those conferred by possession of *B*-properties (see, for example, Kim 1998: ch. 4), then ontological reduction will fail if *A* can be shown to be associated with additional causal powers not exhausted by those associated with *B* (Hendry 2011: 328). In the case of chemistry, it has been suggested that we have precisely this kind of *sui generis* conferral, since the molecular structures in terms of which chemical explanations are given cannot be grounded in the 'exact quantum mechanics of isolated systems of electrons and nuclei' (Hendry 2011: 328).[35] Of course, the reductionist might insist that these structures arise from the interaction of the systems in question with their environment but then the precise nature of these interactions and the associated explanations would need to be articulated. Alternatively, each molecular structure is associated with a *sui generis* law of nature which, although expressible in the language of quantum mechanics, is not reducible to the fundamental laws of the latter (Hendry 2011: 329; also forthcoming: chs 9 and 10).

[34] Of course, such pluralism comes at a cost in terms of overall parsimony and this might be avoided by giving an alternative account, whether reductive or otherwise, of the relationship between theories such as quantum and classical mechanics (see Bokulich 2008 for a discussion of the issues involved in establishing such a relationship).

[35] Ironically, perhaps, the crucial obstacle to such grounding has to do with symmetry: arbitrary solutions to Schrödinger's equation in its exact form should be spherically symmetrical, but molecular structures are clearly not (Hendry 2011: 304; Wooley 1998). Adopting certain approximations—such as the Born–Oppenheimer approximation in which the nuclei are held fixed—does not help since these make a significant difference to molecular symmetry properties (Wooley 1998). We shall return to this in Chapter 12.

In this case we don't have the same tension as in the case of 'horizontal' pluralism. If molecular structures are subject to *sui generis* laws, then, presumably, they have *sui generis* properties.[36] But now, even if these structures are not regarded as intrinsically *sui generis* entities, and are taken to be composed of physical atoms and, proceeding further, the standard array of elementary particles, one can avoid possible attribution of inconsistent properties by insisting that the *sui generis* properties associated with the structures only arise when their constituent particles are in the appropriate arrangement, say. Nevertheless one might feel uncomfortable about this talk of *sui generis* laws holding at different levels. That discomfort might be traced to the thought that laws, on a widespread conception of their role, are supposed to *govern*. But if the *sui generis* laws govern the molecular structures, and the latter are composed, ultimately, of the standard array of elementary particles, then do these laws also govern the latter? And if not, what is it that blocks the transitivity of governance? Now, we'll return to the nature of the governing role of laws later but again one might wonder if some form of object-oriented stance is being assumed here and perhaps a level-dependent form of structuralism would clarify the situation. On such a view, different levels—however they are to be distinguished—manifest different kinds of structure, yielding different putative—and dependent—entities and hence different relations of causation holding between the properties of such. Jumping ahead to our discussion in Chapters 9 and 10, the relevant relation between laws and entities should not be governance but some form of dependence and setting the latter as ontologically primary will allow us to articulate the nature of the structure of the world in general, and to understand the sense in which there might be a further level-dependence manifested in the world. Of course, this is not to say that one cannot be a structuralist and a reductionist but once again, in the spirit of our 'Big Tent' approach, let's keep our options open.

Moving on to causal pluralism at the level of concepts, Hall (2004), for example, has maintained that one can discern two distinct features of causation as this concept is employed in our everyday talk, having to do with difference making and production, respectively. Thus we sometimes characterize a cause as producing or generating the associated effect, but other times we are more interested in those causal factors that are identified as making a difference with regard to the effect. The latter feature forms the core of perhaps the most well-developed incarnation of the causal theory of explanation (Strevens 2008). And, it is argued, some of the confusion that leads to the 'amiable jumble' arises because of a failure to keep separate these features and the associated criteria for what count as a cause. Indeed, Hall insists, they should not be regarded as distinct features at all but rather as manifestations of two different concepts of causation, each with different criteria.

[36] Such as the symmetry properties mentioned previously.

However, this generates obvious concerns (see Godfrey-Smith 2011: 330). In particular, what is the realist about causation to make of such a view? Is it the case that each concept corresponds to a distinct causal relation, so that we have two relations of causation that are manifested in different—difference making and productive—contexts? Can these contexts be appropriately distinguished? Or do we have one relation that manifests different 'modes'? The alternative is to refuse to answer these questions and eschew realism in this case.

Thus Godfrey-Smith suggests that causation is an 'essentially contested concept' in the sense that it is 'reliably subject to dispute with respect to its boundaries and criteria for application' (2011: 336). And this is because although application of the concept has significant consequences, its domain of applicability is sufficiently complex, or multifaceted, that there are no sharp borderlines to act as 'attractors to usage' (2011: 336). Furthermore, recognizing that this is the case will serve us better than simply acknowledging that the concept is disunified in complex ways. Significantly, it is the existence of a set of accepted exemplars which display causation's role in certain important practices—crucially, those involving the assignment of responsibility, in some sense—that prevents its fragmentation into distinct concepts within our discourse (2011). I say significantly because, of course, these exemplars are typically drawn from everyday life, or at best involve the use of 'toy' physics, infected (or some might say, primed) by certain intuitions. But when it comes to science and modern physics we should not expect such exemplars to hold or the associated practices to be displayed. Indeed, as we shall shortly see, although at the level of the everyday, causation might be 'essentially contested', when it comes to physics, there is simply no contest at all!

Perhaps the 'epistemic theory' advocated by Williamson (2009: 204–10) can offer a suitable way forward here. According to this view, causation is not to be analysed in terms of some physical mechanism or relation; rather it is a feature of the way we *represent* the world.[37] This offers a very general framework since causation need not be taken to hold between physical entities, which of course meshes with the Humean view. In particular, if causality is a feature of our representations of the world, rather than of the world itself, then we can say that in those representations that cover the social sciences, say, or folk psychology, or everyday decision making and so on, causation has a place, albeit a contested one perhaps, but when it comes to the representations of current physics, it does not.

Why not? Well, consider one of the standard approaches that has explicitly drawn upon what its proponents regard as key elements of modern physics, namely causal process theory (see Dowe 2009, 2007, and 2000). This is offered as a response to the issue of 'what causation in fact *is* in the actual world' (2000: 3) and has at its core an explication of the notion of causal interaction in terms of the exchange of quantities,

[37] Here we might draw a connection with van Fraassen's view of modality in general as a feature of our models, but not of the world; I shall touch on this again in Chapter 10.

such as energy and momentum, governed by conservation laws. Thus, the notion of *process* is cashed out in terms of the world-lines of objects, and *causal* processes are those in which the object possesses a *conserved quantity*, where this is a quantity, such as mass-energy or charge, that is governed by a conservation law.

Now obviously the central role played by conservation laws will make this view attractive to structuralists in particular with causal processes reconceptualized as simply those that are associated with conservation laws, where the notion of process as world-line has itself to be reconceived in structuralist terms (and will thus depend on a structuralist understanding of space-time). This stance may also help with the question, 'on the process view, *why* are all and only conserved quantities involved in causation?' The answer is that there is a connection between the fact that a putative object possesses conserved quantities and the fact that it has causal powers. This connection is provided in structuralist terms via the central role of conservation laws in tying together the law-like structure and yielding the putative objects that are causally related. Again, the iterative approach comes in handy: if causation is first analysed in terms of the interaction of objects then in so far as, secondly, the latter are dependent upon the relevant symmetries and conservation laws then it should come as no surprise that there is such a connection.

Of course, one might still wonder why it is that that which is engaged in causal relationships—charge, say—is also that which happens to be conserved but such questions only arise because we use our everyday intuitions about causation to open up the gap expressed by 'happens to be'. Causal process theories tell us *what* causation consists in (transmission of conserved quantities) and structuralism can tell us *why*. If this seems to bind causation and conservation so tightly that the former is reduced to only a thin shadow of what we intuitively take it to be, then one can only say that it is only such a thin notion that finds any place, if at all, in modern physics.

Now, it is in the context of Special Relativity that process theories of causation find their natural home, since here the notion of a causal process can be understood in terms of the world-lines of bodies within Minkowski space-time (Hoefer 2009: 697). However, things become problematic when we shift to General Relativity. First of all, there are a huge variety of space-time models that satisfy Einstein's equations and in which causal anomalies can arise (for an overview, see Hoefer 2009: 698–701). Thus, for example, those regions of space-time that are beyond the event horizon of a black hole are, famously, causally disconnected from the rest of the world (Hoefer 2009: 700) and uncaused events can emerge out of 'naked singularities'.[38]

Perhaps one might find a way of keeping such features beyond the remit of one's theory of causation but there is a more fundamental issue that process accounts must face: in General Relativity there is no conservation of energy-momentum (for useful

[38] Of course, one might insist that how one physically and philosophically characterizes singularities is problematic anyway. Lam (2007 and 2008), for example, argues that they should be understood as non-local features, best understood from a structuralist perspective.

discussions, see Rueger 1998 and Hoefer 2000).[39] More specifically, although one can obtain a differential form of the conservation law, which states that no energy-momentum is created in an infinitesimal region of space-time, the integral form, which is equivalent to the differential form in Newtonian and Special Relativistic physics, and which states the same for finite regions, does not hold in general (a useful presentation of this can be found in Baez and Weis undated). In such a finite region, curvature becomes apparent, and hence the gravitational energy within such a region must be taken into account. One might think of an object traversing such a region: as it does so, the space-time curvature itself affects the object's energy-momentum (Rueger 1998: 34). Now the obvious move is to extend the characterization of energy-momentum to be conserved to include the change in this quantity associated with the gravitational field. Typically this can be done via so-called 'pseudo-tensors' but the name gives the game away: they can be non-zero even in flat, empty space-times, depending on the coordinates chosen, and for these and other reasons, are not regarded as yielding a good definition of energy density at the local level (see Baez and Weis undated). Locally, a gravitational field can always be made to vanish by an appropriate coordinate transformation.

In particular, as Rueger notes (1998), one cannot regard gravitational waves as propagating localized energy-momentum through space-time and therefore one cannot capture them via the process account. More generally, 'in spacetimes, or regions of spacetime which lack the requisite symmetries, no process whatsoever will possess conserved quantities and hence qualify as a causal process' (1998: 34; see also Hoefer 2009: 202–3). Where there *are* such symmetries (and the differential and integral forms can be equated with one another), as in the case of asymptotically flat space-times, process theories of causation can get a grip and hence the issue of whether a given process counts as causal or not depends on the nature of the underlying space-time.

Thus it is not the case that 'being causal' is a property that can be considered to be 'local' or intrinsic to a process, in the sense of being determined only by features of that process or those that pertain to its immediate spatio-temporal neighbourhood. Again, however, this conclusion could be accommodated by OSR. Indeed, if one includes space-time structure in one's characterization of the world, then one way of approaching the issue is to say that 'being a causal process', or not, is dependent on the structure of the world, such that under certain conditions, as when the relevant symmetries pertain, the relevant processes may be characterized as causal but in general, they may not. Returning to our general issue of the status of causation in physics, however, Hoefer's blunt conclusion remains: '[t]here is no genuine energy-momentum conservation principle in GTR [General Relativity]' (Hoefer 2000: 195).[40]

[39] General covariance, of course, does not yield invariance under symmetry transformations (Earman 1989: ch. 3).

[40] Hoefer also uses this as a stick to beat space-time substantivalism (2000: 196–7): if one of the 'essential' characteristics of substance is that it possesses energy, or has energy content in some sense

Still, one might wonder whether there is not *something* 'out there' in the world to which the concept of causation in our representations might partially correspond, in at least the sense of some 'thin' relation holding, if not the 'thick' one associated with the contested or, as Healey calls it, 'open textured' concept (Healey 2009). Indeed, we might recall Cassirer's claim that the 'essence' of causality lies precisely with the demand for strict functional dependence that is satisfied even in the context of quantum mechanics. On this basis, I shall suggest that in so far as there is such a 'thin' causal relation, it can be articulated in terms of certain forms of dependencies understood within the structuralist framework.[41]

Of course, we might consider 'thickening' the causal relation by incorporating the fundamental feature of 'difference making' in the sense that causes make a difference to their effects (Strevens 2008). This is broader in applicability than causation, as typically understood, in that non-causal features of the world may also 'make a difference'. But, I shall argue, they do so in virtue of the relevant dependencies that hold and it is these that we should be focusing on. Furthermore, that tightening of our focus will encourage us to understand these dependencies in a structuralist manner.

Consider, for example, the explanation of the halting of the gravitational collapse of white dwarf stars by Pauli's Exclusion Principle (Colyvan unpublished). The core of the explanation as usually given is that the gravitational force is balanced by the difference in what is sometimes called the 'Pauli pressure', or 'degeneracy pressure' created by the occupancy of the relevant energy states. In so far as Pauli's Principle cannot be regarded as a causal law, it has been claimed that this represents an example of an acausal explanation of the behaviour of physical systems.[42] Strevens also considers this example within his general approach to explanation, at the core of which is a difference-making criterion that 'takes as its raw material any dependence relation of the "making it so" variety, including *but not limited to causal influence*' (Strevens 2008: 179; my emphasis). The idea is that once we have established the relevant dependence relation between some state of affairs and some set of 'entities', the criterion will tell us what facts regarding those entities underpin the relation's 'making it so' (Strevens 2008). In the case of white dwarf collapse the relation between Pauli's Principle and the halting of the collapse is 'some kind of metaphysical dependence relation' (Strevens 2008: 178). In this case, the relevant dependence relation appears to be straightforward: the Exclusion Principle drops out of the formalism associated with Permutation Invariance (that is from the anti-symmetric representation of the permutation group) and the latter, as a fundamental symmetry,

(and this was long held as one of the reasons to regard fields as substantival) then the fact that the energy content of empty space-time is so ill defined raises obvious problems for this view.

[41] And where the context is such that these dependencies involve the relevant conservation laws, we may have a sense of causation that matches that articulated by causal processes theories.

[42] This has been argued to open the door to the explanatory role of mathematics.

is a feature of the structure of the world.[43] Of course, how one understands the symmetry is crucial. Huggett draws on the parallel between permutations and covariant spatial transformations and constructs a framework in which quantum statistics (and hence Pauli's Principle) emerge as 'a natural result of the role symmetries play in nature' (Huggett 1999b: 346; for discussion see French and Rickles 2003). Thus he argues that it is implied by the conjunction of a further symmetry principle obeyed also by space-time symmetries together with the formal structure of the permutation group. This further principle is what he calls 'global Hamiltonian symmetry' which implies that the relevant symmetry operator commutes with the relevant Hamiltonian. With regard to the permutation group, of course, permutations of a sub-system are permutations of the whole system and this 'global Hamiltonian symmetry' very straightforwardly implies Permutation Invariance (Huggett, 1999b: 344–5).

Hence, Huggett concludes (1999b: 346), we should view Permutation Invariance as a particular consequence of global Hamiltonian symmetry given the group structure of the permutations (of course, the issue remains as to the status of that group structure; again see French and Rickles 2003). As a result, we recall, the relevant Hilbert space can be thought of as divided up into sub-spaces, corresponding to the different group-theoretic representations and hence different statistics (Fermi–Dirac statistics obtain when the particles 'sit' in the symmetric sub-space, and Bose–Einstein statistics when they sit in the anti-symmetric sub-space). Given the symmetric nature of the appropriate Hamiltonian, once 'in' such a sub-space, particles cannot get out, as it were. From this perspective, the Exclusion Principle can be thought of as an expression of the 'limits' placed on the Hilbert space by Permutation Invariance. The latter, in turn, can be understood as a form of 'constraint', although as we shall see, when it comes to the so-called 'space-time' and internal symmetries (such as those associated with the Poincaré group and the $SU(3)\times SU(2)\times U(1)$ symmetry of the Standard Model respectively) such talk may not be entirely appropriate in the structuralist context where the distinction between symmetries as constraints on and by-products of the relevant laws becomes blurred.

In the case of the white dwarf collapse, the 'metaphysical dependence' to which Strevens alludes can now be understood as underpinning the constraint imposed by Permutation Invariance, where it is this, and hence the associated symmetry, that 'makes things happen' (or not). How one understands both the constraints and the dependence will depend on one's metaphysics but as I've indicated, the structural realist, for example, can take it as holding between the physical structure and the relevant putative entities, processes, and regularities. Taking this general sense of 'making things happen', or 'difference making', as also being applicable to the causal

[43] As I have said, in non-relativistic quantum mechanics the symmetry is imposed on the theory; in algebraic quantum field theory it arises 'naturally' as a result of the imposition of a certain selection criterion on the set of representations of the permutation group.

case, the point is that at the fundamental level it is the relevant dependencies that are doing all the metaphysical work.

Now, what about laws? Aren't these causal—at least those that are, in some sense, 'fundamental'—and don't the relations they express offer specific examples of causal relations? I shall say more about the modal features of laws in Chapters 9 and 10 but here we should at least note the following when it comes to their links to causation. First of all, to say that a law is causal is not usually understood in terms of taking the law itself to be the 'seat' of causation. Rather, the idea is that it represents the causal relations between entities that may act as such 'seats', and typically, these entities are taken to be objects. But even without adopting a structuralist line, this object-oriented understanding may break down in quantum mechanics. For example: Does the Hamiltonian in Schrödinger's equation *cause* the wave-function to evolve in the way it does? Even in Newtonian physics, where the terminology of forces may encourage such causal talk, concerns arise: can Newton's Third Law be regarded as causal? (See, for example, the comment by Tooley 2009: 374.)[44]

Leaving these issues to one side, if one eliminates objects or, at least, no longer takes them to be the 'seat' of causal powers, then in what sense, if any, can laws be understood as causal? As in the case of the Pauli Exclusion Principle/Permutation Invariance, it would seem that, at first glance at least, one can say that it is not the case that Schrödinger's equation *causes* the wave-function to evolve in the way it does, or at least, not in anything like a thick sense. Nevertheless, the relevant dependencies are represented by, or manifested in, the associated laws. So, for example, causal talk of a given charge, say, causing another charge to accelerate (either towards it or away) is to be analysed in terms of the operation of the relevant law—in the classical context, Coulomb's Law—in turn understood as a feature of the (classical) structure of the world and upon which the correspondingly relevant particles, conceived of as putative objects, are dependent. Moving back to the quantum context, one can also talk of the appropriate dependencies in terms of Cassirer's functional coordination as noted before.

Still, as I have said, this is causation only in a thin sense. Even if we restrict our attention to those well-known examples involving forces and related changes of state, as represented via differential equations, such as Newton's Second Law, we do not regain one significant 'thick' feature of the causal relationship, namely the temporal asymmetry between cause and effect. There is no basis for grounding this asymmetry in physics, and indeed the time-symmetry of the laws of classical physics motivated in part the Russellian stance touched on earlier. So, what else might ground this particular 'thickening' of the causal relationship? One option would be to follow a Humean line and take the asymmetry to be imposed by us, or via our representations:

[44] One might try and rule out the Third Law as non-fundamental or as having been superseded since it has no counterpart in Special Relativity. Of course, if one is going to make that sort of move one should carry it through and shift again to General Relativity, but there, as again we have seen, other concerns arise.

in effect we define the cause to be that event that precedes the effect. In that case, the alignment between the causal and temporal 'arrows' becomes a matter of mere 'semantic convention'. However, as Price and Weslake (2009) note, this renders the alignment too tight as it excludes the possibilities of simultaneous and backward causation as simply conceptual confusions. On the other hand, this alignment seems too loose, since it fails to explain the relationship between the temporal asymmetry of causation and the 'fact' that it makes no sense to deliberate with regard to past ends. This brings in a further thick feature of causation, namely that it has to do with deliberation (something that forms the core of interventionist approaches, of course; see Woodward 2003). Given the time-symmetric nature of fundamental physics, the question then is why are our deliberative abilities and associated notion of cause so strongly aligned with the temporal arrow?

The alternative to these attempts to ground the temporal asymmetry in physics is to adopt a subjectivist stance: the asymmetry can be explained in terms of our temporal orientation as '"players" in the dynamical environments in which we live' (Price and Weslake 2009: 436). In other words, the claim is that there is something inherently temporally asymmetric about human agency, in the sense that we deliberate about the future and not the past and this is simply a matter of pragmatics, rather than metaphysics. The price of such a view, of course, is introducing a sizeable subjectivist element into the notion of causation, but perhaps that is unavoidable. On this view, then, what we have is not so much a metaphysical thickening, but the addition of a further pragmatic dimension. Again, this takes us beyond the metaphysical dependencies manifested by the physical structure of the world (at least, at its most fundamental level).

So, what can we conclude? As many commentators have noted, we have to juggle a number of competing demands: we want an analysis of causation that is compatible with current physics, that accommodates our (folk) understanding of deliberation, that meshes with our view of scientific laws, that is metaphysically minimal yet explanatorily productive, and so on... Physics pushes us towards a 'thin' notion of causation; indeed, some might say, one that is so thin as to be not merely a shadow of its former self but empty. Thickness is acquired when deliberative aspects are added to account for the feature of asymmetry, but this comes at a cost. I shall suggest that the costs involved can be reduced by taking laws (and symmetries) as ontologically primary within a structuralist context, with a subset associated with a thin, yet productive, notion of causation at the level of physical interactions. Mapping out such an account involves a complex interplay between conceptual and ontological analyses (see, for example, Paul 2009) and this is what I shall attempt in the next two chapters, where I shall argue for the view that laws and symmetries, as features of the structure of the world, should be regarded as inherently modal.[45]

[45] Thus after insisting that the causal structure of the world is dependency structure, Hall goes on to argue that this dependency structure, in turn, is 'a *counterfactual* dependency structure—a structure

8.9 Conclusion

In his overview of causation in the sciences, Hall writes,

> Looking at the world *just* through the lens of fundamental physics, we won't see the need for any interesting, richly structured concept of causation. True, we could say that each complete physical state 'causes' each later complete physical state—but why bother, once we have the fundamental laws in hand? But the value of a concept of causation derives from details of our actual human predicament. First, we need *control* over our world. Second, we need to *understand* it. Third, while grasping a complete, correct physics would obviously facilitate understanding and control, we can only build up to such a grasp by way of *piecemeal approximations*. The scientific value of causal concepts is precisely that they facilitate control, understanding and piecemeal approximation. (Hall 2011: 97)

That's all well and good, but returning to the topic with which we began this chapter, if there is no thick, or richly structured, concept of causation to be found at the level of physics, how are we to metaphysically distinguish mathematical structure from physical structure?

Of course, the object-oriented realist faces similar problems in coming up with a characterization of her objects that is immune to counterexamples from both the everyday and quantum contexts. If they are conceptualized entirely in terms of properties, as on the 'bundle' view, then when it comes to physical properties such as mass, spin, etc., she is also going to have to deal with the issue whether their group-theoretic description implies their mathematical constitution. But if she adds substance or suchlike to the mix, she is going to have to take this as primitively physical, with no further characterization available. And of course the structuralist can do the same, taking the structures we find in this, the actual, world, as physical in a primitive sense.

Alternatively one might focus on the relevant dependencies, touched on previously and argue that these cannot be conceived of as purely mathematical. However, if these are just the laws in those situations that might be described as putatively causal, then it is difficult to see what renders these physical. One might push further and argue that in this, the actual, world, there is manifested a particular combination of dependencies that together make up the physical.

So, think again of Permutation Invariance: under those circumstances (or, if you prefer, in those worlds) where this applies and the fermionic representation is realized, one obtains the right kind of 'chemical' bonds and hence the property of solidity is manifested. And then further structures yield further properties that we

reflected in the pattern of truth values for counterfactuals relating the state of some localized part of the world at one or more places and times to the state of some localized part of the world at one or more *later* places and times' (2011: 97) and, just to emphasize the similarities with the view I shall defend, he states that this counterfactual structure is 'endowed' by the fundamental laws of the world. In effect what I shall do in Chapter 10 is explore the nature of this endowment.

associate with the physical, including position observables given via imprimitivity and so on. But of course the likes of Tegmark will push back, insisting that this is just more mathematics and that there is nothing here that is distinctively physical.

Perhaps then we simply have to accept that the distinction between the mathematical and the physical has, at the very least, become blurred (French and Ladyman 2003) or that it cannot be drawn at all (Ladyman, Ross, et al. 2007). Perhaps there is no answer to the 'problem of collapse'. Perhaps we should follow Ladyman in dismissing this as a pseudo-problem (Ladyman 2007/2009).[46] Still, one can draw appropriate distinctions between OSR and mathematical structuralism in general, such that criticisms of the latter can be blocked from being imported into considerations of the former. And whether or not object-oriented forms of realism fare any better in this respect by appealing to a primitive notion of objecthood, they are still going to have to accommodate our reflections on the nature of causation in this context and the broader issues associated with distinguishing the abstract and the concrete.

The next two chapters will take up some of these themes and will argue for a view of the structure of the world as inherently modal.

[46] Psillos (2012) objects to this dismissal and presents an answer to the problem in terms of structural universals. Although he himself admits that this answer fails, he takes this to reflect the failure of OSR rather than as indicative of the nature of the problem in the first place. We shall briefly consider this approach in the context of Esfeld's causal structuralism in the next chapter.

9

Modality, Structures, and Dispositions

9.1 Introduction

My primary aim over the next two chapters will be to make good on the promissory note alluded to by Esfeld:

> The structures to which ontic structural realism is committed have been conceived by James Ladyman as including a primitive modality (Ladyman 1998; Ladyman, Ross, et al. 2007: chs 2–5), and Steven French (2006) expresses sympathy with this view. However, it has not been spelled out as yet what exactly that modality consists in. (Esfeld 2009: 179)

Esfeld himself spells it out by means of a form of structuralism that embraces the dispositionalist account of modality. As I shall suggest, however, fundamental problems arise when this sort of account is imported into the context of modern physics; in particular it is unclear how dispositionalism can capture symmetries and conservation laws. Nevertheless, a version due to Chakravartty will provide a useful foil to the structuralist position I shall articulate in the next chapter; indeed, one can reach this position via a kind of 'reverse engineering' of dispositionalism. My central claim will be that we should take laws and symmetries—and hence the structure of which these are features—as inherently, or primitively, modal. As we shall see, this bears certain connections with both the Humean and dispositionalist alternatives and articulating it in the context of OSR will, I hope, reveal the advantages it has over both. Let me begin with the former.

9.2 Humean Structuralism

For much of this chapter, I will assume that modality is 'in' the world, in the sense that the structure of the world is modal, and the issue will be, as Esfeld puts it, how to spell that out. But there is an alternative view that insists that modality is not 'in' the world at all, but lies with the models we deploy to represent the world. Of course, we do not have to be realists about every feature of these models and one may take the modal features to be non-representational in this sense. Indeed, this is the kind of stance the constructive empiricist takes (van Fraassen 1980 and Monton and van

Fraassen 2003; see Dicken 2007), although one does not have to be an anti-realist about theoretical terms or features as well as modal ones (see Psillos 2009).[1] Such a view would be broadly consonant with a Humean account of modality but as is often the case with views that are as historically shaped or conditioned, it is not always straightforward to pin down precisely what is mean by or what counts as 'Humeanism'. However I shall take the essential tenets of a Humean form of structural realism to be the following (Lyre 2010: 10–11; see also Lyre 2011):[2]

i) Humean supervenience: fundamental intrinsic and categorical properties supervene on a micro-physicalist base;
ii) a regularity view of laws (where this is understood as a non-necessitarian view).

Appropriately construed, these ground a view of the structure of the world as fundamentally categorical and allow OSR to resolve a number of critical issues.

Now, it has to be acknowledged that although they might be seen as natural allies, tenet (i) is strictly independent of (ii), so one could give up Humean supervenience while still holding the regularity view. Whether one should or not again depends on how (i) is understood. Typically it is understood as the claim that the world is some kind of 'mosaic of local matters of particular fact' and the incorporation of locality (however understood) is an obvious hostage to fortune, as the non-local features of quantum systems yield well-known counterexamples.[3] However, as expressed here, Humean supervenience is a broader claim and with an appropriate structuralist understanding of intrinsic properties one can maintain that 'whole structures' are taken to be included in the subvenient base. In other words, one gives up the local bits of the mosaic and takes the subvenient base itself to be structural and, crucially, categorical.

The core of this understanding is the claim that in so far as they are structural invariants, in the manner that has already been discussed here (i.e. as arising from the action of the appropriate symmetry group), the putative intrinsic properties of particles are dependent on the structure and should thus be regarded as 'structurally derived intrinsic properties' (Lyre 2010, 2011). They are still to be regarded as intrinsic since they are instantiated, or subsist, independently of the existence of other 'object-like' entities. As a consequence, there are relata on this account (unlike eliminativist OSR) and structurally derived properties, where these are either relational properties or structurally derived intrinsic properties, but there is nothing

[1] Having said that, Berenstain and Ladyman argue that there is a fundamental tension within Humean forms of realism, particularly with regard to explanation: '[it] is unclear how a realist who disavows natural necessity can make sense of the idea that unobservable entities explain the phenomena' (2012: 156).

[2] Maudlin (2007) and Berenstain and Ladyman (2012) ask what motivation there might be for being a Humean in the first place, particularly given the lack of scientific grounds for Humean metaphysics. In terms of what I tried to lay out in Chapter 3, one possible driving force might be the desire to reduce our metaphysical humility to the absolute minimum.

[3] However, see Darby 2010 for a possible response that cleaves to the spirit of Humeanism in this context.

more to the relata than these structurally derived properties (cf. Esfeld and Lam 2011). And the latter distinguish classes of (putative) objects or domains of structure only.

On this account, one can have a physically possible world with one electron only, possessing charge, say (with a relational space-time[4]). Here charge is understood not as a relational property, given by Coulomb's Law, say, but as a symmetry invariant, given by the feature of the structure presented by U(1) gauge symmetry and instantiated by a lone electron (Lyre 2011). Furthermore, this notion of structurally derived intrinsic properties allows us to make sense of the distinction between those symmetries that are instantiated (in this, the actual, world) and those that are not, such as gauge symmetries (see Lyre 2004). In the former case, the ontological commitment to structure amounts to a commitment to the structure invariants, such as mass, spin, charge, etc., taken to be the most fundamental structurally derived intrinsic properties.

This understanding of putative intrinsic properties can be straightforwardly accommodated by eliminativist OSR, but my worry is that, as with moderate OSR (discussed in Chapter 7), the account is not stable (see French 2010a). Again, we have relata, but these are cashed out in terms of either relational properties or structurally derived intrinsic properties. With the former, we have the issue of relations without relata again; with the latter, the relata, as structurally derived properties, are dependent upon the structure. In neither case do we have objects, whether thinly or thickly construed, that may be taken as non-eliminable.

Furthermore, in the case of the lone electron, the structure is going to be pretty attenuated. Consider: the regularities that the Humean sets such store by are going to be limited in scope and content in this case. Those aspects that are encoded in symmetry principles are not going to be instantiated because here all we have are gauge symmetries. This leaves only the law-like features of the structure but that yields precious little to go on. One might talk of the possible behaviour of the electron when a test charge is brought in, but obviously such talk cannot be taken to represent any feature of the actual structure of this particular world. Thus in such worlds we do not have the regularities represented—on the Humean view—by Coulomb's Law, say. But perhaps this is a price the Humean is willing to pay: admitting worlds where the relevant regularities and consequent structure is very 'thin'. The alternative is to rule out such worlds as unphysical, since in effect we do not have the laws (even construed à la Hume) in terms of which our understanding of what is physical is grounded.

How one generates such sparsely populated worlds raises interesting issues, and as a result, their force in undermining the Humean account may depend on the mode of generation. So, they are typically deployed in the following way (see Slater and Haufe

[4] This will obviously have to be spelled out in terms of possible objects (within this possible world) anchoring the spatio-temporal relations; see Belot 2011.

2009): if law statements are merely summaries of regularities, then a world in which there is only a lone electron, moving at constant velocity, say, is compatible with any number of laws, including Newton's laws of motion. In such cases, the laws 'float free' from the Humean base and the nomic scepticism this engenders counts against the Humean account.

Now the Humean can simply insist that such cases beg the question by assuming, for example, a governing conception of laws that she rejects. Lange, however, attempts to block this insistence by appealing to the intuition—that he suggests the Humean should accept—that many facts about the world could have been different without the laws being different (Lange 2000). This motivates the claim that the laws of this, the actual, world, remain the laws of the lone electron world and this form of invariance of laws across such radical changes of non-nomic facts leads one to conclude that the laws cannot be constrained by these facts in the way the Humean maintains. In particular, one can imagine another world where electrons obey very different laws than ours and from that world one would arrive at a lone electron world holding these very different laws invariant; thus, two worlds that contain the same non-nomic facts can differ with regard to their laws, contrary to the Humean view.

However, the manner in which Lange generates such possible worlds offers a way out for the Humean. Basically, Lange adopts a methodology of impoverishment: we arrive at the lone electron world by starting with our world, then 'severely depopulating' it (Lange 2000: 87). Given the intuition that the laws would remain as they are, even through radical non-nomic changes, he concludes that this would hold even when such changes include severe depopulation. And likewise for those other worlds where the laws are different; thus, the same apparent regularity can support very different laws. However, here's the worry about such a move: depopulating is supposed to leave the history of the world intact, so the lone-electron world obtained from this world is actually very different from the lone-electron world generated from another world with very different laws, since the two 'starting' worlds have very different histories (Haufe and Slater 2009: 269). In that case, however, the two lone-electron worlds will contain very different regularities throughout their histories and hence we do not have a case of 'same regularities, different laws' that would undermine Humeanism.

More generally, one might have doubts about generalizing Lange's fundamental intuition (2000: 271–2). That the laws remain stable, or as expressed previously, invariant, under removal of (at least some) 'everyday' objects, seems uncontentious. In a world such as ours in all respects except without the Eiffel Tower (2000), would Newton's laws still hold? It seems hard to find any grounds for claiming they would not. But one does not have to move beyond the everyday to stretch the intuition: what about a world in which everything but the Eiffel Tower has been removed? What history consistent with the relevant laws could produce such an outcome? Or, moving to the micro-level, suppose the world were so impoverished that only a single sodium atom was left (2000)—would this still bond with chlorine to form salt? At such points the intuition seems stretched to breaking point and certainly it is no longer so clear that one can maintain that the laws would be the same.

There is an alternative world-building methodology that one might adopt: one might generate such worlds by 'building from scratch', as it were, a permanently sparse world containing only one electron (Haufe and Slater 2009: 269–70). In this case the history of the world we start from is not carried over, and we lose whatever grounds we had for maintaining the intuition that the laws will remain invariant. Thus the Humean can simply insist that the laws of lonely worlds generated this way correspond to whatever the regularities are in such worlds. At the very least, the burden of argument has shifted away from her (2009: 270).

The choice of method for generating sparse worlds has implications for structuralist claims in this context. In particular, let's recall the previous claim that according to the Humean form of structural realism, one can have a sparse world of one electron, possessing charge, where this is understood as a gauge invariant. Such an understanding might be underpinned by the impoverishment method for generating sparse worlds, in so far as the intuition can be preserved that through such impoverishment the relevant gauge symmetry will be maintained. However, the same concerns as touched on earlier will arise and if the Humean response of taking the laws and associated symmetries as given by the regularity in that world is adopted, then the attenuation is going to be such that it is also unclear whether one can say that gauge symmetry also holds in such a world. On the other hand, if the alternative methodology of generating worlds from scratch is adopted then again it is not clear on what grounds one might build in such symmetries. Of course, one could take them to be features of such worlds, such that the relevant lonely particles would be manifestations of these structural features, but that takes one away from the Humean picture and towards the kind of view I shall be elaborating here.

Moving on to point (ii), a regularity view of laws is, of course, compatible with object-oriented realism. After all, if the latter is articulated in terms of adopting a particular stance towards the unobservable posits, or referents of the theoretical terms, of our theories, then it is not necessary to adopt the same stance towards the law-statements, of those theories. Likewise, then, when it comes to structural realism, the same point holds. Of course one has to take a little care: after all, if one is giving ontological priority to the laws over the objects, then this might naturally incline one to take a realist stance towards the former. But all one has to do is to insist that by 'laws' here one means nothing over and above the relevant regularities; or, better, insist that the issue has to do with what the relevant law statements represent: regularities, rather than relations in the world bound by necessity.

Furthermore, this crucial feature of laws from the Humean perspective appears to be entailed by the relevant structures (Lyre 2004). Thus consider Minkowski space-time: this can be conceived of as a 'global geodesic structure' exemplified by the trajectories of free-falling bodies, where these in turn display regular behaviour.

Furthermore, this behaviour should not be understood as the result of some disposition, whether of the body or the structure itself (we shall consider these options shortly) but rather as an 'exemplification' of the manifest, categorical structure of space-time. So, returning to a (very) sparse world, the movement of a particle along an infinitesimally small path in an infinitesimally small time can be regarded as a 'proper instantiation of the full spacetime structure of that possible world' (2004: 12). If this exemplification of the structure by the behaviour of bodies is understood as arising from the dependence of such bodies on the structure, then one can see how in this respect this view is not going to be so different from the one I defend here (the crucial difference lies with the categorical nature of the structure in the Humean account, as we shall see).[5]

How does this Humean structuralism draw the distinction between laws and accidental regularities? As is well known, the standard move is to appeal to a 'Best System' criterion: those regularities count as law-like that are represented in such a way that they 'best systematize' scientific knowledge, where what counts as 'best' is to be judged in terms of the criteria of simplicity, strength, and so forth (see Callender and Cohen 2009). However, the worry is that striking such a balance between simplicity and strength introduces an element of subjectivity into the analysis. This can be assuaged by taking the 'best' theoretical system to involve appeal to the relevant natural properties and kinds (Psillos 2009). Of course, a possible circularity arises if what is taken for a kind to be natural is that it is part of a nomological pattern (2009: 143). Grounding the natural kind division in symmetry considerations—conceived as distinct from laws—offers a way out of this circle (at least when it comes to physics) but now symmetries must be accommodated within the regularity view.[6]

One option is to take them as constraints on laws (a view that draws support from their heuristic role); another is to regard them as by-products of these laws (with the heuristic role not taken to be indicative of the symmetry's ontological status). Now an obvious way to go would be to regard the notion of constraint as incorporated within that of a 'meta-law', which has as its subject matter the laws themselves, rather than phenomena per se. Then one could presumably run the same analysis on symmetries as one does on laws and incorporate such constraints within the analysis itself, by, for example, taking 'strength' to be dependent upon satisfying such constraints. Thus the

[5] Again, such a structuralist approach offers an alternative response to Brown's demand for an explanation of the behaviour of bodies in following geodesics: rather than the relevant dynamics acting as the explanans, as he argues, it is the structure together with this exemplification relation (see Brown 2005).

[6] However, Callender and Cohen (2009) suggest that the circle cannot be avoided, since in practice scientists devise laws based on their choice of kinds and choose the kinds based on the laws (interestingly they cite the example of the application of SU(3) in particle physics to motivate this claim). They argue for a form of best system analysis according to which laws are relativized to the chosen basic kinds, where the latter remain open to the process of enquiry, rather than being stipulated once and for all.

symmetries would be meta-regularities of which the laws are a subset of (standard) regularities. Such a view could obviously accommodate claims that certain laws, at least, can be derived from the relevant symmetries and also the latter's heuristic role.

Alternatively, one might take the symmetries to be by-products, in the sense that they are features of, or arising from, these laws.[7] In this case they would be subsumed within the analysis of the regularities and again the Humean would have little difficulty in accommodating them. Nor, it seems, would she have much difficulty in adopting the Cassirerian perspective discussed in Chapter 4, according to which law statements describe the network of relations, 'held together' by the symmetry principles which represent what is invariant in the network. Of course one might read this 'holding together' in a strong sense as involving some kind of dependence of the laws on the symmetries. However, one can also read it more weakly as involving a mutual dependence or a kind of 'reciprocal interweaving and bonding'. Again, if this bonding can be explicated in categorical terms, as involving regularities or features thereof, then one could envisage a Cassirer–Hume hybrid form of structuralism.

Still, the sorts of inter-system comparisons that one needs to engage in on the best systems account of laws are problematic. Taking simplicity on its own, for example, a kind of relativity arises since assessments of simplicity are relative to the set of basic natural kinds or basic predicates assumed to apply to the domain or system in question.[8]

Alternatively, by appealing directly to structure, the Humean can account for the difference without having to adopt some form of 'Best System' account. Thus, only structures are law-like, so that,

> the particle following a geodesic is not a subsequence of disparate events which, without further explanation, show a regular behaviour. It is an exemplification of a global regularity itself – the geodesic structure. (Lyre 2004: 12; see also his 2011)

So, the idea is that non-law-like regularities can be considered to be a 'subsequence of disparate events', in some sense. It is the global or holistic nature of the structure that implies that the sequences of events that exemplify it are not disparate. One way of getting a grip on this idea would be to suggest that what marks a difference between a sequence of disparate and non-disparate events is that the former is more easily or readily disruptable, as a change to the surrounding contingent circumstances has more opportunities to have an impact on, and disrupt, a sequence of disparate

[7] Callender and Cohen don't say anything explicitly about how symmetries are to be incorporated into their 'better best system' account. However, they do mention SU(3) and the Eight-fold Way and take the latter to be a law, or the corollary of a law which suggests something like the by-product view.

[8] Callender and Cohen's response (2009) is to bite the bullet and relativize the notion of 'Best System-hood' and hence lawhood; that is, what counts as the simplest, strongest, and thus best system is to be determined relative to a specific choice of basic kinds or predicates. As they note, this is not compatible with standard realism about natural kinds, and instead they favour a form of 'promiscuous realism' (I shall consider a form of this view in Chapter 12). For further criticism see Berenstain and Ladyman 2012.

events than it does on a sequence of non-disparate events. Non-disparate events are more *stable*, a suggestion that I shall return to later.

Now, what grounds this stability? An obvious move is to appeal to the structure, but then what makes a structure stable? Those who hold so-called 'necessitarian' views of laws, according to which, for example, laws are metaphysically conceived of in terms of necessary relations holding between universals, will explicate this stability in terms of such relations. The dispositionalist, as we shall see, appeals to an ontology of dispositions and the modality associated with that, with the idea being that laws are supervenient upon, or 'flow' from, the underlying dispositions, so the stability of laws as compared with non-law-like regularities is to be explained by the stability of the underlying dispositions. Indeed, as a way of accounting for the necessity of laws, this is given as the principle explanatory advantage of dispositionalism. The Humean structuralist, however, has no such resources and has to appeal to some inherent feature of the structure itself that confers this stability. It is not entirely clear how that feature might be spelled out in categorical terms.[9]

Let me now move on to consider structuralist variants of the dispositionalist account, namely Chakravartty's semi-realism and Esfeld's dispositional structuralism. These will not only provide appropriate foils to the alternative 'modal structuralism' that I favour, but by 'reverse engineering' these dispositionalist accounts, we can further motivate the former.

9.3 Doing Away with Dispositions

Characterizing what a disposition *is*, is itself philosophically problematic (see, for example, Fara 2006). Typically, dispositions are associated with subjunctive conditionals as in the standard example: a vase may be ascribed the disposition of being fragile, on the grounds that it would crack or shatter were it to be knocked or dropped (with sufficient force). The characterization of dispositions in terms of the entailment of such conditionals faces well-known problems, not least of which is that by rendering all physical properties dispositional it effects a kind of *reductio* (see the debate between Averill 1990 and Reeder 1995). One response is to strengthen the entailment with conceptual necessity: dispositional properties are those that play, as a matter of conceptual necessity, a certain causal role that is best captured in conditional terms (Mumford 1998).

[9] Lyre notes (2004: 12) that on his view laws are exceptionless, since any instantiation of a structure will display the same regularity as encoded in the structure itself. Alterations to such regularities can only occur if the structure itself changes. On this view, then, 'stability' is not perhaps the most appropriate notion, as it allows for degrees of stability, whereas on Lyre's view, laws are absolutely or ultimately stable in the sense that no exceptions can be tolerated and this is what distinguishes them from non-law-like regularities. As we'll see, the idea of degrees of stability will prove helpful in extending forms of structuralism to non-physical sciences and one might expect tensions in likewise extending the Humean approach.

Standardly, causal properties are taken to be identified with, and hence reduced to, such dispositions. This forms the central claim of the Dispositional Identity Thesis (DIT): the identity of causal properties is given by the dispositions they confer. Characterizing dispositions, or powers in general, via a contrast with categorical properties has been a matter of considerable debate for many years,[10] but however they are conceived, dispositions are taken to possess two crucial distinguishing features (1998: 269): first, they are related to manifestations of a specific kind. More specifically, perhaps, a disposition relates a stimulus and a manifestation such that, if an object *x* instantiates the disposition, it would yield the manifestation in response to the stimulus. Secondly, the disposition and its manifestation are considered to be 'distinct existences' in the sense that the disposition can exist without being manifested, as when there is no appropriate stimulus, for example.

We can call this (in slightly risqué fashion) the 'S&M Characterization of Dispositions':

> Everything that has disposition P and is subject to the relevant stimulus S will yield the appropriate manifestation M; or, more formally (for the hell of it)

$$\forall x((Px \,\&\, Sx) \rightarrow Mx)$$

It is via this characterization that we distinguish and individuate dispositions.

There are well-known concerns that arise with regard to the identification of the relevant manifestations (Molnar 2003), and while some insist that they only become observable under laboratory conditions (Mumford 2009a), others argue that they should be regarded as 'virtual' effects (Wilson 2009), to which we have indirect epistemic access as with unobservable entities in science in general (but see McKitrick 2010).[11] Thus there are 'hidden' costs associated with adopting a dispositionalist ontology that may not be readily apparent when simple or 'everyday' examples are given. When the ontology is conveyed away from the everyday and into the domain of modern physics, however, care must be taken to ensure that a metaphysics appropriate for macroscopic objects is not being illegitimately imported into the micro-realm.

[10] According to Mumford (2011), the 'leading powers theorists' now advocate the view that powers, or dispositions, should not be understood as a kind of property at all but should be regarded as a distinct and irreducible ontological category in their own right. This further raises the ontological price of admission, as it were.

[11] Shifting to the other end of the dispositional sequence, one might also have concerns about what constitutes a stimulus. Some proponents of dispositional analyses have argued that one can have dispositions in the absence of stimuli, giving the Ghirardi–Rimini–Weber 'spontaneous collapse' interpretation of quantum mechanics as an example (Dorato and Esfeld 2010). However, one may fairly wonder if one has any grip on what the relevant disposition is in such examples and whether what is being appealed to is just some partially analysed potentiality–actuality shift.

At first sight, physics might seem to offer a more supportive environment for dispositionalism. Thus consider the S&M characterization and the claim that to say something is disposed to give a response to some stimulus is just to say that it possesses a property that would cause it to give that response if it were to undergo that stimulus. As stated, and in the context of the sorts of 'everyday' examples having to do with vases and such, this is open to well-known objections involving 'finkishness', which hinge on the introduction of non-permanent dispositions such that the relevant counterfactual remains true in the absence of the disposition in question (see Martin 1994). Such moves crucially depend on the intuition that objects may gain or lose their powers but it is, at the very least, not at all clear that this holds for the fundamental objects of physics and their equally fundamental properties. Similar considerations apply to the deployment of so-called deviant processes whereby the relevant response is obtained by means other than those related to the stimulus cited in the conditional (Smith 1977); again there appears to be little scope for such deviance in the micro-realm.

Furthermore, dispositionalism appears to be naturally supported by the way in which we understand objects and properties in physics. Chakravartty asks,

> Why and how do particulars interact? It is in virtue of the fact that they have certain properties that they behave in the ways they do. Properties such as masses, charges, accelerations, volumes and temperatures, all confer on the objects that have them certain abilities or capacities. These capacities are dispositions to behave in certain ways when in the presence or absence of other particulars and their properties. (2007: 41)

Thus the property of mass, for example, confers the disposition on a body to be accelerated under applied forces. And, crucially, it is via the linkage between such dispositions that causal activity is produced. The crucial move here is from the claim that the explanation for the behaviour of particulars is the properties they have, to the assertion that these properties confer upon these particulars certain 'abilities or capacities', which follows from the dispositional identity thesis (DIT).

Now the Humean is going to argue that when it comes to these classical examples, at least, she has a very different but, at least, equally viable take on matters that, importantly, involves no further inflation of our ontology. It is perhaps not so surprising, then, that dispositionalists have turned to modern physics in support of their cause. Thus Molnar asserts,

> Physics tells us what result is apt to be produced by the having of gravitational pull or of electromagnetic charge. It does not tell us anything else about these properties. In the Standard Model the fundamental physical magnitudes are represented as ones whose whole nature is exhausted by their dispositionality: that is, only their dispositionality enters into their definition. Properties of elementary particles are not given to us in experience: they have no accessible qualitative aspect or feature. There is no 'impression corresponding to the idea' here. What these properties *are* is exhausted by what they

have a *potential for doing*, both when they are doing it and when they are not. (Molnar 1999: 13)[12]

There are a number of claims being made here. The first has to do with what physics 'tells us'. This is not straightforward of course, since it is not clear what it is that is doing the 'telling'. As I have already noted, claims that, for example, theories tell us about particulars have been made on the basis of an explicit Quinean reconstruction in which objects emerge 'thinly' as whatever yields the value of the relevant bound variable. It is open to debate whether such reconstructions should be taken as canonical. Secondly, it is not at all clear that even under such a reconstruction the theory could be said to 'tell us' that said objects possess dispositions. Of course, they do not tell us much more about properties than that they are interrelated in certain ways and they certainly do not tell us that they have quiddities or whatever (not least, in both cases, because the language of theories is metaphysically 'thin') but further interpretation is required to go from such interrelationships to the possession of dispositions.

One could always adopt the same perspective in this regard as operationalism: thus one might say that since we can only measure or even detect charge, say, by bringing in a test charge, the most appropriate way of understanding this property is in terms of the disposition to manifest (the Coulomb force) when presented with the appropriate stimulus (the bringing in of the test charge). Indeed, one could try to justify such a move via the tailoring insistence presented in Chapter 3: we should cut our metaphysics to fit our epistemology. However, there's a difference between using this to reduce our overall level of humility and push such items as hidden 'natures' or, indeed, quiddities, out of our ontology and using it to bring in another particular set, such as dispositions. In particular, a structuralist ontology arguably provides a more tight-fitting set of metaphysical clothes.

The second claim made in the quote from Molnar is that the 'Standard Model' of elementary particle physics—that is the currently accepted model of the electroweak and strong nuclear forces—represents the relevant fundamental physical magnitudes in entirely dispositional terms. Again, it is not clear how this might be justified.[13] One thought might be that over and above what was just said about what theories 'tell' us, the Standard Model, by incorporating quark confinement, say, tells us that certain elementary objects and their associated properties cannot be 'observed' or accessed,

[12] Similarly, Mumford insists that '[p]hysics in particular seems to invoke powers, forces and propensities, such as the spin, charge, mass and radioactive decay of subatomic particles' (2011: 267).

[13] Bauer argues against the standard dispositionalist view that mass, for example, is an intrinsic ungrounded disposition on the grounds (ha!) that according to the Standard Model it is grounded in the Higgs field (2011). This offers further grist to the mill that will be ground shortly regarding the status of intrinsic properties in physics and in so far as the moral that Bauer seeks to draw is that we should 'look beyond the objects and particles bearing dispositions, to properties of their environment and of other objects, in exploring the ontological grounds of dispositions' (2011: 98), this feeds into Chakravartty's account and thereby, as we shall see, into an ultimately structuralist ontology.

even in the extended sense in which electrons, etc., might be said to be observed or accessed. Hence, even more so than in the latter case, we can only detect such objects via the effects they produce. But, again, it is not clear what this 'even more so' amounts to here. Quarks, just like electrons, are only accessible via long (defeasible) chains of effects and all that quark confinement tells us is that we cannot detect or access single quarks. It is not at all clear that this provides any additional grounds for dispositionalism.

Furthermore, if one were to let quark confinement tell us something metaphysical, there appear to be just as strong grounds for saying that it tells us to do away with the underlying objects; thus again, an alternative interpretation of this theoretical feature can be given. It is particularly noteworthy that no mention is made of the role of symmetries in the Standard Model and this role of course suggests that it is not straightforwardly the case that the 'whole nature' of the fundamental physical magnitudes is exhausted by their supposed dispositionality. Finally, even granted that properties of elementary particles are, of course, not given to us in experience, further steps are required to warrant the claim that what they *are* is exhausted by what they have a *potential for doing*.

Even if we grant that physics does not lead us directly to dispositionalism, the dispositionalist may argue that the Humean alternative is deeply unsatisfactory, which leaves the question of what accounts for the behaviour of elementary particles? One option is to take this behaviour as a 'brute fact', but then, the argument goes, the properties appear to have instrumental value only (Molnar 1999: 15) and the danger is that 'this instrumentalism about the properties will carry over into anti-realism about the particles themselves' (1999: 15). If we then accept a realist ontology of macroscopic objects and their dispositions, we end up with a strange kind of dualism. That dualism threatens to collapse into incoherence if we further take the history of physics to support the kind of framework for both objects and properties suggested here: physical systems would be dependent on non-existent constituents and their properties would be ontologically dependent on non-existent fundamental properties. To avoid this, the realist must embrace dispositions all the way down, as it were.

However, why should one accept that without dispositions, the behaviour of objects must be taken as brute? One could of course explain this behaviour in terms of the relevant laws, taken as 'governing' in the appropriate sense and understood in a structuralist context, as explicated shortly. In particular, on such a view we could avoid the inference 'If there is no electric charge (but only "electric behaviour"), then there is no electron' (Molnar 1999: 15) and in general the worry that instrumentalism about properties carries us to anti-realism about the objects since, according to OSR, we can still be scientific realists about both but deny that they constitute metaphysical entities in any robust, non-dependent sense.

Nevertheless, we have already seen that the Humean framework of monadic, categorical properties is problematic, particularly when it comes to establishing their intrinsicality.[14] We recall that we are invited to get an initial handle on what is meant here by considering a given object and then conceiving of a world in which there is just that object; the properties of the object that are retained in such a possible world where there are no other objects are those that we might take to be monadic and 'intrinsic'. Whether such possible worlds are constructed via impoverishment or 'from scratch', such exercises are problematic in so far as they abstract away the relevant physics[15] in order to achieve an appropriate state of affairs on the basis of which metaphysical judgement can be passed on the status of certain properties, this judgement then being held to remain in force when we shift back from the aforementioned state of affairs to that which is fully clothed with the relevant physics.

When it comes to such conceptions, as Hacking said in a related context, 'bland assertion is not enough' (Hacking 1975: 251). What else one needs to specify obviously depends on the project one is engaged in; at the very least some form of spatio-temporal background might be introduced (French 1995). Likewise, if the project is to establish whether a property such as mass, for example, should be counted as intrinsic, then abstracting away the framework of General Relativity is tantamount to metaphysical bland assertion and leaves the project open to the sorts of charges levelled by Ladyman, Ross, et al. (2007). (And of course how we should understand mass in the context of General Relativity is a delicate issue.)

The point, again, is that conceiving of a possible world in which there is a 'lonely' object and simply asserting that such an object has any of the standard physical properties is not enough to establish that such properties as they feature in *this* world are intrinsic, monadic, or whatever. Indeed, the attribution of these standard physical properties in an appropriate manner—that is, one that respects their role within our best theories—may undermine this notion of 'lonely objects' and in general the possible-world conceptions it is associated with. The question now is, if one is going to carry over the relevant panoply of physical theory into one's possible-world conceptions, why even bother trying to abstract out the supposed intrinsic properties? Why not simply 'read off' the metaphysics of properties from the theoretical context?

Thus consider charge again: it is obviously completely straightforward to 'conceive' of a possible world in which there is a single charged particle and no other objects. So, we have a situation of metaphysical loneliness and we may be tempted to conclude that charge can thereby be considered an intrinsic and monadic property. But in this context, what features does this property have? In particular, does it have

[14] Dorato also argues that quantum mechanics supports dispositionalism because quantum states are best seen as relational and indefinite (Dorato 2007). However, being relational does not imply being dispositional and equating the disposition–manifestation relationship with that which holds between being indefinite and being definite, at the very least, extends the meaning of dispositional beyond that which is being considered in this chapter (see McKitrick 2008).

[15] Where such abstraction proceeds either by stripping away or positing without including the relevant context.

any that we standardly attribute to charge? Can we say, in this physically attenuated situation, that the charge on this lonely object is either positive or negative, or that it obeys Coulomb's Law, or that it is the generator of the U(1) symmetry of electromagnetism, with an associated conserved current?[16] Of course, we *could* say, were we to bring in a test charge (suitably designated as either positive or negative) from infinity, but then we are no longer so lonely.[17] If we want to ascribe to charge those features that it is standardly taken to have, and that 'make' it *charge*, at least as far as the physics is concerned (irrespective of whether we take properties to have quiddities or not), then we need to consider it in the appropriate theoretical context. But if we do that then it seems much harder to maintain that charge is intrinsic or monadic in the relevant sense; indeed, if we are undertaking such a consideration, then we might just as well 'read off' the metaphysics from the appropriate theory.[18]

Related concerns arise with regard to the motivation for introducing dispositions, and not just because they are typically taken to be intrinsic themselves. Thus charge is argued to be dispositional because it entails certain subjunctive conditionals; that is, an object is taken to be charged only if the object would produce the appropriate manifestation—such as an acceleration in the appropriate direction—when subject to the appropriate stimulus conditions—namely in the presence of another charged body (see Mellor 1974: 171). This accommodates the kind of scenario articulated here but it invites us to ascribe charge on the basis of considering what an isolated object would do, if under the appropriate conditions, and constructs a metaphysics on that basis. We recall the characterization of dispositions as playing, as a matter of conceptual necessity, a certain causal role that is best captured in conditional terms (Mumford 1998). Thus charge is dispositional because as a matter of conceptual necessity it plays the causal role it does in repelling or attracting other charges in accordance with Coulomb's Law. The obvious question now concerns the grounds for asserting such conceptual necessity and this pulls us back, once more, into the broader theoretical context (and again the issue is whether we need to advance beyond that context to metaphysically construct isolated dispositions).

[16] I am grateful to Kerry McKenzie for discussions on this and related points.

[17] Unless we were to take a further step up the ladder of modality and insist that the introduction of such a test charge is not to introduce a further feature of the possible world we have conceived but is merely a further modal exploration of that world that allows us to add to our conception of it; so the idea would be that the $1/r^2$ feature associated with Coulomb's Law can be held to be a feature of our lonely situation but that as a matter of epistemology (in that possible world) it cannot be detected. At this point one might counter-insist on bringing one's metaphysics into line with the relevant epistemology and that if our only access to the features of charge is through its effects on other charged bodies, then ascribing such features to lonely charges that by stipulation or bare assertion are not able to have any such effects amounts to the elaboration of illegitimate metaphysics.

[18] All this is not to say that physicists should not consider lonely scenarios involving, say, universes with single masses, for example (think of the Schwarzchild solution of Einstein's equations). These are fine for helping to pin down and explore certain physical features but the point, again, is that this does not amount to an appropriate (meta-metaphysical) methodology for establishing the metaphysical nature of properties that feature in this, the actual, world.

More importantly, for my purposes, however, this characterization is still too broad and fails to distinguish between a dispositional analysis of physical properties and the kind of structuralist conception that I favour. We recall the claim that dispositions are conferred by causal properties and manifest, continuously, as causal processes. As I said previously, how one cashes out the contrast with categorical properties—whether dispositions reduce to the latter, whether this distinction even applies to properties, and so on—involves a number of open issues, but I don't think what I have to say hangs on that, as it is the nature of the distinction itself that is crucial here. As Chakravartty, for example, notes (and we shall consider the specifics of Chakravartty's own view shortly), it is usually explicated in terms of the manner in which dispositions and categorical properties are described: 'the former in terms of what happens to objects under certain conditions, and the latter without reference to any happenings or conditions' (2007: 123). It is this distinction which undergirds the metaphysical picture and perhaps the fundamental difference between dispositional accounts in general and OSR lies with the issue of the extent to which that most basic metaphysical picture needs to make reference to these 'happenings and conditions'.

It is not just that the underlying metaphysics of the dispositionalist is typically object-oriented and particularist, but that it is one in which we are invited to conceive of such particulars and their properties as being 'disposed to behave in certain ways in the presence and absence of other particulars and properties' (2007: 120), where it is these properties and resulting causal processes that scientific theories describe. Thus underlying this account is something akin to the 'loneliness' assumption previously discussed, namely that particulars may exist in the absence of others and hence may not manifest the requisite behaviour, but by conceiving of what that behaviour would be in the presence of the further particular we come to ascribe a disposition to behave in such a way. As should be clear, this is an assumption and picture fundamentally at odds with OSR (cf. Ladyman, Ross, et al. 2007: 3) and, as indicated already, is not pressed upon us by the fundamental physics.

Having softened up the terrain, as it were, let me now present three crucial problems for this picture.

9.4 S&M and Laws

Consider Coulomb's Law, again as usually and simply stated:

$$F = Cq_1q_2/r_{12}^2$$

As so stated, the various quantities involved in this law—the charges, the distance between them, and the force—can all be regarded as *determinable* quantities. They are all such that they can take on specific values and become *determinate*. We shall return to the distinction between determinables and determinates and what it implies in the next chapter but for now note that an electron, for example, is not merely charged; it has a determinate charge.

Now, returning to the stimulus and manifestation conditions that characterize dispositions, these also involve determinates. Given any such determinate charge, Coulomb's Law states the mathematical relation between any determinates of the stimulus and manifestation condition. The issue now is how one gets from these determinates to the *determinables* related in the law. As we have noted, one of the chief advantages of the dispositionalist approach is the way it can account for laws and their necessity but as we shall now see, with the issue I am about to present and the next, there are significant problems that must be faced (see Vetter 2009).

So, according to the dispositionalist, if we retain our 'fundamental base' across possible worlds, so we have the same set of particulars in any possible world as we have in this world, then we will have the same set of properties, and these will confer the same set of dispositions, which in turn yield the same laws; hence the laws must be necessary. But how do we get the laws from the dispositions in the first place? Bird's treatment is elegant and instructive.

Thus he takes properties to essentially and hence necessarily confer the relevant dispositions (2007).[19] If P is the property concerned, S the stimulus, M the manifestation, and $D_{(S,M)}$ the associated disposition, then we have:

$$\Box(Px \rightarrow D_{(S,M)}x)$$

If we substitute the associated conditional for the disposition, then we obtain:

$$\Box(Px \rightarrow (Sx \Box\!\!\rightarrow Mx))$$

It then follows, by elementary modal logic, that

$$\forall x((Px \,\&\, Sx) \rightarrow Mx)$$

With P as charge again, it is of P's essence that like charges repel, unlike charges attract (Bird 2007: 45); S would be the presence of another charge at a distance r from the given charge (2007: 22); and M would be the acceleration exhibited by the charges. Now, given all that, is the schematic derivation here enough to give us Coulomb's Law?

Vetter 2009 thinks not, for the reason that S and M are both expressed as determinates and what Coulomb's Law expresses is a relationship between *determinables*. Given a determinate charge, such as that on an electron, and a determinate distance from another determinate charge, the law states not merely *that* a force will be experienced and an acceleration manifested; it states *how much* force and *what* acceleration (2009). In other words, it 'acts' at the level of determinables, yielding determinate manifestations when determinate stimuli are 'plugged in', as it were.

[19] Thus as Vetter notes 2009, this view is modal 'twice over': first because properties have essences in the sense of certain characteristics that they possess in every possible world, in virtue of being the property that they are; and secondly, because these properties have a dispositional character, involving some kind of relation to a manifestation that need not be actual.

Now, the question is, can Bird's derivation capture the appropriate relationship between determinables and determinates encapsulated in Coulomb's Law (2009)? It doesn't appear that it can.

We recall the S&M characterization. Bird notes that in the case of charge, or more specifically, that of having a determinate charge, what this characterization yields is the following: the stimulus of a charge q_1 brought to distance r_1 from the given determinate charge yields as the manifestation a force and associated acceleration a_1; the stimulus of a charge q_2 brought to distance r_2 from the given determinate charge yields as the manifestation a force and associated acceleration a_2; and so on. Thus charge must be understood as a 'multi-track' disposition (2007: 21–4) in the sense of being characterized in terms of the conjunction of different stimulus and manifestation conditions, corresponding to the many different determinates of the determinable. Now the next question is, which is more fundamental, the conjunction or the distinct conjuncts? For the dispositionalist, it is the conjuncts since these correspond to the distinct stimulus and manifestation conditions.

Now although this conclusion—that having determinate charge e, say, is not fundamental—seems surprising, it fits nicely with our derivation (Vetter 2009: 9). Substituting the specific values given in one of the distinct conjuncts yields,

$$\forall x((x \text{ has charge e \& is at a distance of } r_1 \text{ from charge } q_1) \rightarrow \\ x \text{ exerts a force resulting in acceleration } a_1)$$

But this is not Coulomb's Law. Thus the dispositional analysis yields not the laws but only instances of them. Each of these instances can explain something, namely the regularity associated with the specific stimulus and manifestation conditions, but clearly their explanatory power is much reduced compared to that of Coulomb's Law itself. Indeed, the crucial (meta-) regularity is now the similarity between these instances, which remains inexplicable if the only resources one has are the distinct dispositions constituting the 'multi-track' (Vetter 2009).[20]

How then is this meta-regularity to be explained? At the very least, it is unclear what the dispositionalist can appeal to. Certainly, she cannot appeal to Coulomb's Law itself, since this was supposed to be obtained by the derivation. But, of course, this is the easiest way of answering the question so we could take the connection between the instances of the law 'on board' and regard the conjunction as more fundamental than the distinct conjuncts (2009).[21] Now, what grounds could we have for doing so? Certainly if we follow the S&M characterization, there appears to be none. Alternatively we might seek such grounds in a view of determinables as just as fundamental as determinates and thus as yielding the requisite connection between the

[20] We might also recall again Katzav's argument that the Principle of Least Action cannot be obtained from a dispositionalist analysis of properties (Katzav 2004).

[21] As she says, this would be to take Bird's suggestion that the conjunction is 'natural' more seriously than he himself does.

instances of a law by taking these (qua determinates) to be entailed by the law itself. Although at this point, '[i]t is not at all clear what the metaphysical picture here is supposed to be' (Vetter 2009: 10), I shall shortly suggest one way of clarifying it via OSR.

Let's now move on to the second problem, which is that if one takes seriously what dispositionalism says about the fundamental nature of the relevant dispositions, then laws themselves cannot be fundamental elements of our metaphysical pantheon and should be eliminated. Indeed, we can see how this move might go from reflecting on the derivation given previously and the focus on specific dispositions as fundamental. Mumford has expressed the concern more generally as a dilemma arising from the role of laws as 'governing' the relevant properties (Mumford 2004).

9.5 Mumford's Dilemma[22]

Mumford couches the dilemma in the following terms: Laws are either external to, or independent from, the properties they supposedly govern or they are internal to, or dependent upon, these properties. In the former case, the identity of the properties that participate in the relevant law cannot be given by the role they play in that participation; that is, it cannot be given by their nomic role. This raises the issue of what then grounds the identity of properties. Introducing quiddities is both ontologically inflationary and increases the level of humility, as noted in Chapter 3. If, however, laws are understood to be dependent upon these properties in the manner implied by dispositionalism then they cannot be said to govern them. In that case the metaphysical status and role of laws becomes, at the very least, unclear. Hence the dilemma: if we introduce laws as metaphysically substantive and with a supposed governing role, then we must either accept quiddities or drop the governing role; neither option is palatable to the realist about laws.

Mumford's own resolution is to bite the bullet and take laws out of the picture altogether.[23] The regularities of the world are determined by modally informed properties that can be conceived of as bundles of dispositional powers, understood as ontologically fundamental and it is these that provide the relevant necessary connections. In a sense, the requisite modality 'flows up' from the properties, rather than down from the laws, which are hence not needed as metaphysically substantive entities and can be eliminated.[24] It hardly needs saying, this is quite a radical conclusion to draw!

Let us move on to the final problem that dispositionalism must face. As we shall now see, although Bird faces up to it squarely, his 'solution' is also radical and

[22] The following is taken from an early and extended draft of Cei and French (2014). I am grateful to Angelo Cei for agreeing to let me use this material.

[23] A move that he further justifies by pointing out the lack of a unitary conception of laws within science itself. A more nuanced approach to scientific practice undermines this claim (see Chakravartty 2007).

[24] For a response see Bird 2006; and for a counter-response, see Mumford 2006.

together with Mumford's, casts doubt on the viability of the stance as a whole. This is the problem of accommodating symmetry.

9.6 Dispositions and Symmetries

As Bird himself notes, from the perspective of the dispositionalist, it is mysterious why, in the manifestation of charge, for example, the total charge should remain constant. Given the S&M characterization, it is hard to see what it is about the disposition itself that ensures that charge is conserved. More acutely, perhaps, on the view that symmetry principles and conservation laws play a constraining role with regard to the standard or regular laws, such principles and laws raise an obvious problem for the dispositional essentialist. Put bluntly, she cannot accept such constraints, since she holds that the laws being constrained owe their necessity to the dispositional properties that ground them and so there is simply no metaphysical room for further constraints:

> Properties are already constrained by their own essences and so there is neither need nor opportunity for higher-order properties to direct which relations they can engage in. (Bird 2007: 214)

Bird's solution to this problem is as drastic as Mumford's with regard to the dilemma over governance: we should regard symmetry principles as 'pseudo-laws', that will eventually be written out of our scientific worldview. His argument is as follows: symmetries involve invariant quantities (that are conserved); the latter can be regarded as part of our theoretical background structures; the dispositionalist should be committed to the elimination of such structures, a stance that chimes with that of modern science;[25] hence, she should regard symmetry principles as eliminable features of our theoretical representations.

Given the significance of symmetries (and conservation laws) in modern physics, some might take this conclusion as a form of *reductio* of the whole dispositional essentialist enterprise. Furthermore, it creates a tension with Bird's own attempt to regard laws as metaphysically substantive, if symmetries are associated with laws (consider, for instance, the time reversal invariance of laws of motion, which is not displayed by phenomena such as billiard ball collisions). One might wonder how the

[25] Thus Bird takes space-time to constitute a form of background structure and suggests that General Relativity effectively dispositionalizes and eliminates it (2007: 161–6). Leaving aside the issue of whether dispositional accounts can appropriately capture spatio-temporal features of the world, one might question this relegation of symmetries to background structures on a par with space-time. Certainly, one could argue that it might be plausible to suppose that not all features of the world can or should be subject to a dispositional analysis. Consider the various physical constants, for example: these might simply be regarded as initial conditions that help define the kind of world we exist in. Having said that, as we shall now see, Bigelow et al. take these conditions to define the essence of 'the world', taken as an entity in itself.

dismissal of a substantive nature for symmetries can be reconciled with the attempt to retain such a nature for laws.[26]

And of course, if one were sympathetic to the motivation for a realist account of laws (via a form of Inference to the Best Explanation, say), and in particular as laws as governing, then it is hard to see how one could resist a similar motivation when it comes to symmetries. Just as with laws, we encounter certain regularities in nature—think of the hadron 'zoo' in elementary particle physics, for example—and the best explanation is given in terms of certain fundamental symmetry principles—as represented via group theory. Here too one can claim that the relevant symmetry governs the classification in elementary particle physics and if one were to hold that the identity of certain properties is given in terms of such symmetry principles, then one could mount a similar 'Mumford-type' dilemma for symmetries. In this case, Mumford himself would be on shakier ground in rejecting symmetries as metaphysically substantive, since, for example, there is not the diversity in such principles as he claims to find in laws (at least not at first glance). And the dispositionalist would presumably have to say that the symmetries supervene on the relevant powers in such cases and hence are internal to the relevant *property*, since they 'flow' from the latter's essence.

Alternatively, she might adopt the view of symmetries and conservation laws as 'by-products' of the 'regular' laws, arguing that the role of such principles as constraints reflects their heuristic role only and not their metaphysical status. In this case, however, she is going to have to show how such by-products arise from the dispositional grounding of the 'regular laws', by, for example, demonstrating how they can be accommodated within the derivation in section 9.4. However, it is not at all clear how that will be possible: what could it be about the S&M characterization that would allow

$$\forall x((\mathrm{P}x \,\&\, \mathrm{S}x) \rightarrow \mathrm{M}x)$$

to entail the relevant symmetry principle or conservation law? Indeed, we would again face the Vetter problem since at best all that the S&M characterization is going to give are the instances of the law, and the relevant symmetry will be a further meta-regularity going beyond that covered or described by the law itself!

Although the significance of symmetries has been noted by some dispositionalists, it remains a challenge to be taken up (Livanios 2010). Certainly, the prospects for some kind of accommodation look dim. Perhaps, however, this is going about the issue in the wrong way. Rather than seeking to derive symmetries from laws, taken to

[26] Indeed, one can turn Bird's strategy for defending dispositional essentialism against him: just as alternative views to dispositional essentialism are to be seen as deficient for failing to appropriately accommodate the nature of laws, their necessity, etc., as well as the overall success of science, so the dispositional essentialist account can be viewed and dismissed likewise for failing to appropriately accommodate the nature and role of symmetries in science.

be grounded in dispositions, perhaps we should try to obtain them directly, as it were, from some other kind of essential feature.

In this vein, Bigelow et al. (1992) have suggested that symmetry principles and conservation laws might be seen as deriving from the essence of the world as a whole, regarded as a kind (the only one of its kind, of course). Now on their view, laws derive from the essences of particular natural kinds, so regarding conservation laws as deriving from the essence of a very general kind might be seen as a 'natural' extension of this line. Furthermore, the particular essences may then be seen as contributing to the world-essence and in so far as they do that, laws derive from them. However, explaining regular (i.e. non-conservation) laws in terms of a world-essence would then be redundant (Livanios 2010), given that they are already explained in terms of particular essences. But if the latter are taken out of the explanatory picture, then the positing of such a world-essence seems acutely ad hoc.

It also takes this approach a significant step away from any grounding in science. This is made obvious in Bird's recasting of the proposal in terms of properties:[27] there is a property of 'being a world' and this has as its essence the disposition to conserve charge, etc. (2007: 213). Bird himself sees this as still ad hoc and furthermore notes that it does not account for local as well as global conservation (see also Livanios 2010).[28]

In effect then, the dispositionalist faces a further dilemma: if she adopts the first option and takes symmetry principles and conservation laws to be 'super-laws' or higher-order constraints on 'regular laws', then she must deny their significance in current physics. If, on the other hand, she takes the alternative, and regards 'the world' as a kind or property to which these symmetry properties can be referred, then she is committed to an explanation of the regular laws that is coarse and ad hoc.

In both cases, I suggest, the root problem lies with the adoption of an object-oriented metaphysics. Thus, from the ontological perspective, conservation laws must be treated on a par with the regular laws. Since these are understood and explained in terms of essences associated with kinds of objects, so the conservation laws must be explained as deriving from the essence of a kind, where there is only one entity of this kind. Or, on Bird's suggestion, just as the property of charge has as its essence certain dispositions and is instantiated by certain objects, so the property 'being a world' has the disposition to conserve charge and is instantiated by 'the world'. Taking the alternative, it is because the essence of a given property fully constrains that property that there is no metaphysical room to manoeuvre for higher-order constraints; if there were, then the relevant essence would not be

[27] Since he rejects kinds as the sources of laws and takes the latter to derive from the essences of particular properties.

[28] Furthermore, one might wonder whether 'the world' could be said to constitute an entity such that it forms a unique kind or possesses the property 'being a world' (or better, the property 'being the world' such that this has an essence); see van Fraassen 1995.

'particularized' but would depend on features beyond that property and its instantiation. In effect, allowing such constraints would introduce a holistic element and it is not clear why this should be barred, unless it is because of an underlying commitment to a particularized, object-oriented metaphysics. Dropping this bar would allow the dispositionalist to better capture the role and significance of symmetries in physics but, of course, at the cost of a major reorientation of her position.

Someone who is willing to pay that cost is Esfeld, who offers a kind of dispositional structuralism that by incorporating primitive modality in this form bears some resemblance to the view I shall ultimately defend. However, before presenting the latter, I shall also consider a less radical approach that also brings together dispositions and structures, but retains objects. This is Chakravartty's 'semi-realism' and it will provide a useful bridge to my view.

9.7 Dispositional Structuralism: Causal Structures

Esfeld explicitly takes himself to be spelling out the French–Ladyman idea that the fundamental structures of OSR should incorporate a form of primitive modality (2009). He does this by proposing that these structures should be understood as causal, building on a suggestion of French (2006: 181–2) which then gives rise to necessary connections in nature and accounts for the necessity of laws. Thus he argues that if the fundamental properties are conceived of as categorical, in the sense that their identity is not given by their causal, or more broadly, nomological, role, then one can obtain possible worlds that differ in the distribution of such properties instantiated in them but do not differ with regard to the relevant causal, or, again, more broadly, nomological relations. Given that our epistemic access to such properties is obtained via their causal role, there would be no epistemic—that is, discernible—difference between these worlds. Thus in order to account for the identity of properties we must invoke something other than their causal role, such as quiddity, again, but given the constraints of our epistemic access we cannot know that feature of the properties, requiring the adoption of a deep form of humility. A gap arises between metaphysics and epistemology which should be closed.

This is what the Dispositional Identity Thesis does, with the identity of the relevant properties tied to certain causal relations. Again, the laws 'flow' from the nature of the properties, and their necessity is grounded as before (Esfeld 2009: 5). However, the twin moves of avoiding quiddities and reducing humility lead to a radical shift in the dispositionalist stance: if the properties are regarded as needing external stimuli for exercising their powers, then we could obtain a situation in which we have properties of two different types being instantiated in the world but with that difference never being made manifest because the relevant stimuli never occur. To avoid such situations, it has been suggested that we drop the S&M Characterization and allow dispositions without stimuli (see also Esfeld and Dorato 2010).[29] In this case the relevant powers are no longer mere potentialities but actual properties[30] and the model to be thought of is not that of fragile vases or the dissolution of sugar in water but rather the disposition of radioactive atoms to undergo spontaneous decay or that of charged particles to generate electromagnetic fields (Esfeld 2009: 5).

However, it remains unclear how such cases illuminate what the dispositions are in such cases, or how we are to understand the sense in which such powers are modally informed. Furthermore, once one gives up the S&M Characterization it is also not clear whether one is entitled to avail oneself of the dispositional analysis or even call these powers 'dispositions'. At best it would seem that all one can say is that the relevant properties are modally informed.

Hence, it is not clear to me what thinking of structures as powers or dispositions brings to the metaphysical table. Certainly if the S&M Characterization is abandoned then it seems to me that one is left with a pretty minimal view that cleaves to a form of causal structuralism (see Hawthorne 2001) in holding that concrete structures are to be identified via their causal profile but does not expand on the 'source' of the relevant modality. For a radical dispositionalist like Mumford, that source lies in powers understood as the fundamental ontological category; for a more moderate dispositionalist like Chakravartty, that source lies in the properties, understood as dispositional as characterized by S&M.

Thus, Anjum and Mumford (2011) argue that dispositional modality should be thought of as involving a primitive, *sui generis* modality rather than the necessity usually associated with dispositional essentialism, so that causes, for example, have a disposition to produce their effects that lies somewhere in the metaphysical space between the contingent and the (physically) necessary (2011: 10).[31] Properties are still clusters of powers on this view, and it is their manifestations that grant powers their identity, although they may exist unmanifested. A given effect may be the result of many powers operating at once, and this encourages Anjum and Mumford to draw an analogy between powers and classical forces, to the point that they suggest that the vectorial representation of the latter can be employed with regard to the former. Thus they represent powers within a 'quality space' in terms of which they may additively combine (2011: 28). Furthermore, and crucially, it is due to this additivity that we

[29] A claim that is supported by the Ghirardi–Rimini–Weber interpretation of quantum mechanics according to which the wave-function undergoes spontaneous collapse; see Esfeld 2009: 10–12.

[30] In response to the concern that such properties need to be 'seated' in objects, one can repeat the line taken in the previous chapter and insist that there appears to be no obstacle to structures taking on this seating role (particularly given that when objects are put forward for such a role, they are typically conceived of in bundle or cluster terms).

[31] Their argument is basically that for causation to be necessary it should survive what they call 'antecedent strengthening', whereby the effect should still be produced when further factors are added to the cause; but of course, in many cases such further factors interfere with or otherwise offset the cause; for criticism of this feature of their account in particular, see Lowe 2012.

may derive explanations and predictions of the effect of several powers acting together.

However, as a recent commentator has noted, although this vectorial representation plays a 'key heuristic role' in the development of Anjum and Mumford's position (Glynn 2012), it is actually misleading since, unlike the case of forces, there is no common metric in terms of which the powers can be represented. In the absence of such, the proposed representation makes little sense. Furthermore, powers are typically interactive in ways that forces are not, in the sense that their contribution to a given effect may be determined by their interaction with other causal powers, so the overall contribution is not factorizable into distinct contributions attributable to different powers (Glynn 2012).

Can we strip away this unfortunate analogy? Certainly, but then it is unclear to me what remains of Anjum and Mumford's view except their primitivism with respect to powers and their claim that the latter's modality sits somewhere between the contingent and the necessary. Now, I am sympathetic to both these features of their view and it may be that it could be adapted to the structuralist stance, by drawing on the same sorts of considerations that I will present in Chapter 10.[32] But in that case, I suspect it will inevitably look more like the more moderate dispositionalist account, particularly if instead of just appropriating vector analysis, the results of modern physics are seriously considered.

Let me then turn to Chakravartty's position which, on the one hand, retains a robust notion of object, while on the other, incorporates the relational features of modern science in a way that not only situates his 'semi-realism' within the structuralist camp but also allows me to appropriate and invert the relationship between laws and properties in a useful and (hopefully) illuminating manner.

9.8 Semi-Realism and Sociability

'Semi-realism' takes the kinds of structures we should be realist about as concrete and conceives of them in terms of relations holding between first-order, causal properties of objects (Chakravartty 2007: 41):

> Causal properties are the fulcrum of semirealism. Their relations compose the concrete structures that are the primary subject matters of a tenable scientific realism. They regularly cohere to form interesting units, and these groupings make up the particulars investigated by the sciences and described by scientific theories. The continuous manifestations of the dispositions they confer constitute the causal processes to which empirical investigations become connected, so as to produce knowledge of the things they study. (Chakravartty 2007: 119)

[32] Perhaps yielding another example of going 'a-Viking' the other way, with metaphysicians improving their positions by drawing on some relevant physics.

Hence, it is not just that we can infer the natures of things from the structure but that the latter is encoded in the former (2007: 43). This is the case because the first-order causal detection properties should be understood in terms of dispositions for specific relations which comprise and are recognized as the concrete structures alluded to in Chakravartty's quotation.

On this account, laws and properties are just 'flipsides of the same coin' (2007: 147): the laws are effectively encoded in the dispositions conferred by the properties and the identity of the latter is given in terms of the laws they participate in. Mumford, of course, would insist that if the laws really are encoded in the dispositions then, qua metaphysically substantive entities, they should be dispensed with. Alternatively, if they are retained, as metaphysically substantive, over and above the properties and the dispositions they confer, then the metaphysical gap reappears and one may wonder what the relationship is between the laws and the latter properties and dispositions.

Now, it is not necessarily a feature of semi-realism that laws have a governing role and so Mumford's dilemma may be avoided (Chakravartty 2013). On this view, laws are simply relations between properties and convey no further modal force beyond that conveyed by the dispositions in terms of which the properties are to be identified. Nevertheless, the laws are 'distinct things' since a disposition for a relation is not the same thing as a relation. However, this seems to reinstate the metaphysical gap, with the attendant worries. More generally, it raises the concern that this form of dispositionalism inflates our ontology by including both dispositions and laws as distinct things. Mumford avoids this, as we noted, by eliminating laws and taking the dispositions, or powers, to be fundamental; I shall do the reverse: drop the dispositions and take the laws to be fundamental.

Furthermore, reinstating this gap illuminates the difference between modally informed structuralism and dispositional-based semi-realism. Consider again Coulomb's Law (again, we'll restrict ourselves to the classical context, since any shift to quantum mechanics or relativity theory will only serve to aid the structuralist further[33]): both positions can agree that it expresses a set of relations between properties. However, the structuralist will take these relations to be always there, in the world, as it were, or always and continually 'manifested', to adopt dispositional language, whereas the dispositionalist sees charge as having the disposition to enter into these relations. If we remain at the level of the phenomenology of this, the actual, world, there appears to be no grounds on which such a difference could be conclusively established. It emerges when, again, we shift to a sparse world: when the dispositionalist conceives of a world containing a single charged entity she takes that property, charge, to be identified in terms of the dispositions that would be

[33] So consider mass. If, in the classical context, it at least makes preliminary sense to regard a mass as having the disposition to yield the relations expressed by the Law of Universal Gravitation, it is not clear that it does in the General Relativistic context where we have the tight association with space-time.

manifested were an appropriate stimulus to be present, namely another charge. These dispositions are dispositions *for* certain relations that do not exist as 'distinct things' in that single-object world. The structuralist, on the other hand, either rejects such a world as a mere chimera of sorts or adopts something like the Humean structuralist's manoeuvre. In the latter case, the structure is there 'in the world' as it were, and the charge is an instantiation of it, although problems arise as we have seen. Of course, if we move to a two-charge world, then the relevant relations are manifest in either case, but again there is nothing in the phenomenology of such a world that would ground the aforementioned difference.

Again, we might recall our concerns over how the notion of intrinsic property is tied to a conception of lonely objects existing in such sparse worlds and the worry whether the property that is called 'charge' in this one-object sparse world would have any of the features that we standardly attribute to charge (in particular such as its role as the generator of the U(1) symmetry of electromagnetism, with an associated conserved current).

Perhaps this conflates epistemology with metaphysics again:

> On the dispositional view, a particle's charge is something it possesses independently of its interactions with test charges—that is a metaphysical proposal. *How one comes to know* its charge is another matter, and may well require experiments (either real or in thought). To think that the relations manifested in such experiments somehow 'make' charge the property that it is, however, is once again to beg the question. (Chakravartty 2013)

However, one person's conflation is another's tailoring and at this level, it may become unclear to third parties who is begging what against whom! But the crucial issue, for me, concerns the justification for the metaphysical methodology that underpins the dispositionalist stance: we have epistemic access to charge in this, the actual, world via various interactions and on that basis (mediated by theory and experiment in the 'usual' way, whatever that is), we attribute certain features to it. The dispositionalist (and others) constructs a sparse world with a lonely charged object and assumes that the property has those same features. But recall our previous considerations: if the world is constructed by stripping away properties and features of this world, then it is not clear that one will automatically retain all the features of charge that one wants through such a process. On the other hand, if these relevant features are simply stipulated, then it is also not clear on what basis one can simply assert that charge is a generator of the U(1) symmetry for example. But the dispositionalist needs some such basis if she is to be able to assert that the property she is attributing in the lonely world is indeed that which we call charge. And it is only if she can do that, that we can establish the difference between her and the structuralist.

The latter of course insists that we do not have a conflation of epistemology with metaphysics when she insists that one cannot posit charge in the absence of that context in terms of which we can attribute to it those features that she will regard as ultimately structural. Instead, she claims, we are bringing our metaphysics into line with our epistemology and given the lack of justification for the kind of metaphysical methodology that underpins the dispositionalists' position, together with the requirement to reduce our level of metaphysical humility as much as possible, it is this form of tailoring we should adopt.[34]

Returning to Mumford's dilemma, one could mind the gap between laws and dispositions as follows: epistemically we identify the relevant relations via the laws but metaphysically such relations are to be conceived of as arising from the dispositions in terms of which we identify the relevant properties. However, it is hard to see how laws can be conceived of as distinct things on such a view and the difference between this and Mumford's view may seem tissue-paper thin. Another way of avoiding this kind of dilemma would be to adopt the kind of structuralist perspective on laws that sets them as primary and the relevant properties as emergent, in a sense to be explicated shortly, and semi-realism can be metaphysically 'reverse engineered': if what makes a causal property the property that it is, are the relations that it enters into with other such properties, with the conjunction of the laws comprised by these relations specifying the natures of all the causal properties there are, then reading this identity chain from right to left, as it were, and ontologically, we can take the laws (understood structurally of course) as fundamental and the powers and properties as emergent from the relevant relations, with no need for dispositions.

Indeed, the kind of holism that semi-realism entails with regard to the natures of causal properties meshes well with a structuralist stance (Chakravartty 2007: 140). As in the case of the natures of objects presupposed by ESR, the natures of properties get cashed out in relational (and hence structural) terms. And the network of properties and relations comes as a package or not at all (2007: 147).[35] If properties and laws are merely flipsides of the same coin, then we can take the law side as ontologically basic, as a straightforward reading of the physics would suggest anyway.[36]

[34] Chakravartty draws a contrast between the kinds of relations involved with charge and those we have with our neighbours, say. In the latter case, if we were to move houses, the relations would change or disappear altogether, but physical relations are different. So, he asks, 'If one were to take the test particle away (from the world, even), would the subject particle no longer have charge?' (2013). But there is a difference between changing things within a world and changing worlds. In the former case, moving the test charge a long way away does not remove the relation with the given charged object, because of the nature of the electromagnetic interaction (as expressed classically by Coulomb's Law). It is the latter that justifies the claim that the relation does not 'disappear' in some sense. However, when we shift to a different world, one in which the test charge does not exist, it is, again, not clear what justification there could be for maintaining that the property that we are attributing to the lonely object still counts as 'charge'. Again, the dispositionalist is appealing to an intuition that may not be so well founded.

[35] Psillos (2013) worries that this holism is pernicious in that no property can be identified until all the properties it is related to are identified but then no properties can be identified at all. Chakravartty (2013) points out that we can draw on the distinction between epistemology and metaphysics that Psillos himself accepts and take a given property to be distinguished epistemically, while accepting that metaphysically the identity conditions for such properties are holistic.

[36] In addition, given that the dispositions are genuinely occurrent, if their manifestations are causal processes, then as with gravitational capacity, when it comes to the most fundamental dispositions, the question arises: do we really need both the causal power *and* the process it manifests?

What about 'ceteris paribus' laws? Being able to give an account that sets these on a par with 'traditional' laws is typically taken to be one of the advantages of the dispositional view.[37] If ceteris paribus law statements are accurately formulated to describe causal laws, then they can be understood as 'partial maps' of relations (2007: 149). This can also be made consonant with the structuralist perspective, although instead of saying that they hold partially because in formulating them one does not specify all of the potentially relevant dispositions, one would have to say that—ultimately—they so hold because in formulating them one does not specify all the relevant features of the fundamental structure. Thus both the semi-realist and the structuralist can agree that, if correct, ceteris paribus law statements are accurate descriptions of possible relations (2007), but whereas the former takes these relations to be between causal properties, the latter takes them ontologically *simpliciter*, and the sense of possibility here is grounded, not in dispositions and their stimulation under different conditions, but in the modal abstraction of certain aspects of structure.

The crucial difference between semi-realism and OSR remains the status of objects (Chakravartty 2003b). For the semi-realist, objects fill the fundamental role of acting as the 'seat' of causal powers. In the previous chapter I have tried to respond but here I would like to suggest that, again, certain particularist assumptions might be at work behind the scenes: if one thinks of powers as the sorts of things that can be brought into play, or manifested, under certain conditions, where those conditions are typically articulated in particularist terms—that is, in terms of objects interacting with other objects—then it becomes natural to see these powers as grounded in the objects. Natural, but not necessary.

In addition, as we discussed in Chapter 7, properties cluster: we see a certain charge, observed at a particular spatio-temporal location, always associated with a certain (rest) mass and a certain spin and we infer that we have observed an electron, for example (Chakravartty 2007). The explanation of this coherent clustering in terms of the 'sociability' of properties offers the possibility of accommodating symmetries within the dispositionalist account.

Thus, consider natural kinds, which we shall take to cover both what are traditionally seen as 'essence' kinds, such as those we find in elementary particle physics, and also so-called 'cluster' kinds, associated with biological species, for example. Instances of the relevant properties are not randomly distributed across space-time; rather, they cluster together, a phenomenon that is described in terms of 'sociability'. Furthermore, sociability comes in degrees:

[37] Drewery argues that certain ceteris paribus laws are not subject to this type of account, namely those that state that other things being equal a member of a kind is like other members in possessing a certain property (Drewery 2001). However, Chakravartty understands the traditional counterparts to such statements as 'definitional generalisations' and one could extend this to treat Drewery's examples as ceteris paribus descriptions of objects (which as such would not apply to the fundamental objects of physics anyway).

The highest degree of sociability is evidenced by essence kinds, where specific sets of properties are always found together. In other cases, lesser degrees of sociability are evidenced by the somewhat looser associations that make up cluster kinds. (2007: 170)[38]

A crucial question is whether this notion of 'sociability' should be taken as primitive or not. It can certainly be analysed further in the case of cluster kinds, through Boyd's 'homeostatic clustering', for example. However, such mechanisms are not robust enough to account for the co-instantiation of the so-called intrinsic properties of elementary particles and in the case of such 'essence' kinds, sociability must be admitted as a 'brute fact' (2007: 171).

Now clearly this notion captures an important feature of the physical world and by introducing it semi-realism scores over Mumford's account, for example, which allows for entirely promiscuous property clustering. Crucially, it allows the dispositionalist to accommodate symmetry, at least in so far as this contributes to the grouping of entities into kinds, such as fermions and bosons, for example, as well as the sub-kinds of leptons and quarks, with the former including electrons, muons, tauons and their associated neutrinos, and so on, all captured via symmetry considerations, as represented by group theory and, in particular, the (restricted) Poincaré group or, better, the associated Lie algebra.

Nevertheless, it remains unclear what explanatory work the notion of sociability is doing (a similar concern arises with regard to 'compresence' and 'foundation' as outlined in Chapter 7). If the explanandum is the clustering of certain fundamental physical properties (forming both kinds and objects), then what the brief outline here suggests is that the relevant explanans is the appropriate symmetry consideration, as represented group-theoretically. One might argue that this gives us only the physical explanans as it were, and that what we need is the appropriate metaphysical correlate. But then 'sociability' seems merely to label the phenomenon itself,[39] taking it close to functioning along the lines of 'dormative virtue'. But perhaps that is unfair: the semi-realist could legitimately respond that both sides agree that the explanans of the physical explanation is the relevant symmetry principle. What is at issue is the further metaphysical explication of that explanans; for the semi-realist it is sociability, for the structuralist it is taking the symmetry to be a fundamental, perhaps primitive, feature of the structure of the world. Put like that, it is not clear which account has the advantage.

One might claim that structuralism extends further than an approach based on sociability, since the latter is primarily concerned with the sociable nature of *properties*, and that's not what we seem to have in the case of fermions and bosons, at least not the so-called intrinsic or state-independent properties such as mass, charge, and spin. If we can be said to have properties at all, they are those of the relevant

[38] In either case, it is the sharing of causal properties that underwrites the relevant inductive generalizations and predictions.

[39] Like its trope-theoretic kin of 'founding' or 'saturation', perhaps.

aggregates, reflected by the appropriate quantum statistics. However, the spin-statistics theorem does relate these latter properties to one of the intrinsic set just listed; the issue then is whether one can take the statistics and related distinction into kinds as a result of a purportedly more fundamental distinction between half-integral and integral spin (see French 2000c). There are continuing debates over what counts as an adequate proof of this theorem (see Berry and Robbins 1997; Sudarshan and Duck 2003), but the semi-realist can certainly use it to argue that the reach of symmetry does not extend further than that of causally empowered properties.

Nevertheless, the advocate of OSR can insist that the explanatory arrow runs the other way, and that the aforementioned symmetry considerations can be said to explain sociability, in the sense that embedding this notion in a structuralist metaphysics provides a 'deeper' metaphysical explanation—not least because these considerations reflect a fundamental aspect of the structure of the world. Thus OSR yields a unified account of the sociability of particulars and kinds, citing the relevant symmetries and group-theoretic features as explanans in both cases. In this sense, OSR offers a stronger explanatory framework than that based on object-oriented metaphysics (French 2013; cf. Chakravartty 2003b).

There is yet a further, broader, dimension to this issue of explanation, however. Here I am concerned with physics, but semi-realism is explicitly intended to apply to other scientific fields, including, significantly, biology. In such fields we do not find anything like symmetry or its correlates in terms of which we can articulate a structuralist grounding of kinds. Now one could argue that what this shows is precisely how fluid and ungrounded such putative kinds are in these fields but Chakravartty can press the case that here the notion of sociability can do some useful work, not simply in grounding whatever kinds are appropriate for a given domain—since then one could simply insist, again, that it is some feature of the entities of that domain that provides the relevant grounding for the grouping into kinds and hence that sociability is again redundant—but rather in providing a trans-domain understanding, the details of which are domain-specific.

Thus in considering this issue of explanation, Chakravartty notes that there are two explananda one might consider. The first has to do with the grounds for kinds grouping within a domain, and here he admits sociability does no work (2013). But the other concerns the success of the inductive practices we find in science, particularly concerning generalization and prediction, and in explaining this success, sociability serves as part of the explanans:

> [E]ntities behave in certain ways in certain circumstances as a function of the (causal) properties they possess; therefore, the greater the extent to which the members of a class of entities share (causal) properties, the greater the success one should expect of inductive generalizations and projections over their members. (2013: 50)

Since sociability offers a measure of the degree to which properties are shared, it is directly correlated with the relevant measure of success and thus helps explain it.

In the latter sense, sociability functions in the way we might expect metaphysical or more broadly, philosophical, explanans to do—indeed, perhaps it is the only way they can function—namely as umbrella terms that span different scientific domains. As such they may offer a useful framework for comparing the accounts specific to the separate domains. So, in the case of explaining the clustering of properties into kinds, one might claim that the degree of sociability is higher in physics than in biology, say, where we don't have the symmetries we find in the former to underpin the kind classifications. When it comes to the former, at least, sociability needs to be supplemented (if not supplanted) by a structural understanding (French 2013).[40]

Whatever stance one adopts on the status of sociability, introducing these considerations allows us to respond to a problem that arises for those views that take particulars to be 'bundles' of properties (as semi-realism does; see Chapter 7): this is the problem of 'free mass' (Schaffer 2003). On such views, the question arises as to what prevents one or more of the properties—mass, for example—'breaking free', as it were, from the others, so that we have a one-property 'bundle' of, in this case, just mass. The possibility of such a 'free mass' has been taken as a *reductio* of the bundle view of particulars and a typical response is to introduce certain property interdependencies such as 'founding' or 'saturation' relations. To these one could add sociability. Schaffer rejects such metaphysical interdependencies on the grounds that they involve 'occult' and 'brute' necessities; that there is no plausible way to specify exactly *which* interdependencies hold, and that it seems possible to obtain respectable property-clusters in ways that preclude interdependence (2003: 132). I am not particularly bothered by the first or third issues, since it is not clear to me how a metaphysical necessity that is underpinned by its physical counterpart counts as 'occult' or 'brute'; nor, relatedly, do I think the kinds of combinations Schaffer constructs correspond to anything we find physically. But we can specify which interdependencies hold, at least at the level of physical relationships, in precisely the ways indicated earlier.

The conclusion reached is:

> An explanation for why properties cluster remains elusive. All attempts to explain the impossibility of free masses, whether in terms of the relation between *object* and *property,* or in terms of principles internal to *property,* look to fail. Perhaps it was all along a mistake to think of free masses as impossible. (2003: 133)

Again the issue of what counts as an adequate explanation arises. Perhaps the difficulty here is that if one looks only to metaphysics for such an explanation then there is nothing for whatever principle one proposes—founding, saturation, sociability, etc.—to get any purchase on. And insisting that free properties are impossible because of some object–property or property–property relation that ties them into a cluster looks like little more than repeating the insistence. The alternative is to

[40] For an attempt to extend OSR into biology, see Chapter 12.

articulate an explanation on physical grounds and then embed the explanans in an appropriate metaphysical framework.

This then allows us to account for the contingent lack of free properties (2003: 134), and in a clearer way than the analogy suggested with quark confinement. It also allows a response to the 'positive' argument for free masses, based on subtraction: for any particular, we can construct a 'near twin' sub-duplicate of it by subtracting any property, following which 'it seems clear' that we still have a particular.[41] Hence we obtain the 'generalized subtraction premise': for any n-propertied object, it is possible for there to be an n-1 propertied sub-duplicate (2003: 136). This is taken to garner further support from the apparent lack of necessity of any particular property and we are now merely iterations away from a free mass (or any other property). The problem of course is that although this argument applies to particulars regarded in an abstract, metaphysical sense—namely as clusters of properties—it does not apply to the contingent particulars of the actual world. Here, if one strips away charge, for example, from a certain cluster, one no longer has an electron;[42] and more importantly, recalling what was said previously, only certain clusters make up the kinds and particulars we observe around us, as determined by the relevant symmetry considerations. Perhaps one might counter that what we are concerned with here is the notion of particular in the abstract or metaphysical sense and there is nothing in that notion that blocks the subtraction argument. Well of course, but that is because of the kinds of considerations already noted: at this level of abstraction there is nothing that could act as such a block that doesn't have the look and feel of something akin to a 'dormative virtue'.

9.9 Conclusion

To finish, then: dispositions, as articulated in the context of modern physics, are tied to a problematic metaphysical picture and I hope to have indicated how robust disposition talk is at best unmotivated and at worst undermined by considerations from this context. More positively, semi-realism's metaphysics is in better shape than other forms of dispositionalism, when it comes to the accommodation of symmetry principles in science. Nevertheless, if 'sociability' is to function as less of a metaphor and more as a metaphysical explanans, then it needs to be further articulated, and the most appropriate way of doing that, I suggest, is through structural considerations. Certain of these already lie at the heart of semi-realism and what I shall suggest is that this commitment be thought of as a bridge to OSR, with the relationship between properties and laws inverted, as indicated earlier, to give a modally informed structural realism. That will be the topic of the next chapter.

[41] As Efird and Stoneham note, this 'clear' intuition is surely question-begging (2010).

[42] Efird and Stoneham make a similar point: if we take a certain object such as a post box then it is not at all intuitive that there could be a 'near twin' *of a post box*, yet lacking colour and shape. Their conclusion is that Schaffer's argument is either invalid or fails to be independently suasive (2010).

10

The Might of Modal Structuralism

10.1 Introduction

Having looked at some of the more prominent alternatives for accommodating causation, in particular, and modality, in general, within a structuralist framework, let me finally turn to the kind of account I favour. In short, it takes the structure to be 'inherently' modal. Much of this chapter will be spent trying to spell this out, in terms of how we should understand laws and symmetry principles and with due regard to how the structuralist should view 'the actual' and its relationship to 'the possible'. I shall also consider how this account compares with Lange's view of modality as manifested in primitive subjunctive facts and with Maudlin's primitivist account of laws.[1] But let me begin with some motivation for my account.

Modality has to 'fit' somewhere in our metaphysical picture, unless you're a Humean of course, in which case I refer you to the concerns outlined previously. As Vetter nicely puts it:

> Anyone who does not either deny modally loaded facts about the world, or outsource them to real other worlds—anyone, that is, who thinks that counterfactual or law-like, counterfactual-supporting statements are true in virtue of something in the actual world—has to include unrealized possibilities in actuality. (2009: 6)

The issue then is where to place this modality; or, alternatively, what is this 'something' in the actual world in virtue of which counterfactual or counterfactual-supporting statements are true? Dispositionalism offers one answer but I have already noted the problems it faces. Given that, one should certainly not take it for granted that accepting possibility in actuality means accepting that our fundamental properties are endowed with modality, or that that endowment is dispositional in character (cf. Vetter 2009: 5).

Nevertheless, I am sympathetic to the dispositionalist's overall strategy: select that which is taken to be fundamental in one's ontology—dispositions or powers in this case—and take them to be inherently modal in such a manner that they can effectively 'endow' the relevant properties with that modality. Indeed, I think that

[1] For further argument that realism involves a commitment to 'objective modality', see Berenstain and Ladyman (2012).

by reverse-engineering dispositionalism, we are led naturally to the structuralist view. So, whereas the dispositionalist takes the laws to arise from or be dependent in some way upon the properties (giving rise to Mumford's dilemma, of course), I shall invert that order, taking the properties to be dependent upon the laws and symmetries. Because of this inversion, I *have* to relocate the modality, shifting it along the line of dependence from the properties to the laws and symmetries themselves. Furthermore, instead of creating a new fundamental ontology of dispositions, thereby inflating our commitments and increasing our level of humility, I shall retain that which we read off our theories—laws and symmetries—and invest *those* with the requisite modality.

So, if you are already inclined towards that dispositionalist strategy, then given the role of symmetries in physics, and the arguments motivating OSR, in particular when it comes to keeping your level of humility low, I would hope that you would follow the inversion and shift the modality to the laws and symmetries. Of course, you might still have qualms about regarding these as 'inherently' modal, but I hope to dispel these by the end of this chapter.

10.2 Laws, Symmetries, and Primitive Modality

Let's recall the central issue:

> The structures to which ontic structural realism is committed have been conceived...as including a primitive modality...However, it has not been spelled out as yet what exactly that modality consists in. (Esfeld 2009: 179)

So, let me try and spell it out. I'll begin with the Cassirerian vision sketched in Chapter 4: law statements express the network of relations, 'held together' by the symmetry principles which represent what is invariant in the network. How this holding together is effected obviously depends on how one conceives of the relationship between laws and symmetries. It might seem that views according to which symmetry principles act as constraints on laws, or, at least in some cases, directly yield laws, are more amenable to this idea of holding together than the view which holds that symmetries are merely 'by-products' of laws. However, even in the latter case, one might argue that symmetries can fulfil the same role even as by-products, or more precisely, higher-order features of laws, since, as such, they are able to span, as it were, the laws and thus function to 'tie' the structure together. Attempts to adjudicate between these views typically involve appeal to features of scientific practice which are not, in fact, decisive. Given this, and for other reasons already discussed, I shall continue to maintain the Cassirerian line that there is a kind of 'reciprocal interweaving and bonding' between laws and symmetries and that it is this that yields those entities formerly known as objects.

We also recall how laws and symmetries do this. Kinds of (putative) objects are given to us by the relevant symmetries in terms of what Chakravartty calls 'sociable' clusters of properties. Thus, beginning with the fundamental (or close to fundamental) kinds associated with quantum statistics, namely bosons and fermions, this 'natural' distinction is cashed out structurally, in terms of the symmetry features of the relevant wave-functions. These are encoded group-theoretically in the principle of Permutation Invariance (PI) and we have already considered how this might be conceived as a kind of initial condition, imposed on the structure of Hilbert space, or as a consequence of global Hamiltonian symmetry given the group structure of the particle permutations.[2]

Moving down the hierarchy of fundamentality, as it were, the symmetries associated with the Poincaré group yield further properties such as mass, charge, and spin (via the associated sub-groups), forcing a review of their metaphysical status as 'intrinsic'. And with the laws encapsulating the relations between tokens of these properties we have the kind of dependence of the latter on structure that I discussed previously. We can illustrate the kind of interweaving and bonding Cassirer had in mind with the case of spin which effectively 'drops out' of the Dirac equation, in the sense that its existence is required by both the mathematical formalism of the Dirac equation and group theory in order to guarantee conservation of angular momentum and to construct the generators of the rotation group (cf. Morrison 2007: 546). The equation itself can be understood as a consequence of the existence of spinor representations of the orthogonal group SO(4, C). I shall discuss the relationship between groups and their representations in this context, shortly.

The question now is how do modality/possibility and actuality enter this picture? And furthermore, does the accommodation of actuality require the introduction of a non-structural element that is fatal to the project?

10.3 Symmetries and Modality

10.3.1 Kind distinction$_1$: Bosons and fermions

So, as already noted, this kind distinction emerges as a consequence of PI, and the particle statistics are given by the action of the permutation group on Hilbert space, dividing it up into sub-spaces corresponding to the irreducible representations of the group: the symmetric bosonic representation, the anti-symmetric fermionic and the plethora of paraparticle representations.

It is in that last feature that the presence of possibility lies: all the kinds of possible particle statistics, and therefore all the possible kinds of particles (at this level of the kinds hierarchy) are encoded in PI. Of course, it appears to be a fact about the actual world that only the bosonic and fermionic representations are needed and here we

[2] As already noted, and as we shall consider in more detail in the next chapter, in the context of quantum field theory (as conceived of from a certain perspective), Permutation Invariance arises naturally as a kind of gauge symmetry (Halvorson and Müger 2006).

can recall Weyl's famous statement (made before the relevance of the bosonic representation was fully appreciated):

> The various primitive sub-spaces are, so to speak, worlds which are fully isolated from one another. But such a situation is repugnant to Nature, who wishes to relate everything with everything. She has accordingly avoided this distressing situation by annihilating all these possible worlds except one—or better, she has never allowed them to come into existence! The one which she has spared is that one which is represented by anti-symmetric tensors, and this is the content of Pauli's exclusion principle. (1950: 288)[3]

Thus actuality is also included, as it were. I'll come back to the issue of how we can understand the relationship between possibility and actuality in this specific context.

One could regard the further possibilities represented by paraparticles as just so much surplus structure and attempt to exclude them on various grounds but this may not be methodologically prudent (recall the discussion in Chapter 3): parastatistics played an important role in the early history of quantum chromodynamics (French 1995) and anyon statistics have proved useful in getting a theoretical grip on the quantum Hall Effect (Stern 2008).[4] Excising these possibilities as surplus to requirements and adopting a theoretical framework in which they fail to appear may then reduce the heuristic resources available to us, to the detriment of scientific progress.

Now, it is too quick to say that the kind structure of the actual world and the 'space' of physical possibilities is given simply by PI. We need both the group representations[5] and the dynamics. The former underlie the division of Hilbert space into the relevant sub-spaces. Formally a representation of a group G is any group composed of mathematical entities which is homomorphic to G (or, a representation of G is a homomorphism from G to the automorphism group of G on some object). When those entities are linear operators in a (n-dimensional) vector space (the object of the automorphism group) we obtain the linear representations that are so useful in quantum mechanics. In terms of this vector space, a representation is said to be fully reducible if this space can be decomposed into the direct sum of sub-spaces (of dimension less than n) which are invariant under all transformations of the relevant group. A basis for the entire space can then be formed from the sets of basis vectors spanning each sub-space. If the space cannot be decomposed into sub-spaces of representations with lower dimensionality then the representation is said to be irreducible (see, for example, Hamermesh 1962[6]). The representations corresponding

[3] See also Weyl 1931: 238 and 347.

[4] Here the relevant group is the braid group; see French 2000c.

[5] As McKenzie has argued, OSR has been identified with a kind of group-theoretic structuralism that, at the very least, seems to downplay the role of the representations; here I want to correct that impression.

[6] In the case of the symmetric group associated with PI, various features of these irreducible representations can be obtained and presented in graphical form by the method of Young's Tableaux. A useful (well, I think it is) introduction to these can be found in French 1985 (appendix) or more recently in Butterfield and Caulton 2012.

to Bose–Einstein and Fermi–Dirac statistics respectively are just such irreducible representations.

One might be tempted to see in this 'realization' of the group by representations a way of getting a handle on the distinction we've been touching on in the past few chapters, namely that between the abstract and the concrete. Indeed, these are terms that are often deployed in this context, with G taken to be abstract and the homomorphism group as concrete. But nothing massively metaphysically significant, I think, hangs on that deployment, for when it comes to the representations we're still talking about mathematical entities—vectors in Hilbert space, etc.—and the best we can say, perhaps, is that they are closer to the concrete, in Chakravartty's sense (see the previous chapter). Of course, the critic may not be assuaged, pointing out that in so far as the structures that are the ontological focus of OSR incorporate the group G (as they must, else one can't make sense of the group representations) then they are still significantly abstract and in a way that raises the concerns outlined in Chapter 8.

There are two ways one can go here. One could accept that last point and insist that the structures one is realist about are those presented via the representations and that the group G is just so much mathematical descriptive resource to which no ontological significance should be given. This will require some deft manoeuvring when it comes to drawing a clear line between the group G and its automorphism group, as we saw in Chapter 6. Alternatively, one could remain committed to 'the structure of the world' as informed by, or presented via, the group and the associated representations, taken together as providing the presentational resources as discussed in Chapter 5.[7] This would mean biting the bullet when it comes to the inclusion within the ontology of OSR of something that *appears* abstract, but I think that particular bullet has to be bitten when it comes to the first option as well, and, furthermore, it's not clear to me that biting it will cause much pain anyway, given what was said in Chapter 8.

Also, in so far as the structure we are discussing is that of the actual world—so, the representations are those of Bose–Einstein and Fermi–Dirac statistics—there is no shift in modality when considering representations. As I have just said, these only make sense in the context of the relevant group G, so it is not as if the actual is solely covered by the representations, or equivalently, that the latter fully cover the extent of the former. Nevertheless, there is of course a sense in which G yields the possible—precisely that sense indicated already, where possible representations can be yielded via G, or possible particle statistics are 'encoded' in the symmetry PI. Later on I shall try to articulate this sense in terms of the actual as being constituted by determinables (represented in this case by PI, or presented by the group) and determinates

[7] Bryan Roberts has suggested in discussion that the representations can be thought of as properties of the group, which would mesh with the line taken in this chapter.

(represented by the specific kinds, bosons and fermions, or presented via the associated irreducible representations).

Of course, this still leaves the question of why we only observe or encounter bosons and fermions in this, the actual, world, when there are all these other possibilities encoded in PI. This is typically answered within the context of consideration of quantum mechanics by imposing the so-called Symmetrization Postulate (SP) which restricts the state vector for the given assembly of particles to the bosonic or fermionic sub-spaces (see Messiah and Greenberg 1964, for the classic discussion of the relationship between PI and SP; see also French and Rickles 2003). But this is ad hoc and simply pushes the explanatory task back one step—how, then, are we to understand such an imposition?

Previously I have suggested that PI can be understood as a kind of initial condition that determines the accessibility of certain states (French 1989; French and Rickles 2003).[8] To see this, we recall that, with the relevant Hilbert space divided up under the action of PI, the symmetry of the relevant Hamiltonian ensures that once a particle is 'in' a sub-space associated with an irreducible representation, it cannot get out of it. So, bosons will always be bosons, fermions will stay fermions.[9] Without this, the distinction between these kinds would not be so robust and the relevant structure would not be 'fixed' in the way it is via the action of PI. Hence PI imposes a restriction on the states of the assembly such that once a particle is in a given sub-space, the others—corresponding to other symmetry types—are inaccessible to it.

This idea of restrictions being imposed on the set of states accessible to a system is nothing new of course: it can also be found in classical statistical mechanics, where it is the energy integral which imposes the most important restriction. What PI represents is an additional constraint or initial condition and in particular, the symmetry type of any suitably specified set of states is an absolute constant of motion equivalent to an exact uniform integral in classical terms (see Dirac 1958/1978: 213–16). Of course, some may wonder whether this actually sheds much light on the status of PI, since it appears to leave it standing as a kind of 'brute fact' but at least it is no more brutish than the other, classical, constraints. Certainly, however, when it comes to the question why we only see bosons and fermions in this world, perhaps the only answer we can give is, that's the way the world is!

Does either that answer or the broader claim that PI acts as a kind of initial accessibility condition on the set of states introduce a non-structural element sufficient to undermine OSR? Clearly not. When it comes to the fact that this world is a bosonic and fermionic one (and not a para-bosonic and para-fermionic one, say), this fact itself can be articulated in structuralist terms, as I am indicating here. Both the very distinction between bosons and fermions and the manner in which this

[8] This characterization was originally given in the context of accommodating the view of quantum particles as individuals; see French and Krause 2006.

[9] This may not be the case for paraparticles under certain circumstances (French 1987).

relates to the relevant representations of the permutation group can be accommodated within the structuralist programme. As for the role of PI, the principle itself is regarded as a feature of the structure of the world and describing this role in terms of its acting as a kind of initial condition imposed on the world is simply a way of saying that the structure of the world is like *this*, rather than *that*, where 'that' will refer to worlds (not, presumably, physically possible) where the states of an assembly of particles are not so constrained.

Still, one might wonder how PI compares to other symmetries and also and relatedly, still have concerns that it does not arise from or relate to other features of the structure of the world in a smooth or 'natural' way. In response, recall Huggett's proposal (1999b) that it be regarded straightforwardly as on a par with rotational symmetry, for example. Now, of course, as Huggett acknowledges, the two symmetries are very different,[10] but, nevertheless, PI is implied by the conjunction of 'global Hamiltonian symmetry' which holds that the relevant symmetry operator commutes with the relevant Hamiltonian[11] and which space-time symmetries also obey, together with the formal structure of the permutation group. With regard to the latter, permutations of a sub-system are permutations of the whole system and this 'global Hamiltonian symmetry' very straightforwardly implies PI, without any additional assumptions concerning the structure of state space (Huggett 1999b: 344–5).[12]

Hence, Huggett concludes,

> we should view permutations in a similar light to rotations: we should not take [Permutation Invariance] as a fundamental symmetry principle in order to explain quantum statistics. Instead we should recognize that it is a particular consequence of global Hamiltonian symmetry given the group structure of the permutations. Further, if we accept the similarity of permutation and rotation symmetry, it becomes natural to see quantum statistics as a natural result of the role symmetries play in nature. (1999b: 346)

However, as he acknowledges, Permutation Invariance only follows from his general symmetry principle given the particular structure of the permutation group. So the issue of the status of PI is pushed back a step: what is the status of the structure of the permutation group? Or, to put it another way, why should that particular group structure be applicable?

[10] A quantum system of the kind we have been considering is not just covariant with respect to permutations but invariant: permutations are not just indistinguishable to appropriately transformed observers but to all observers.

[11] What we take the relevant Hamiltonian to cover is crucial here because, as Huggett acknowledges, the principle would appear to be violated in the case where, for example, we have a non-central potential term in the Hamiltonian of an atomic system, but, he insists, the symmetry is restored if we consider the 'full' Hamiltonian of system plus field, which does commute with the operators of the rotation group. As he points out (1999: 345), if observers are taken to be systems too, this symmetry principle is equivalent to covariance for space-time symmetries.

[12] It does, however, assume that the system being measured and the measurement apparatus are composed of the same indistinguishable particles, otherwise the Hamiltonian will not remain unchanged. Thanks to Nick Huggett for pointing this out.

One answer would be to follow Poincaré and Weyl and appeal to an understanding of it as a priori. As is well known, Weyl (1952: 126) insisted that 'all a priori statements in physics have their origin in symmetry'. Not surprisingly, empiricists such as van Fraassen have tended to resist this line (van Fraassen 1989) and move to the other end of the spectrum, offering a broadly pragmatic answer to our question. From this perspective, PI comes to be seen as nothing more than a problem-solving device (see Bueno 2006). Occupying the middle ground between these extremes is the understanding of PI as reflecting a feature of the structure of the world in the sense that, together with the other fundamental symmetry principles, it effectively binds the 'web of relations' constituting that structure into the relevant kinds. Furthermore, PI can be considered to be inherently modal by virtue of encoding these possibilities represented by the varieties of quantum statistics, including para-statistics. Let me now briefly consider other kinds of symmetries, namely the space-time symmetries represented by the Poincaré group and the 'internal' symmetries that feature so prominently in elementary particle physics.

10.3.2 Kind distinction$_2$: particle/property classification[13]

The general strategy here is to classify states of elementary particles in terms of unitary irreducible[14] representations of a group that can be represented thus: $G \times P$, where G is the 'internal symmetry group' which depends on the theory we happen to be concerned with. In the case of the Standard Model we have: $G = \mathrm{U}(1) \times \mathrm{SU}(2) \times \mathrm{SU}(3)$, where SU(3) represents the symmetry associated with quark flavours. P stands for the Poincaré group, which represents the symmetries of Minkowski space-time. The (unitary) irreducible representations of $G \times P$ can then be constructed by taking the tensor product of an irreducible representation of G with one of P. This construction thus yields a twofold strategy; let us begin with P.

Described by Mirman as the '(necessary) transformation group of any geometry that allows a universe in which physics is possible' (2005: 110), the Poincaré group can be resolved into the semi-direct product of the translation group and the group of all linear transformations that preserve the Minkowski metric, or, in other words, that preserve the speed of light (aka the Lorentz group[15]). In quantum mechanics a symmetry operation is represented by a 'Wigner automorphism' and according to Wigner's Theorem, every Wigner automorphism is induced by a unitary or anti-unitary transformation, which is unique up to an overall phase (for details see

[13] Special thanks here go to Kerry McKenzie, although of course she is not to be held in any way responsible for anything I say.

[14] It is the irreducibility of the representation that represents the elementarity of the system (Wigner 1939; Newton and Wigner 1949).

[15] Strictly speaking it is the universal cover of this group that is isomorphic to the group of 2×2 complex matrices with determinant 1 and it is the latter which forms the Poincaré group via the semi-direct product with the group of translations. This double cover of the (proper) Lorentz group is required in order to account for spin.

Straumann 2008). So, any symmetry corresponds to a unitary or anti-unitary operator in Hilbert space (Lorentz transformations are associated with a unitary operator; time-reversal symmetry with an anti-unitary operator).

What this strategy yields is a classification of the non-negative energy irreducible unitary representations of the Poincaré group that have sharp mass eigenvalues.[16] These correspond to the possible elementary states of particles, given any theory for which the associated symmetries hold. They can be labelled by the Casimir invariants (see again Straumann 2008), one of which gives the total mass (squared) and the other yields the spin, although further labels associated with the energy and helicity may also be required (Mirman 2005: 110). Various classes can then be obtained: those representations for which the mass is real; those for which it is imaginary; those for which it is zero; and those for which the momentum is zero.[17] The last do not correspond to physical objects (2005).[18] The massless representations cover objects that exist on the light cone, such as photons and gravitons[19] and also the vacuum state.[20] Those corresponding to imaginary mass cover tachyons, which travel faster than the speed of light and thus enter into 'non-standard' causal relationships. Of those for which the mass is real, those representations that are labelled with spin 0 correspond to the Higgs boson, those with spin 1/2 correspond to electrons, neutrinos, and quarks, those with spin 1 to the W and Z bosons and those with spin 3/2 to the W baryon.

Let us turn now to the 'internal symmetries' and the example of quarks. The so-called 'quark model' was originally introduced as another kind of classification scheme in order to make sense of the hadron 'zoo'—that is, the large number of different kinds of hadrons.[21] Here it was another group—SU(3)—that made sense of this zoo by underpinning the division of hadrons into multiplets containing 1, 3, 6, 8, 10, 27 . . . members, associated with representations of the group, with the triplet containing (of course) 3 members as the fundamental representation from which the others can be obtained. The history of the application of SU(3) to physics is interesting in itself[22] and it begins with the introduction of isospin, represented by

[16] An interesting discussion (bordering on, and sometimes plunging head-long into, the arcane) of the details of Wigner's classification can be found on the blog 'n-Category Café', at: <http://golem.ph.utexas.edu/category/2009/03/unitary_representations_of_the.html>.

[17] As Mirman also notes (2005: 2), inhomogenous groups like the Poincaré group have a range of different types of representations and, in effect, there is a considerable amount of surplus structure here, corresponding to possibilities that may or may not be physically applicable.

[18] Neither do those representations for which the spin is continuous.

[19] See Mirman (2005: 3–4) for the properties of the massless representations and the relationship to gauge transformations.

[20] Drake et al. (2009) take this to correspond to particles 'not moving through time'. But if the vacuum is understood to be devoid of particles, it is not clear what this means. Hopkins (2009) uses this confusion as the basis for his claim that we shouldn't take these representations as describing particle states at all but rather as 'media'. Despite Baez's enthusiastic reception for this idea, I'm none the wiser!

[21] A useful account of the history can be found in Pickering (1984); fortunately the sociological conclusions are detachable.

[22] And by paying attention to the details of this history, the applicability of this piece of mathematics becomes less surprising than certain commentators feign to believe.

SU(2) (the double covering group of SO(3), the rotation group), itself the result of an analogy between atomic and nuclear structure (as we noted in Chapter 5). Efforts to combine SU(2) and the U(1) group appropriate for the strangeness or hypercharge led Gell-Mann and Ne'eman, independently, to propose SU(3) as the group of the quark model.[23] Within this model, in which SU(3) was viewed as the symmetry of three quarks, two of them (the 'up' and 'down' quarks) generate the isospin symmetry, which came to be regarded as 'a sort of accident' (comments by Fritzsch in Doncel et al. 1987: 633). And of course with the development of quantum chromodynamics and its incorporation of gauge symmetry,[24] SU(3) came to be seen as the appropriate gauge group, with colour emerging as the value of a quadratic Casimir operator in the relevant representation (see Pickering 1984).

Up, down, and strange quarks were thus each considered to be both massive spin-1/2 irreducible representations of the Poincaré group and states in the fundamental representation of the SU(3) group. As McKenzie has pointed out, since the properties that label the representation are common to all the quarks in a given multiplet, it is the states the quarks are in that serve to distinguish the quarks from one another. These latter properties do not remain invariant under the transformations represented by SU(3), because these transformations map between the states and hence map to different properties (McKenzie 2012). As she notes, this appears to raise a problem for the structuralist claim that 'only those properties that are invariant under the symmetries of the theory are real'.

However, as she also acknowledges, although such a claim makes sense in the context of the Poincaré group, it is not clear that it does in the quark 'flavour space', where there is nothing equivalent to an 'observer' in terms of which the claim can be understood. A plausible option is to suggest that 'we treat the distinct states of the multiplet as physically distinct because the symmetry is broken; otherwise, physical meaning would not be attributed to a quantity corresponding to a state in the relevant representation' (2012). The symmetry that is broken is associated with SU(2) and isospin and it is broken in two ways: by the electromagnetic interaction (obviously, because the proton is charged and the neutron is not, and they also possess different magnetic moments) and by the different masses of the up and down quarks. Fortunately the effect is very small in most cases (less than 1 per cent), which means that isospin symmetry remains useful for the practice of elementary particle physics (and so many of the properties of nuclei can be attributed to isospin symmetry).

[23] For the history of these developments see Gell-Mann 1987 and Ne'eman 1987 and for the role of group theory in particular, see Doncel et al. 1987: 485–90 and 512–14, respectively; the comments by Speiser in Doncel et al. 1987: 552–3 are also useful.

[24] It was in this context that parastatistics was applied, with quarks regarded as parafermions of order 3. This proposal was subsequently rejected in favour of the introduction of the new property of colour; see French 1995.

Of course, even at the level of less than 1 per cent, the symmetry is still broken and the existence of symmetry breaking has been presented as an obstacle to the structuralist programme.[25] The concern seems to be simply something like the following: according to OSR, symmetries are an important feature of the structure of the world; but symmetries are broken; hence, symmetries cannot be an important feature of the structure of the world.

In response, one should distinguish between 'explicit' and 'spontaneous' symmetry breaking (see Brading and Castellani 2008): the former can be said to occur where the relevant laws do not possess or manifest the symmetry in question; in the case of the latter, the laws themselves have the symmetry but the vacuum background does not. In the case of explicit symmetry breaking the structuralist simply needs to adopt a fallibilist attitude and urge that the ontological focus should shift to the fundamental laws and associated symmetries. Indeed, in this particular case of isospin, it was acknowledged from the get-go that the symmetry was broken once the charge of the proton was taken into account and of course that the masses of the proton and neutron, although very close, were not exactly the same.[26]

Shifting to quantum chromodynamics, SU(2) is then treated as a sub-group of the larger 'chiral'[27] symmetry group associated with that theory. Here too we have explicit symmetry breaking, due to the different masses of the quarks. It is because the up and down quarks are both comparatively 'light' and have almost equal masses and because the gluons do not couple to their flavour that the interactions are independent of whether they are 'up' or 'down'. Under these circumstances the masses of the proton and neutron can be treated as equal and SU(2) applied. Again, of course, chiral symmetry can be broken, both explicitly via the 'naked' quark masses, or spontaneously, via the 'chiral (or quark) condensate'[28] that forms in the vacuum state in low-energy QCD and which 'gives' mass to the hadrons. With regard to the former, there appears to be no obstacle to the structuralist insisting that we move to the more fundamental level, or, at the very least, that we adopt the same fallibilist attitude that all realists have to maintain.

As for the latter, the concern is that if a symmetry is broken, then it cannot be invested with the significance that the structural realist wishes to attach to it, since the symmetry is not manifested in the relevant domain. However, instead of thinking of the symmetry as somehow 'lost', the situation is better understood as one where the

[25] In discussion following my presentation of the ideas contained in this chapter.

[26] For a discussion of the idealizations involved in the introduction of isospin with an argument that these can be nicely accommodated by the partial structures variant of the semantic approach, see French 2000b.

[27] Chiral symmetry involves invariance under left-handed and right-handed rotations, taken independently, and is related to parity.

[28] Chiral condensates are fermionic condensates, which, like their Bose–Einstein counterparts, are types of 'superfluids'. A well-known example of a fermionic condensate is associated with superconductivity where, according to the 'BCS' model, electrons with opposite spins form bound states called 'Cooper pairs' which then form the condensate.

relevant phenomena is characterized by a symmetry that is 'lower' than the unbroken symmetry (see, for example, Castellani 2003; Brading and Castellani 2008). And expressed in group-theoretic terms, this means that the group characterizing the latter is 'broken' into one of its sub-groups. In the case of the Standard Model, for example, the fundamental symmetry of SU(3)×SU(2)×U(1) is spontaneously broken down to SU(3)×U(1) at the electroweak energy (about 100GeV). Thus, symmetry breaking can be described in terms of relations between transformation groups, something the structural realist can easily accommodate. We shall return to this issue in Chapter 11, in the context of considering the role of unitarily inequivalent representations in understanding spontaneous symmetry breaking and the further challenge these representations pose to OSR.

Returning to the central point: what we obtain in the case of these 'internal' symmetries is, again, a kind classification of the associated quantum numbers (such as mass, charge, spin, isospin, strangeness, baryon number, and so forth). These are conserved under transformations associated with the relevant symmetry and this, in turn, determines the particle's location within the multiplet (here, octuplet or decuplet) that forms the basis of the irreducible representation of the relevant group. And as is well known, this classification has led to novel predictions, notably that of the Ω^- particle (see Bangu 2008 and 2012), which was proposed to fill the gap in the spin 3/2 baryon decuplet in terms of which the properties (and significantly, for detection purposes, the mass) of the new particle could be determined.[29]

Let us now move on to consider how modality is encoded in the laws, via the models of the theory.[30]

10.4 Laws, Models, and Modality

Let me begin with the following question: On what basis can we ascribe lawhood to the relevant propositions of the theory and thereby, via truth or representational faithfulness, to the world?[31]

[29] The structural realist can take this example to be on a par with the more familiar ones, such as the predictions of the existence of Neptune and neutrinos, and argue that just as these feed into 'No Miracle Type' arguments for the truth of the associated theories (Newtonian mechanics and the theory of beta decay respectively), so a similar argument can be run in this case, concluding with the 'truth' or, better perhaps, representational faithfulness of SU(3).

[30] Before we leave the internal symmetries it is worth noting that in the case of the symmetries associated with the Standard Model, the role of the dynamics is further diminished. It is not the case that the symmetry is imposed and then the dynamics ensures that the resultant classification does not break down; rather, as McKenzie has emphasized, once we 'fix' the relevant symmetries, we also fix the dynamics, at least to a certain extent.

[31] Roberts argues that there is no such basis and defends a meta-theoretic conception of laws (2008); for a response in the current context see French 2011d.

Here we need to articulate the ascription via two stages within the structuralist framework: first of all, there is the attribution of laws, as features of structure, to the world. The grounds for that are whatever the grounds are for attributing charge to the world, or features thereof, involving Inference to the Best Explanation (IBE), for example.[32] The structuralist can follow this line and maintain a form of the No Miracles Argument, concluding that the best explanation for the success of a given theory is that 'its' laws are 'out there' in the world, as features of the structure of the world (cf. Ladyman, Ross, et al. 2007).

So, we attribute laws to the world, as it were (as features of structure) on whatever grounds we attribute (as realists) charge or 'up-ness' to the world. Indeed, given the dependence that the structuralist insists holds between charge, say, and the relevant laws, or between 'up-ness' and the relevant symmetry principles, then one cannot attribute the properties without attributing the laws and symmetries. We can even piggyback the structuralist's attribution of laws and symmetries on those grounds the object-oriented realist invokes to justify the positing of electrons, say, but once the positing is done, where the object-oriented realist sees a metaphysical robust object, the structural realist sees, at best, a thin stand-in or no object at all, but rather manifestations of structure via the dependence relations that hold between the laws and symmetries and the relevant properties.

But of course this is not to attribute *lawhood* to the world, at least not in the sense of a property with modal features. The Humean structuralist can agree with all that has been said so far but insist that what we are attributing to the world are certain regularities, structurally conceived in the way indicated in Chapter 9. What are the grounds for attributing lawhood qua modally informed property? Answering this takes us to the second stage and here the grounds must be broadly metaphysical, having to do with (non-Humean) reasons for taking modality to be 'in' the world rather than a feature of our theories and models, say. We have already mentioned these reasons in the discussion of Humean structuralism and dispositionalism, and these the 'modal structuralist' can also appropriate. But what about the specific reasons why we should attribute lawhood in the sense of *laws qua features of structure as inherently modal*, rather than in the sense of laws as supervening on dispositions?

In the next section I will present one way we can understand this idea of structure as being inherently modal—namely via the models the theory presents (cf. Brading 2011)—and I will then return to the shift of modality from those models to the world.

[32] Where the invocation of IBE is local and domain-specific, just as is its deployment in support of attributions of charge or 'up-ness'.

10.5 Modality 'in' the Theory

Brading presents the accommodation of modality as an obstacle to OSR, in so far as it cleaves to a conception of scientific representation according to which 'a theory successfully represents a physical system if there is a model of the theory that "sufficiently resembles" the system in question' (2011: 58). In terms of the partial structures account, this idea of 'sufficiently resembling' is cashed out in terms of partial isomorphisms holding between the relevant partial structures (in terms of which the theories are represented, at the meta-level of the philosophy of science, as discussed in Chapter 5). The worry is that if by model here we mean a particular model, then within this framework we cannot accommodate modality since that will involve a consideration of the range of models a theory makes available. If we broaden our focus to accommodate this range, then we must, at the very least, appropriately modify this conception of representation.

Thus, '[w]hen we ask whether [a] model contains modal information, a great deal hinges—for the structuralist—on whether by "model" we mean generic or particular model' (2011: 58). The latter incorporates a particular solution of the equations of the theory—Newtonian mechanics, say—whereas the former covers a range of solutions, effectively telling us how the system in question will behave under a range of initial and boundary conditions.

So, option 1 would be to focus on the structure of one particular model of the theory, in accordance with our characterization of representation. However, '*this structure, in and of itself, contains no modal information*' (2011: 59; Brading's emphasis). The information carried by this model concerns a particular solution to the equations, or one particular trajectory of the system through the relevant phase space. Here the structural realist may appear to suffer in comparison to the object-oriented realist who can accommodate the relevant modal commitments within a single model by virtue of the fact that the properties of the objects concerned—electrons, say—are given not by the model alone but by the theory as a whole (2011: 58). However, I think there is no real suffering involved, since the structural realist can also 'load' the modality onto the properties, either by regarding the latter as clusters of dispositions, or taking them to be dependent on the laws, in the way I indicate, and simply leave out of the picture any 'thick' notion of object.

Option 1, then, yields a form of 'structural actualism' in which the model describes only the actual structure of the physical system (2011: 59). The Humean would of course be content with this and will either insist that any modality remains 'in' the theory or is outsourced (to use Vetter's handy phrase) to possible worlds. The dispositionalist, on the other hand, will see structural actualism as incorporating only the 'manifest' relations and thereby leaving out those features that help us get a grip on the relevant counterfactual scenarios.

Option 2, advocated by Brading herself, is to understand 'model' in the generic sense and take the shared structure of the collection of models made available by the theory as capturing the relevant modal features. As we saw in Chapter 5, this shared structure is taken to present the 'kinds' of objects[33] that the theory talks about and by virtue of stipulating the possible trajectories for entities instantiating the kind this framework lays down modal constraints on those entities (2011: 60–1).[34] Invoking the Newtonian example again, the solutions to the generic two-body problem can be regarded as a family of models that prescribe all and only the possible trajectories for the relevant entities (Newtonian inertial-gravitational entities) that are in two-body motion (2011: 61). With respect to each solution or model, there will be features that are peculiar to that model and that can be considered to be 'contingent'. There will also be features that are due to the entities being instantiations of the kind and thus are features of the shared structure of the family of models as a whole; these can be regarded as 'necessary'. Thus, 'modality is a feature of the *collection* of models of the theory, and not of any particular model of the theory' and, drawing on the distinction between representation and presentation again, 'modality is *presented* through the *shared structure* of the models of the theory' (2011: 61).[35]

Now, of course, I understand 'shared structure' in terms of partial structures, interrelated via partial isomorphisms (and with mathematical structures via partial homomorphisms) and, as I indicated in Chapter 5, where Brading and others see 'mediating models' lying between 'theoretical' and 'phenomenological' models, I see partial structures related, 'from the top', to theoretical models via partial isomorphisms and, from the bottom, to data models likewise (da Costa and French 2003).[36] Whichever account one prefers doesn't matter so much when it comes to accommodating the modal features but the issue will have some bearing on how we might overcome the further obstacle that Brading sees as lying in the structural realist's path, as we shall now see.

10.6 Representation, Modality, and Structure

So, OSR can accommodate modality via the following claims (Brading 2011):

1. An individual model of the theory represents the particular physical system in question through a relationship of shared structure.
2. Modality is a feature of a collection of models, deriving from their shared structure and is not a feature of any individual model.

However, the desired conclusion is

3. Our theories represent the modal properties of the world.

[33] We also recall that Brading and Landry urge that we should then refrain from adopting any ontological commitments with regard to these putative objects.

[34] cf. 'Laws are the patterns that nature respects; to say what is physically possible is to say what the constraint of those patterns allows' (Maudlin 2007: 15).

[35] cf. Maudlin (1994: 131): 'the models depict the possible worlds allowed by the laws'.

[36] Brading herself, we recall, advocates minimal 'methodological' structuralism.

As given, the argument is clearly invalid (2011: 62).

Of course, we could render it valid by changing premise 2, so that modality is understood as a feature of an individual model. That obviously presents some difficulties. The alternative is to keep premise 2 and drop 1, but that means that the structuralist must take representation to involve more than a relationship between a particular model and the given system. Alternatively, we might take the representational relationship to hold between the system and a generic model; that is, drawing on the Newtonian example, a structure that presents a range of possible trajectories of the system (Brading 2011: 62–3). So, on this view, we would not only have to claim that when a particular trajectory is realized by a system, that trajectory shares structure with a particular model—so we are realists about that structure—but also acknowledge that the system shares structure with the other non-realized but possible trajectories—so we are realists about this structure too.

Now, some might find this odd, but let me try to dispel that sense of oddness. Consider the following claim that any realist will surely agree with: our best theories represent how the world is.[37] The object-oriented realist takes the representational elements to be objects and properties and can then cash out the representation relation in terms of denotation and reference and so on. If she is a dispositionalist, it is these representational elements that will be 'imbued' with modality and if she is a structuralist it will be the laws and symmetries. Furthermore, in so far as the laws yield a range of models (or more generally, contribute to that 'yielding' or construction of models[38]), this view is compatible with the account of representation I set out in Chapter 5, with the models represented, at the meta-level, in terms of partial structures, and the representational relationship, at the 'object' level—that is, between the models and the system—captured via partial isomorphisms. Given that these models are all interrelated both hierarchically, or vertically, as we move from theoretical models to models of the phenomena, data models, and so forth, and horizontally, or inter-theoretically, it should not come as a surprise that representation 'spreads out', as it were, through these interrelationships. When we think of specific examples, as in the case of the particular trajectory realized by a Newtonian system, we tend to think of the system as represented by a particular model as indicated previously. But that model is interrelated to others, either directly, as it were, if it is sufficiently complex and draws on other theoretical resources, or indirectly, via Newton's laws that, as we've

[37] Of course, by best I mean approximately true, so by represent I mean approximately, or faithfully represent to a certain degree, or whatever.

[38] Here I am of course aware of and agree with claims that models may be built from the 'bottom up' or from some level labelled 'the phenomenological', rather than from the 'top down' or theoretical. Where I differ from some advocates of such claims (see, for example, Cartwright, Shomar, and Suarez 1996) is in the further insistence on their part that such construction proceeds independently of theory, in methods and aims (see, for example, French and Ladyman 1997).

seen, encode the range of possibilities. In so far as these laws play a role in representation, so does modality.[39]

Or think of the example of superfluidity, touched on in Chapter 5 (Bueno, French, and Ladyman 2002). There the 'fountain effect' exhibited by liquid helium is accounted for through models that reach up, as it were, to Bose–Einstein statistics and hence Permutation Invariance. Here we have a nice example of a form of symmetry playing a role in representation and as far as the structural realist is concerned, the fountain effect and other phenomena should be seen as a manifestation of the (bosonic) structure of the world. One can climb further of course, following the relationships with mathematical structures, represented—again at the meta-level—by partial homomorphisms and here one can appreciate the manner in which the mathematics—group theory in this case—can also play a role in representation.[40]

What this all suggests is that we should be careful not to take the particular model that is invoked to describe a system as encompassing all there is to the representation of that system. In so far as this model has relations to others, via which further resources can be drawn upon, this representation can be extended and, in particular, in so far as it involves the kinds of laws and symmetries we are concerned with here, it will involve modality.

Moving on, there is a further obstacle to be overcome: laws (and presumably symmetries) have to do with, or are, depending on one's views, *determinables*; but (it has been argued) only *determinates* can be fundamental; hence structure, in so far as it is constituted by laws and symmetries, cannot be fundamental.

I shall clamber over this obstacle by drawing on recent work by Wilson (2012) that argues that determinables can in fact be fundamental. In effect this will be my response to the 'Vetter Problem' presented in the previous chapter and it will allow us to get a better grip on the way in which actuality and modality together inform the structure of the world.

10.7 Determinables, Determinates, and Fundamentality[41]

That the entities in our 'fundamental base' must be determinate is a common assumption made by many contemporary philosophers who,

[39] One of the consequences of those views that seek to render models independent or autonomous from theory is that modality cannot then feature in the way I have indicated here. Perhaps the advocates of such views would not regard that as so unfortunate. However, the sense of autonomy that is typically invoked is either problematic or can be straightforwardly accommodated by the sort of approach I favour.

[40] This is unproblematic of course. What is problematic is whether it can also play an explanatory role; for further discussion see Bueno and French 2012.

[41] Once again I am tremendously indebted to Kerry McKenzie for helping me with a number of issues in this section. I am also grateful to Jessica Wilson for sharing a preprint of her paper (Wilson 2012)—the influence of which should be clear—and for discussion of the issues.

commonly suppose that any fundamental entities there may be are *maximally determinate* [and] that, whether or not there are fundamental entities, any determinable entities there may be are grounded in, hence less fundamental than, more determinate entities. (Wilson 2012: 1)[42]

Let us consider the nature of the relationship between a determinable and 'its' determinates, using the standard examples of 'coloured' and 'red' (see Sanford 2011).[43] The latter is obviously more specific in some sense than the former but the two are tied together in a way that may cast doubt on the insistence that only determinates can be fundamental: something cannot be red without it being coloured, so if 'redness' were in our fundamental base, one could argue there is a sense in which being coloured is too. On the other hand, it is difficult to conceive of a fundamental base solely constituted by determinables (could something be coloured without being a particular colour? It would seem not) so the determinates have a crucial role to play also, one that might be characterized in terms of specificity[44] (I'll return to this shortly). And this specificity is, in certain respects, exclusive: if some element of the fundamental base is coloured, it cannot be both red and green.[45]

These features of the relationship can be expressed as follows:

(1) if a determinate concept (e.g. red) can be predicated of something, then at least one determinable concept (e.g. coloured) must also be predicable of that thing;
(2) if a determinable concept (e.g. coloured) can be predicated of something, then there must be some (unspecified) determinate concept (e.g. red) that is also predicable of that thing;
(3) two determinates (e.g. green and red) of the same determinable (coloured) cannot characterize something at the same time (Johnson 1921, 1922, 1924; for the hypertext version see <http://www.ditext.com/johnson/toc.html>).[46]

[42] See, for example, Armstrong 1961: 59; Geach 1979: 55; Lewis 1986: 60.
[43] For a formal characterization, see Fine (2011).
[44] Thus Searle called the relation between determinable and determinate the 'specifier' relation (1959).
[45] I'm going to assume that beach balls are not elements of the fundamental base. Of course this last feature rests on some form of the Principle of Non-Contradiction, which has, at least in some cases, been expressed in terms of some fundamental object, or some appropriate piece of a non-fundamental object, not possessing more than one determinate property associated with a given determinable (so something can have both mass and spin but it can't have mass x and mass y—leaving quantum considerations to one side for the moment). For more on this and inconsistency-tolerating formalisms in the philosophy of science see da Costa and French 2003: ch. 5.
[46] Johnson is typically relegated to the 'old guard' of pre-*Principia Mathematica* logical thinking, perhaps a little unjustly (although Armstrong 1978: 111 acknowledges that modern discussion of the determinable–determinate relationship has been dominated by Johnson's principles); for a useful attempt to reassess Johnson's contributions see Poli 2004. Johnson expressed these features in terms of his '*Principles of Adjectival Determination*', where, as the name indicates, these are presented in the context of the distinction between substantives and adjectives. Johnson also adds to this the 'Principle of Alternation' which he held to supplement principle (2) in the main text (called, significantly, as we shall see, the 'Principle of Counterimplication') in postulating that 'the range of possible variation of the determinable can be apprehended in its completeness'. As he then went on to note, whether we can so apprehend this range is a further issue. He maintained that we can for any determinable (continuous or

The last will turn out to be crucial for our discussion (for already obvious reasons) and might be expressed as: two determinates cannot exist 'in' the same particular at one and the same time. Johansson expresses this constraint as,

The Principle of Determinate Exclusion: for *some* ontological determinables it is true that two determinates cannot possibly exist in the same spatiotemporal region. (Johansson 2000)[47]

More generally, and perhaps more usefully for our purposes, Searle nicely expressed the idea as follows:

Genuine determinates under a determinable compete with each other for position within the same area, they are, as it were, in the same line of business, and for this reason they will stand in certain logical relations to each other. (Searle 1959: 148)[48]

As we'll see, this notion of 'competition' between determinates nicely fits with the inherently modal nature of structure.

So far we have focused on *properties* but the framework articulated here can be extended. Johansson in particular (2000; see also his 1989) embraces universals as ontological determinables in general,[49] analysing the internal structure of universal determinables, for example[50] (for a brief summary, see Poli 2004: 189–90). Of particular interest here is his claim that such universal determinables may be related to further higher-order universals, such as patterns. These, he maintains, involve two different types of determinables, namely shapes and some other property (1989: 85). Thus, a pattern of colour presupposes both determinates of shape and determinates of colour, although the converse does not hold. This framework then allows us to get the beginnings of a metaphysical grip on the notion of pattern. So, Johansson argues that a colour pattern is more than an aggregate of colours that happen to be situated close to each other (2000). The *components* of such a pattern are 'spatial unities' of the *determinates* of colours and shapes. And this follows from the claim that any finite uni-coloured colour spot must have a border, whether sharp or fuzzy, and if it has a border then it has a shape. If the border is sharp, the border has a determinate shape; if the border is blurred, it has a blurred shape. But there is nonetheless a border and a shape.

discrete) 'whose determinates have an order of betweenness and can therefore be serially arranged'. This would certainly be so for the determinates we find in science, given their quantitative representation.

[47] For further discussion of these principles see also Poli (2004: 172–6) who presents them in the context of a 'field structure' (as compared with the 'tree' structure of classes, sub-classes, etc., applicable to substantives).

[48] Poli (2004: 179–90), again, gives a nice account of the reception and subsequent impact of Johnson's work and notes Searle's organization of determinates into levels.

[49] Interestingly, given what we covered in Chapter 7, according to Johansson, the relations between determinables are relations of existential dependence.

[50] So the determinates of volume are related in a part–whole way that the determinates of colour are not. Furthermore, with regard to the competition mentioned previously, whereas one determinate of colour excludes another, a determinate of volume may include another such determinate as its part.

Now it may stretch things too far to insist that *every* kind of property pattern has properties-with-shapes as components,[51] but perhaps this offers a further way of conceptualizing Ladyman and Ross's talk of patterns, with physical structures in the 'material mode' understood in terms of such patterns, and represented, in the formal mode, by mathematical (set-theoretical) structures. From Johansson's perspective, such a pattern should be seen as a unity of the relevant determinates (we'll drop the spatial aspect for obvious reasons), themselves related to the appropriate determinables. The take-away message is that patterns, and hence structures, involve both determinables and determinates.

Now, Johansson notes that much of the discussion of the determinable–determinate relationship until now has focused on monadic properties, but that it can straightforwardly be extended to relations (2000). Indeed, if one is resistant to the reduction of relations, then, he maintains, we have two ontological (higher-order) determinables: a property-determinable and a relation-determinable.[52] Of course, within that subset of properties and relations that might be designated as 'physical', these two are going to be interrelated in my account, since the relevant properties are dependent on the associated relations, as expressed via the laws. But Johansson's extension offers a bridge over which we might carry this framework to embrace laws. Indeed, Armstrong appears to have come to believe that laws are, or at least involve, determinables (1997: 246–8). Thus think of a (functional) law like

$$F = G\{m_1 m_2\}/\{r^2\}$$

If we insert specific values for m_1, m_2, and r, we obtain a determinate instance of the law, expressing the relation that holds between determinate properties. So, the law as a whole can be regarded as expressing a determinable relation, which can be made determinate in specific situations.

More generally, these determinate instances correspond to solutions that can be represented by (more) specific models of the theory, and it is the shared structure between these models that yields the relevant physical modality. The relationship between the more generic and more specific models can be captured by that between determinables and determinates.

Vetter's demand—introduced in the last chapter—for an explanation of the (meta-) regularity of similarity that holds between instances of Coulomb's Law can now be

[51] I'll also pass over Johansson's *reductio ad absurdum* argument that concludes that colour and shape must be regarded as ontological determinables: if they were not—if they were constructed in some fashion—then it would be possible to think of components of colour patterns in which a determinate colour is united with a determinate of something other than shape; but that is impossible (here I think the notion of the relevant unity being spatial plays a crucial role in the argument); hence colour and shape cannot be merely conceptual but must be ontological determinables.

[52] As he notes, when it comes to these higher-order determinables, at least, one must be careful with regard to the 'competition' between associated determinates, since two determinates of the property-determinable can obviously be associated with the same thing; hence he prefers his Principle of Determinate Exclusion to Johnson's third principle.

met: these are determinates of the determinable that is Coulomb's Law itself and indeed, the kind of picture that Vetter appears to have in mind can be nicely articulated in these terms. Furthermore, laws, qua relation-determinables in Johansson's terminology, yield properties, qua property-determinables, via a kind of dependence relationship (to which I shall return shortly) and the latter yield determinate instances of these properties. Thus Coulomb's Law as a relation-determinable both has determinate instances and yields charge as a property-determinable, which has determinate instances such as *e*, the charge on the electron.

It hardly needs spelling out that the competitive aspect between determinates expressed by Searle is manifested by the relationship between different models of Newton's and Coulomb's theories and different solutions of the respective equations. And as we have already indicated, this can be extended to symmetries. So, for example, the permutation group can be conceived of as an 'interweaving of relations'-determinable—to bring Johansson's and Eddington's terminologies into unholy matrimony—which has as determinates, the symmetric and anti-symmetric (and 'mixed' symmetry) representations. Likewise, the Poincaré group is a determinable which also yields spin as a property-determinable, which has spin 1/2 as a determinate. And in the case of both laws and symmetries it is through the determinable (with the emphasis on the '-*able*') that the relevant possibilities are encoded.

Now this is all well and good but in extending the determinable–determinate framework to laws and symmetries, I still haven't dealt with the insistence that determinables cannot be included in the fundamental base. Wilson considers a variety of objections to the claim that determinables can be fundamental, of which the most significant for my purposes are the following (2012: 9):

> Objection 1: determinable entities are metaphysically grounded in more determinate entities, rather than vice versa; hence, determinables are less fundamental than determinates.
>
> Objection 2: fixing a given determinate fixes the associated determinable but not vice versa; hence, determinables are less fundamental than determinates.

Let us consider each in turn. With regard to Objection 1, if we cash out 'grounding' (see Rosen 2010) in terms of entailment, the claim is that it is because determinates entail their associated determinables, but not vice versa, that determinables are grounded in and less fundamental than determinates. Here the notion of entailment is being taken quite loosely but we can follow Searle (1959) in extending it to predicates and hence properties.[53] So the idea, using the standard example, is that 'being red' entails 'being coloured' but not vice versa and so the former grounds and is more fundamental than the latter. Putting it another way, the asymmetry that we

[53] For some considerations of taking conjunctions as determinates and disjunctions as determinables, see Zimmerman 1997.

associate with grounding (so the less fundamental is grounded in the more, and not vice versa) is matched by the asymmetry of entailment.

However, Wilson argues that, although it is true that instances of determinables do not entail any particular instances of their associated determinate instances, there is nevertheless a kind of 'two-way' entailment in so far as every instance of a determinate entails an instance of an associated determinable, and every instance of a determinable entails an instance of some associated determinate. Thus there is no 'deep' asymmetry of entailment here and the motivation for taking determinates rather than determinables as fundamental evaporates. Of course, one may still feel there is a residue of asymmetry, as indicated by the use of the word 'some' in expressing the entailment between determinable and determinate. But, as Wilson suggests, one can take this as merely a reflection of the distinct nature of determinables and determinates whereby the former are less specific than the latter. This has to be acknowledged but then those who exclude determinables from the fundamental base need to show what specificity has to do with fundamentality.

Turning now to Objection 2, this might appear to have some initial plausibility, since if we have the given determinate in our fundamental base, then we also have the determinable, but not vice versa. So, fixing the determinate fixes the determinable but not conversely; again, we have an asymmetry. But now, consider a little more closely the sense in which the determinate fixes the determinable. The latter is 'modally flexible' in a way that the former is not, and given that, fixing the determinate cannot fix this flexibility (how could it?! The flexibility is by its very nature unfixable). As Wilson puts it, 'it is a constitutive modal fact about every determinable instance that it is of a type whose instances might be differently determined. But no specific determinate instance seems suited to ground this fact about its associated determinable instance' (2012: 12). What the determinate can fix regarding the determinable are certain non-modal facts about it, such that it is in this case determined in the way that it is. But the constituent modal flexibility of the determinable is left unfixed.

So how can it be claimed, as it often is, that determinates fix determinables? There is a tacit assumption here to the effect that the fundamental base need only ground the non-modal facts at a world. Given this, the fact that determinates cannot fix the modal flexibility of determinables does not undermine the claim. But the assumption can be questioned. In particular, it arises from the broader presumption that in our metaphysical inventory of 'the world' we include only such non-modal facts. This is a presumption that the Humean would obviously be happy with, but it begs the question, of course. On my view, fixing the non-modal facts and leaving modal flexibility out of the picture would yield an incomplete inventory. And the most obvious way of fixing the modal facts would be to include the determinables themselves in the fundamental base.[54]

[54] Wilson considers other ways of fixing the modal facts via complex determinate entities, such as disjunctive entities, but rules these out on grounds of 'naturalness' or 'objective similarity'. Interestingly, she also considers the suggestion—which may be amenable to the Humean—that modal facts might be

This still leaves a number of important points to clarify. First of all, admitting determinables into one's fundamental base should not be taken to imply that these are the only entities in such a base. Given their modal flexibility, a possible world with only determinables in its fundamental base would not be a very specific world. To obtain that specificity we need to incorporate determinates as well. Here one can think of the determinate entities as acting as 'existential witnesses' (Wilson 2012: 15). So, using the familiar examples again, 'red' is a determinate witness to 'coloured' (but the latter cannot act as such a witness to the former).[55]

Furthermore, as a way of understanding how determinables and determinates might together form one's fundamental base think of how scientific laws combine with initial or boundary conditions (Wilson 2012: 15–16). So, again, consider Newtonian mechanics and the way in which the relevant laws are applied to a given system: determinate values must be assigned to the relevant variable quantities in order to get the requisite results. Likewise in quantum mechanics, Schrödinger's equation does not specify the form of the Hamiltonian, the details of which must be 'filled in' when the equation is applied.[56] Thus, it is only when the relevant determinables are given determinate values that we can describe the 'concrete' world, just as the laws plus determinate initial or boundary conditions appropriately characterize the relevant phenomena.

Indeed, if we agree that physics is in the business of identifying entities in the fundamental base, then we can take Wilson's picture as representing how scientific determinables and determinates jointly enter into the fundamental base for everything else. Indeed, I think we can go further in not just admitting properties, such as charge or spin, into our fundamental base as determinables, but the laws on which I take these properties to depend. The ultimate ground for doing so is that whereas the likes of Lewis see physics as having undertaken 'an inventory of the sparse properties of this-worldly things' (Lewis 1986: 60), with Wilson taking these 'sparse' properties as determinables, I see physics as having provided an inventory of laws and symmetries—that is, features of the structure of the world—with properties, sparse or otherwise, as a kind of metaphysical by-product, and take these laws and symmetries as both determinables in themselves and as elements of the fundamental base of the world.[57]

fixed by global patterns of local determinate facts. As in the case of the possible dispositional accommodation of symmetries canvassed in Chapter 9, this amounts to the invocation of a highly conjunctive 'world property' corresponding to this global pattern, which Wilson again rejects as unnatural (2012: 14).

[55] So, of course we still have an asymmetry in the relationship but it expresses the difference in the natures of determinates and determinables rather than something fundamental about their relationship in terms of grounding or fixing.

[56] One might recall Cartwright's well-known point about effectively taking down specific Hamiltonians 'off the shelf', as it were (1983).

[57] One might worry that including both determinables and determinates in this way is ontologically inflationary and broadens the fundamental base beyond necessity. After all, if we have 'red' then we already have, in a sense, 'coloured', so why inflate the base by including both? Of course, this presumes a certain

The role of determinates as existential witnesses is nicely exemplified in the case of groups and their representations. So, consider yet again the example of Permutation Invariance. Here we can take the permutation group as the determinable and its representations as the corresponding determinates. Just as the determinable 'coloured' can yield the range of determinate instances, from 'red' to 'green' and so forth, so the permutation group yields its representations, through the 'action' of the group on, say, Hilbert space. And likewise, the modal flexibility of the symmetry is exhibited via this relationship between the determinable and its determinates. Furthermore, in so far as not all of the representations are manifested in this, the actual, world, so those that are—the bosonic and fermionic ones—stand as existential witnesses to the specific features of this world, as contrasted with one that has paraparticles, for example.

Likewise in the case of laws and specific values of certain properties, such as the mass of quarks or the charge on the electron. These can be thought of as akin to initial conditions that specify the nature of this world as contrasted with other possible worlds in which the relevant laws (but not, perhaps, all of them or all of them plus certain symmetries) hold. As such they also act as 'existential witnesses' to the specific 'goings on' of the world (cf. Wilson 2012: 8).[58]

One might be tempted to raise the criticism, again, that allowing for determinates to act as existential witnesses in this way is to introduce non-structural elements into the picture, and thus sully the purity of the structuralist project. My response is going to follow the same line as before: such purity was always made of straw and I do not see these elements as inherently non-structural in a way that undermines structuralism.[59] Indeed, presented in the way I've tried earlier, such elements are required if we are to distinguish the actual world from all the other physically possible worlds. If such a distinction is taken to undermine structuralism, then the project is doomed from the start, along with many others. But of course, to admit that the actual world is different from other possibilities is not to admit anything non-structural in the way

asymmetric relationship that Wilson has already cast doubt on but also, in the case of laws, we've already effectively surveyed other possible components of our fundamental base—such as dispositions, and regularities—and found them wanting. If we are to capture the modal features of the world, a little ontological inflation is necessary.

[58] Johansson (2000) suggests that there is a sense in which when one observes a given determinate, one also 'observes' the relevant determinable. Now there are arguments to the effect that when one observes, or more generally perceives, a particular, one also perceives the kind. Equally plausible, however, are arguments that this cannot be the case and Logue has used this conflict to lever the view that it is metaphysically indeterminate whether one can perceive kinds or not (Logue 2013). I shall not discuss Johansson's suggestion here, except to say that if I am 'observing' an electron *as a fermion*, then it does seem plausible to say that I am observing the kind, in a sense, but this is different of course from observing an electron (the distinction between seeing and seeing-as has, of course, played an important role in discussions of the empiricist's understanding of observability within the philosophy of science).

[59] The critic might latch onto Wilson's characterization of determinables as abstractions from the corresponding determinates, but I have already dealt with that way of viewing things when it comes to OSR.

that critics conceive of it. Consider: it is the Poincaré group, via the sub-group U(1), that yields the property of charge, as itself a determinable, and without this, and the group and the relevant laws, the specific electromagnetic phenomena would not be what they are, but equally, without the specific, determinate charge on the electron, or quarks, the world would not be the way it is. I agree that this initial condition is not structural but it is not clear how it could ever be conceived of as such.[60]

Two options then present themselves: if both the group and certain specific representations (or the law and specific values of the associated properties) are included in our fundamental base, we can either take the latter as strictly non-structural, but in an unproblematic way as indicated previously, or take the whole package of determinables and determinates as what we mean by 'the structure of the world'. My own inclination is towards the latter option. After all, what we are really interested in here is not 'the structure' but 'the structure of the world', and to cash out that definite article we have no option but to bring in specific, determinate elements.

We can draw again on Cassirer's work to help us here. We recall again his picture of the relationship between statements of measurement, statements of laws, and statements of principle, where the first are regarded as individual, the second as general, and the third as universal. We also recall that these are not to be conceived of as related via a hierarchy but as in a Parmidean well-rounded sphere in which the elements can be conceptually distinguished but not ascribed an independent existence and by which these statements all mutually condition and support one another in a kind of 'reciprocal interweaving and bonding'. If we generalize a little, perhaps, and take statements of measurement to embody or at least include determinate elements, then with laws and symmetries as determinables, adopting the Cassirerian framework yields a picture of the structure of the world in which we have a reciprocal interweaving and bonding between the determinate and determinable elements. Furthermore, although these can be conceptually distinguished—so we can distinguish the charge on the electron from charge in general, or the symmetric representation from the permutation group as a whole—they cannot be ascribed an independent existence—it makes no sense to ascribe an independent existence to the charge on the electron or the bosonic representation in the absence of charge or the permutation group in general, respectively (cf. Wilson 2012).[61]

With this framework in hand, and recalling our previous discussion, we can now understand the form of dependence that holds between structures and putative objects, such as quantum particles, in terms of the determinable–determinate relationship.

[60] We recall, of course, that Eddington did try to obtain such determinates on broadly structural considerations but this was doomed to failure.

[61] Wilson notes that this goes against the grain of that feature of Humeanism, which holds that fundamental entities can be freely recombined (2012: 13 n. 20). Clearly a given determinate cannot be 'freely combined' with any old determinable. McKenzie has explored less general ways in which this feature may fail in the context of modern physics (2012).

10.8 Dependence and Determinables: Delineating the Relationship between Structure and Object

However, there is a further problem that must be faced: Armstrong (1997) has argued that because determinates entail the corresponding determinable, it is the determinable that is dependent upon the determinate.[62] However, I am arguing that properties and hence putative 'objects' are dependent upon laws, and so I need the determinable–determinate relationship to hold in the other direction.

Ellis, on the other hand, takes the direction of existential entailment to be the same as that of ontological dependence and argues that determinables, and in particular, determinable natural kinds are ontologically more fundamental than any of their species (1999). Consider the example of methane molecules which depend ontologically on hydrogen and carbon atoms: the molecules could not exist if these atoms did not exist; but the atoms could exist, even though the molecules did not exist.[63] So, here, the direction of ontological dependence is the same as the direction of existential entailment, where, putting it in general terms, the As are ontologically dependent upon the Bs iff the existence of the As entails the existence of the Bs, but the existence of the Bs does not entail the existence of the As.

Now Ellis applies this to property kinds, concluding that '[t]he more general property kinds (namely, quantities and other determinables) are ontologically more fundamental than the more specific' (1999: 67). Think again of Permutation Invariance and the kinds represented by Bose–Einstein, Fermi–Dirac, and parastatistics: here I want to say that the former, qua feature of the structure of the world, is more fundamental than the latter, where certain of the latter may be instantiated in some possible worlds but not others. As Tobin has shown, Ellis' account yields a useful hierarchy of determinables (albeit with some modification; cf. Tobin 2012).

This framework offers a useful way of capturing the view I have in mind and allows laws and symmetries, as determinables, to be fundamental, and not to be dependent on sub-determinables, such as properties (e.g. charge) or determinates, such as particular values of properties (e.g. the charge on the electron).[64] In particular, in the case of OSR it's precisely the lack of independent existence of the relevant entities and kinds that acts as one of the motivations for this position. So, again, you can't have bosons or fermions qua kinds without Permutation Invariance (because it's the

[62] Rosen also invokes dependence in this context, suggesting that it explains the determinable–determinate link (2010: 128–30) and yields an account of why a thing possesses a determinable property in virtue of possessing some determinate thereof.

[63] As Ellis notes, '[t]he paradigm of ontological dependence is to be found in the theory of micro-reduction' (Ellis 1999).

[64] Of course, the alignment of dependence and entailment may depend on the particularities of the case under consideration (see Wilson 2012). And it is not true in general that determinables can exist independently of the relevant determinates.

action of the permutation group that yields the representations in terms of which the distinction makes any sense); likewise you can't have spin or charge without the Poincaré group and, pushing back a level, the symmetry that expresses; and, the claim is, you can't have the relevant physical properties without the laws. Furthermore, in so far as the particles of modern physics are given and thus determined by the relevant group-theoretic structures, you can't have these particles without these structures.

Likewise, Brading argues that the kinds that a theory as a whole talks about are encoded within the shared structure of the models of the theory (2011). So, Newton's laws plus the Law of Universal Gravitation yield the relevant kind-properties, in Brading's account, or more generally, are what those properties are dependent upon, and which feature in the collection of models and thereby encode/capture the relevant modality. And as she says, this opens a door for the structural realist:

> if the *kinds* that the theory presents are also what the theory *represents*, then the way is clear for the objects instantiating a given kind to inherit the modal properties associated with that kind. (2011: 63)[65]

Thus, the kinds, properties, and particles qua putative objects are all instanced or manifested within the relevant structure, and with that regarded as a determinable we have the situation envisaged by Ellis whereby the relevant determinable is not only part of our fundamental base but is that on which less determinable entities are dependent.[66]

So, just to sum up: focusing first on the symmetry features, the group fixes or grounds the relevant representations, and in those terms encodes the relevant possibilities.[67] Thus, using the example of the Poincaré group again, the representations are classified according to the (eigenvalues of the) Casimir operators, and certain of these get ruled out as 'non-physical' (since they involve tachyons or yield continuous spin values). These operators then give us the properties of mass and spin which are thus (metaphysically) dependent on this feature of the structure of the world. In this case, we also get the relevant differential equation (namely the Dirac equation) and in general the laws also encode the relevant modality via the suite of models that they make available. In this sense both the laws and symmetries can be regarded as determinables. The laws also fix or ground the properties—reversing the dispositionalist relationship discussed in the previous chapter—and again, the

[65] By object here she means the entities presented by the theory; what I would call the pre-reconceptualized entities, like elementary particles.

[66] I should emphasize that I have no truck with Ellis' essentialism, in terms of which his account is articulated.

[67] cf. Ryckman (2005), who notes that for both Eddington and Weyl, the actual world is to be reconstructed as a selection from a wider conceptual space of possibilities that delimit the notion of physical object. And, of course, this selection involved an ineluctably subjective component.

latter, as determinables themselves, can be understood as dependent upon the former.

But given the inherent modal nature of the laws and symmetries it cannot be the case that the fundamentality base of this, the actual, world is entirely determinable. Here the determinates act as 'existential witnesses' for the determinables in that they are indicative of the non-modal aspect of the latter. Thus it is determinables plus determinates that form the fundamentality base and in this world that base is composed of groups and the relevant representations together with laws plus the relevant initial conditions.

With that picture in mind, then, I claim that the nature of the dependence between the structure and kinds, properties, and putative 'objects' (e.g. elementary particles) is shaped, or fleshed out, by the relationship between determinables and determinates. So, if we think of a particle such as an electron, and characterize it in terms of the kind, fermion, with the properties of spin 1/2, charge *e*, (rest) mass 9.11×10^{-31} kg, etc., then it is the permutation group, as a determinable, that encodes the possibilities in the representations, one of which gives us Fermi–Dirac statistics; it is the Poincaré group, also qua determinable, that yields spin as a sub-determinable, with spin 1/2 as the determinate (and a basic fact or initial condition of this, the actual, world). Likewise, the laws and symmetries (gauge invariance) of quantum electrodynamics as determinables yield charge as a sub-determinable with the charge *e* of the electron as a determinate.

However, I have not yet explained how certain crucial characteristics of laws, associated with their supposed necessity, can be articulated from the structuralist perspective. These are characteristics that dispositionalism handles very nicely but I shall also draw on comparisons with two other approaches: first, Lange's recent account of that relationship in terms of primitive subjunctive facts and secondly, Maudlin's 'primitivist' account of laws.

10.9 Structure, Counterfactuals, and Necessity

It is the supposed necessity of laws that

i) distinguishes them from accidents
ii) explains their non-violation
iii) explains why events obtain; and
iv) grounds counterfactuals.

As Lange puts it,

> A law's necessity gives it explanatory power. That like charges must repel, for example, explains why in fact, every pair of like charges does. It is no accident or coincidence that in every such case, there is a repulsive force; it does not reflect some special condition that just happened to prevail each time. The reason for this regularity *p* is that *p* is required by law; even

if charge pairs had existed under different conditions, it would still have been the case that *p*. By entailing that *p* was unavoidable (that *p* would still have obtained, under every naturally possible circumstance), *p*'s natural necessity explains why *p* obtains. (Lange 2007: 472–3)

Now, I used the modifier 'supposed' when referring to the necessity of laws because this is typically cashed out in a way that is inimical to the structuralist project. One first supposes the same fundamental base in each possible world and then considers whether a particular putative law also holds in those possible worlds. Again typically, it is assumed that the fundamental base is entirely composed of non-nomic elements: so, for example, it may be assumed that this base is composed of all the known elementary particles, which, also typically, are conceived of in classical or at best semi-classical fashion, and then the question may be asked whether the laws that hold for such particles in this world must also hold in all possible worlds in which there is the same fundamental base as in this one. If they do, then their necessity is established and we can draw a sharp distinction between such laws and accidental generalizations (the latter do not hold in all possible worlds given the same fundamental base), explain why laws are not violated, explain why events obtain according to them, and ground the relevant counterfactuals.

Of course, some account must be given that explains why, if the fundamental base is the same, the laws *must* be the same. In effect the role of this consideration of possible worlds is to reveal a gap between the fundamental base and the laws that must then be closed. On the Humean view, this base is conceived of in terms of objects and their properties, understood as categorical, and the gap obviously cannot be closed using any features of the latter (as explanans, as it were). The obvious option then is to insist that there is no such gap, since there is no necessity 'in' the world, but only in the relevant models and theories. The explanation of (ii) to (iv) must then be sought somewhere else, namely in the 'best system' approach, as discussed in Chapter 9. Alternatively, one might seek to close the gap by re-conceiving the entities in the fundamental base. This is what the dispositionalist does, in effect, by regarding the properties of the objects not as categorical but as themselves inherently modal, that is, as dispositional. As we saw, it is then argued that the laws supervene upon or flow from such properties (generating Mumford's dilemma). Thus, if we have the same particles (as objects) in all possible worlds we will get the same laws in those worlds. Indeed, this ability to close the explanatory gap and thereby account for (i)–(iv) is seen as one of the primary virtues of the dispositionalist account.

Now it might seem that structuralism is at a disadvantage when it comes to this general methodology for establishing the necessity of laws since it denies the basic presupposition of a fundamental base that is distinct from the very laws whose modal status is being investigated. Indeed, one does not have to be an advocate of OSR to have doubts whether we can distinguish such a base as consisting of elementary particles distinct from the relevant laws and symmetry principles (see McKenzie

2011). Still, before we consider an alternative methodology, it is worth noting the following: first of all, like the Humean, the structuralist denies that there is an explanatory gap to begin with, since if we have the same fundamental base across all the (physically) possible worlds, we obviously get the same laws! Secondly, like the dispositionalist, the structuralist 'encodes' the modality into the fundamental base, although whereas the former packs it into the properties, as it were, the latter takes it to be inherent to the laws. There is certainly a sense, as noted before, that the difference between the two is just a matter of where the modal bump in the carpet is pushed to: properties or laws. However, whereas the dispositionalist closes the explanatory gap via the device of the laws supervening on or flowing from the properties, the structuralist denies there is any gap to begin with. In both cases the outcome is the same: in each of the relevant physically possible worlds we get the same combination of laws/symmetries and elementary particles. And although the dispositionalist must rely on some metaphysical notion of supervenience (or some appropriate understanding of 'flows from') to close the gap, the structuralist must likewise invoke some metaphysical notion of dependence holding between laws and properties and objects in order to deny that the gap is there to begin with. Presented like this, it may seem that, having dismissed the Humean stance, there is little to choose between dispositionalism and structuralism. However, the tie-breaker(s) lie in the incompatibility of dispositionalism with current physics, with its inability to appropriately handle the symmetries of the latter, and, ultimately, with the naturalism of the structural realist's 'reading off' our ontological commitments from the relevant theories.

That still leaves an issue for the structural realist, however. If she is not to adopt the standard methodology, with its reliance on an understanding of the fundamental base in terms of particles-as-objects, then how is she to establish the supposed necessity of laws, and explain (i)–(iv)? Now, it should come as no surprise that if we adopt an alternative methodology, we are going to arrive at a different understanding of what it is that sets laws apart from accidental generalizations, say. Concerns that the standard methodology correctly captures our intuitions about what sets laws apart are going to be swept aside on the grounds that these intuitions have no place in a realism grounded in modern physics. And indeed, the alternative nicely captures the way in which we think about laws, and their difference from accidents, in this context.

This takes laws to possess a certain 'counterfactual stability', where this is to be understood in terms of the law-like generalization remaining true under logically independent counterfactual circumstances that are accidental (Lange 2007). This can be articulated as follows: Take those propositions that do not contain the phrase 'it is a law that' or any modal operator; the set of such 'sub-nomic' propositions can then be defined as stable if the members of the set remain true under every sub-nomic supposition consistent with the set; a generalization is then regarded as lawful if and only if it belongs to the largest non-maximal stable set of true propositions; or, in other words, 'necessity involves a kind of maximal persistence under counterfactual suppositions' (2007: 472).

As it stands, of course, this says nothing about which are more fundamental, laws or counterfactuals (see Loewer 2011: 36). The dispositionalist takes both to flow from the modal nature of dispositions; whereas the Humean takes both to supervene on the set of all categorical propositions (cf. Callender 2011).[68] As we shall see shortly, Lange argues that what makes the set stable is the truth of certain counterfactuals and what makes these in turn true are certain subjunctive facts, regarded as primitive. Ultimately, then, it is the primitive subjunctive-ness in the world that accounts for the necessity of laws and explains (i)–(iv). The structuralist, on the other hand, inverts this relationship between the laws and the relevant counterfactuals and shifts the primitive modality into the former.

Thus, if we drop an account of necessity based on the methodology of using possible worlds to open a gap that must then be closed, and adopt Lange's alternative methodology, our understanding of what it is that is in need of explanation must change. We now have a characteristic feature of laws, namely their modal stability, which can be explained by: modally informed dispositions (that is, properties that are modally primitive); or primitive subjunctive facts (that is, facts that are modally primitive); or the inherently modal nature of laws (that is, laws that are modally primitive). Having eschewed Humeanism, one has to accept some primitive or inherent modality somewhere in one's picture, and I hope to show here that placing it in the structure offers an account that meshes best with how we should understand laws and symmetries in physics.

Of course, I still have to account for (ii)–(iv). (ii) and (iii) can be dealt with in a prima facie straightforward manner: laws are not violated because they are part of the fundamental base given by the very structure of the world; likewise, events obtain because they follow from or are determined by the laws, as features of the structure of the world. Thinking of Lange's quote, they are 'unavoidable' because they follow from the structure of the world. Likewise, it is 'no accident' that like charges repel because this is simply a manifestation of the structure of the world as represented classically by Coulomb's Law or, in the quantum context, by the equations of quantum electrodynamics. But what about the structuralist's response to (iv)? Here we need to give some further explication of the relationship between the laws and the relevant counterfactuals and once again it is useful to take dispositionalism as our foil.

So, as we have already noted, it is one of the perceived advantages of the dispositionalist stance that it yields both the apparent necessity of laws and the

[68] Although as Loewer notes (2011: 36 n. 12), for the Lewisian Humean laws are conceptually more fundamental than counterfactuals, since the latter are analysed in terms of similarity of possible worlds and Lewis' account of similarity involves laws.

appropriate relationship with the relevant counterfactuals.[69] Thus dispositions act as the truthmakers of law statements and the necessity associated with the latter is now taken to derive from that which is tied to the former. Of course, this just shifts the issue of the grounds of this necessity,[70] or the bump in the carpet, as we noted earlier. The nature of the modal informing, then, as itself a metaphysical feature, is not substantially illuminated by such a shift.[71] However, if, on the dispositional account, the laws are taken to function as partial identities for the relevant properties, then we can read that identity relation in reverse, as it were, and identify properties in terms of the laws. Extracting powers from the laws presented by scientific theories and taking the former as fundamental is dependent on reading those identity relations a certain way only and in this sense the dispositionalist feeling that one is getting at the ground for such laws and their associated necessity, is in fact illusory (cf. Drewery 2005: 386).

In fact we may be able to go further, since one can argue that referring to laws in a non-eliminable manner may be unavoidable within the dispositionalist stance.[72] Thus if the totality of all the behaviour of electrons, again, is determined by the relevant dispositional properties, then this must include their interactions with other kinds of particles (such as protons). But then these dispositions must encode these interactions as well, and hence there could be no conception of such properties or powers, regarded as that which determines (or governs) the totality of behaviour, that does not make reference to these other kinds of particles (Drewery 2005).[73] The

[69] Mumford retains the idea that there are relations of necessary connection between properties, and that relevant counterfactuals can be asserted but insists that the latter supervene on the appropriate potencies.

[70] Blackburn (1999: 635) presents the following dilemma to those who would pin down the 'source' of necessity: suppose this source is some truth F. Then either F is contingent or necessary. If the latter, then it seems we have not really located the source, since this option leaves us with a 'bad residual must' (1999: 635). If the former, then it would seem that far from explaining the original necessity, we have actually undermined it. Cameron (2010) questions Blackburn's presupposition that the source of necessity is to be found in the truth of some proposition. Rather, he suggests, one might think that the source of necessity is some 'thing', in virtue of which there are necessary truths. As he says, this meshes with the truthmaker theory that I deployed in Chapter 7. However, whereas the modal truthmaker theorist is concerned with pinning down 'the necessity maker', such that for every necessary truth p, p is necessary in virtue of the existence of that necessity maker (2010: 137), we might adapt the theory to talk about the necessity makers for different kinds of necessary truth, such as logical, metaphysical, physical, and so on. Thus for the dispositionalist the necessity maker for those physically necessary truths expressed by laws would be the relevant dispositions, understood as 'seated' in the appropriate objects. For the structuralist the necessity maker would be, of course, the structure of the world, although given our adoption of the alternative Langian methodology this might be better referred to as the 'stabilizer'.

[71] And hence an empiricist who adheres to a 'de-modalized' view of properties and laws may remain unimpressed by such a move. Indeed, for the Humean, the 'source' of modality is being connected to a good systemization of the world. As Callender (2011) has put it, the modality 'flows' from the systematization.

[72] And of course, if it is maintained that all that there 'really' are, are dispositions, so that laws can be eliminated, then Mumford will claim victory (2006).

[73] Drewery presses this point against the kind essentialist but insists that it can be made against the dispositionalist view of properties as well.

behaviour of the latter will in turn be determined by the relevant set of dispositional properties and thus we seem to be driven towards the kind of holistic dispositionalism that Chakravartty favours (2007). But in so far as this signals a shift from the particularist picture that appeared to undergird the dispositionalist stance, it brings this stance closer to structuralism.[74]

Furthermore, the latter resolves a further problem that can be raised in this context (Drewery 2005). Thus, according to the dispositionalist, leptons and quarks are individuated purely by the relevant properties (rest mass, charge, spin, etc.), conceived of in dispositional terms. However, the relevant kinds are empirically discovered and the possession of these properties is obviously not an analytic matter. Thus, from this standpoint, it seems epistemically possible that the charge on the electron, say, could be slightly different from what it is. In such a case, the dispositionalist seems forced to admit that we would then have a different kind of particle but as Drewery notes, whether there could be no other grouping of fundamental particles with similar but slightly different properties is a matter for science to determine:[75]

> The fact that we are discovering the fundamental particles empirically means we can sensibly ask the question: must electrons have such-and-such a mass, charge, or spin? In our world, we individuate them by these properties: we see that the fundamental particles are divided into similar groups which share properties in this way. We label the groups 'electron', 'up quark', 'muon', and so on. But this does not tell us that basically the same groupings could occur but where the different particles possessed slightly different properties. (Drewery 2005: 390)

However, science does 'tell' us this via the sorts of symmetry-based considerations alluded to here; indeed, the case of the Ω^- particle demonstrates that these considerations allow us to predict what the relevant properties of new kinds of particle must be. Of course this offers little in the way of support for the dispositionalist for the kinds of reasons already discussed. And as I have argued, the metaphysical necessity typically ascribed to laws cannot be simply grounded in dispositions conceived of as distinct; at the very least we must consider them as holistically interrelated if we are to accommodate the interactions between different kinds of particles.

How then does this feature of necessity look from the structuralist perspective? Note, first of all, that where the dispositionalist insists that an electron, say, could not continue to be the same sort of entity should the laws be different because changing the laws would be tantamount to changing the relevant cluster of dispositions

[74] cf. Drewery who writes, with regard to kind essentialism in this context, 'the fact that the so-called essences are co-dependent vitiates their claim to be essences rather than laws' (Drewery 2005: 388).

[75] The source of this feeling lies in a comparison with the kind of essentialism Kripke and Putnam espoused with regard to water, where she follows Psillos in suggesting that if we still had molecules consisting of two hydrogen and one oxygen atom but that did not bond to form the structure that water actually has, we would still call the substance consisting of such molecules 'water'. However, the concern here is that without that structure the properties of this new substance would be very different from (actual) water indeed.

(cf. Ghins 2007: 142), the structuralist can agree with the first part of this claim, but assert that it is because of the relevant ontological dependence between the electron, qua object, and the laws, qua relevant features of the structure, together with the relevant symmetries of course. Furthermore, where the dispositionalist also insists that electrons are the same in all possible worlds, because they must interact according to Maxwell's laws or otherwise they would be another type of particle (Ghins 2007), the structuralist would also agree. However, the dispositionalist takes this as a basic presupposition and then argues that with the introduction of dispositions, powers, etc., the laws would also be the same, whereas we take it the other way round: where the laws, or more generally, the relevant structure (since we need the symmetries as well) are the same, then we will obtain the same kinds of particles. And finally, where the dispositionalist takes the necessity of a law to be 'rooted' in the dispositions of the associated entities (Ghins 2007), the structuralist takes this statement of the relationship between laws and entities and reads it in reverse, as it were, taking the 'rooting' to proceed in the opposite direction.

Of course, just as from the dispositionalist perspective we may imagine a world with different objects and properties, and hence different dispositions, and then consider what the laws would be like, so the structuralist can conceive of worlds in which different laws hold, and then consider what different kinds of particles would result. One way of doing this is to conceive of worlds in which different fundamental constants apply, yielding a form of necessity in this context:

> Our laws are physically necessary in that in any world where the fundamental constants have the same values, the laws are the same. (Drewery 2005: 392)

I shall return to this point and the associated issue of how we might understand so-called 'counterlegal' statements shortly.

Having grounded the necessity of laws, dispositionalists—along with others who take laws to be necessary—can then explain why laws support counterfactuals: they do so in the same way that other necessary truths do. Thus consider the assertion, 'if an electron were to fall under the influence of an electromagnetic field it would experience an appropriate force and associated acceleration'. The truth of this counterfactual is taken to follow from the truth of Maxwell's laws and it is this legitimacy of such inferences that supports the view that laws must be modally informed, where, of course, the modality is grounded as indicated previously.

Now as we have seen, although the structuralist may also take laws to be modally informed, she cannot support such claims of necessity with the same methodology nor can she take them to be necessary in the same sense as the dispositionalist. This may seem to present a problem, except of course that the very same reason why she does not take them so leads her to re-conceive these counterfactuals. If the 'electron' in the antecedent is understood to be an object ontologically distinct from the relevant structure, then the counterfactual is true by default because the antecedent is false—the structuralist denies that there are electrons in that (metaphysical) sense.

If the 'electron' is understood in a broader, phenomenological sense as that kind of particle the behaviour of which is investigated through the theories and experiments of physics, then the antecedent and counterfactual as a whole can be taken to be true (within the domain of classical Maxwellian electrodynamics). And the truth of the counterfactual can be appropriately grounded, even adopting an eliminativist stance. The counterfactual's truth follows from that of claims about the relevant features of the structure of the world and the support given to that counterfactual by the relevant laws is also explained by the modal nature of the latter.

In the absence of such a reconceptualization, of course, the counterfactual refers to a picture—of electrons interacting with a field and 'experiencing' a force—that the structuralist is simply going to reject as out of step with current physics. Recall how particle interactions are understood according to the Standard Model: there we have forces mediated by 'particles' (gluons, photons, etc.) that are themselves understood field-theoretically with symmetry playing a crucial role, since $SU(3) \times SU(2) \times U(1)$ yields the relevant interactions (strong and electroweak). Attempts to recover a concept of 'particle' qua object that might fit with the aforementioned picture of particles 'banging into one another' within the space-time arena are of course notoriously problematic (French and Krause 2006: ch. 9). In the absence of an interpretation that is at least more consonant with current physics, the structuralist will be inclined to dismiss the counterfactual as a potential explanandum (or to be blunt, as not the sort of thing worth explaining). As reconceptualized, it can be explained in structuralist terms, just as the form of the counterfactual that the dispositionalist takes as requiring explanation can be explained in dispositional terms.[76]

Furthermore, let us consider again claims involving relationships between kinds of particles. Thus consider the counterfactual, 'Had an electron been present at a certain spatio-temporal location, then all protons would have possessed a rest mass of 1.63×10^{-24} g' (Lange 2004). They possess this mass out of metaphysical necessity; so, how is the following counterfactual to be explained, on the dispositionalist stance: 'Had an electron been present at that location, atomic nuclei would have still contained protons, rather than schprotons (stipulated to have half the mass of protons)'? (2004: 224). Of course, one could appeal to a law that rules out schprotons (by setting out all the acceptable natural kinds), but then this too would have to be grounded in an appropriate set of dispositions and it is not clear how the relevant set of objects could be introduced in a non-ad hoc manner.[77] Again the structuralist, by appealing

[76] In so far as anti-Humeans take the relevant counterfactuals to articulate that (determinable) structure of the world that goes beyond the (determinate) regularities, such reconceptualization is more or less forced upon the structuralist, given how she sees this structure (as compared to the dispositionalist). Of course, for the Humean, there is no such further structure. As Hall puts it (2011: 107), '[t]here is just a decision that, when evaluating counterfactuals, we must always hold fixed those de facto claims about the world that she counts as "fundamental physical laws"'.

[77] Again, one possibility would be to appeal to 'the world' as the relevant kind or 'being the world' as the relevant property but this will face the same objections as indicated earlier (cf. Lange 2004: 230–2).

to the kinds of symmetry considerations touched on earlier (and incorporated into the 'Eight-Fold Way', for example), and effectively extending the relevant sense of metaphysical necessity to cover these symmetries, can provide the appropriate explanation.

Such explanatory power also extends to and could be used to supplement a similar account, which takes lawhood to be ontologically primitive (Maudlin 2007). This can be motivated through reflection on the practice of science: since this practice (at least as it is manifested in physics) takes laws to be 'further unanalyzable primitives' (2007: 105), so should we as philosophers. And once we have laws as primitive, handling possibility, causation, and counterfactuals becomes a comparatively straightforward matter (2007: 66). So, for example, take a standard kind of counterfactual conditional: consider the time at which the counterfactual is to be evaluated;[78] adjust the relevant physical magnitudes so as to make the antecedent true;[79] then consider the future physical magnitudes generated by the relevant laws.[80] If the consequent holds across all the states so generated then the counterfactual is true; if it holds in none, then the counterfactual is false; and if it holds in some, then the counterfactual has an indeterminate truth value.

Now, it may appear that the relevant laws only enter at that final stage (2007: 23) but when we adjust the relevant physical magnitudes to make the antecedent true, we presumably keep the relevant laws fixed (see Lange 2009b: 199). That then raises the crucial question: 'what are laws such that we must be so mindful of them?' (2009b). By virtue of refusing to offer any analysis of laws, the primitivist cannot account for why we take them, and the form of necessity they are associated with, as special or significant. In particular, we cannot explain why a given fact acquires a kind of necessity by virtue of following from the relevant laws (Lange 2009b: 198).

Here it seems the structuralist can step in: what are laws that we must be so mindful of them? Why, they are features of the fundamental ontology of the world! Why does a given fact acquire a kind of necessity by virtue of following from the relevant laws? Because that fact drops out of, and is dependent on, the fundamental ontology of the world. Of course, this is not a *reductive* analysis, but to insist on that would be to beg the question against primitivism. Nevertheless it is an analysis of sorts, since by situating primitivism in a structuralist context, it acquires a certain ontological 'force', that allows a response to the kinds of concerns expressed previously.[81] The upshot, then, is that the primitivist should be a structuralist!

[78] Strictly, consider the Cauchy surface running across the world at that time.

[79] Strictly, adjust these magnitudes on the Cauchy surface.

[80] Strictly, 'fundamental laws of temporal evolution'. Strictly... come dancing.

[81] Having said that, I tend to agree that we should treat the insistence on regarding the practice of physics as our metaphysical touchstone with some care; after all, physicists do not typically analyse counterfactuals in the way that the primitivist about laws does (Lange 2009b: 200); rather, the crucial question is what is the most appropriate way of interpreting our best theories? (2009b: 199).

10.10 Counterlegals and Structuralism

Of course, there are some important kinds of counterfactuals that simply cannot have their truth values fixed by taking laws as fundamental or primitive because they involve consideration of changes to the laws themselves. These are 'counterlegal' statements, such as 'if the fundamental forces had been different, the relevant conservation laws and fundamental dynamical law would still have held', and 'if the fundamental dynamical laws had been different, the relevant symmetry principles would still have held' (Lange 2009a: 140–1). First of all, let me note that even before one considers what it is that fixes the truth values of such claims, one might wonder on what grounds they might be held as true in the first place. In the case of the first sort of claim, it might seem obvious that if the 'grubby force laws' (Lange 2009b: 141) of classical mechanics, such as Hooke's Law, for example ($F = -kx$, in case anyone has forgotten their high school physics), had been different, $F = ma$ would still have held. Indeed, this obviousness seems to follow from the practice of 'plugging' in different force laws into Newton's Second Law in order to obtain the relevant state evolution in different situations. But now consider the 'truly' fundamental forces—electroweak, strong, gravitational. If there had been different such forces, would conservation of energy have held? Here, the answer is not so obvious. Indeed, the history and practice of physics itself suggests we might exercise caution in asserting such counterlegals. Consider, for example, the Bohr–Kramers–Slater (BKS) theory of the emission and absorption of radiation by atoms, which implied that energy and momentum were only conserved statistically overall and not necessarily in each interaction; or Pauli's account of β-decay which retained conservation of energy—that appeared to be experimentally violated—by introducing a new particle, subsequently discovered and dubbed the neutrino. Now suppose Pauli (or anyone else) had been unable to posit the neutrino as emitted in β-decay and, as in the BKS case, physicists had entertained the possibility of a failure of energy conservation. If that was a possibility for electroweak interactions, is it obvious that if there were different fundamental forces energy conservation would still hold?

And as for symmetries, we know that the possibility that one symmetry, namely parity, may not hold for certain forces, is an actuality—it is violated in the case of the weak interactions. So again, can we be so sure of the truth of our second counterlegal? In particular, do we have a sufficiently strong grip on the kinds of forces that would have existed under various counterfactual suppositions? Consider: prior to 1956 we might have entertained with some degree of confidence the claim that all forces obey parity symmetry. This itself should give us pause in taking such a claim to 'confirm' further claims involving even more recondite counterfactuals than those entertained by Lee and Yang et al.

It is important to acknowledge that physicists do not take the relevant symmetry claims as *evidence* for such counterlegals, but rather take them as heuristic principles which may serve to construct new theories, both within the same domain or in new ones (Post 1971). If such new theories are empirically successful then we may take *that* as further evidence for the universality of the relevant symmetry, as in the extension of gauge invariance to the strong nuclear force (see French 1995). We may then further speculate as to the structural similarities between the relevant domains that this common symmetry reveals. However, as the case of parity violation indicates, this is a fallible procedure and hence the idea that claims regarding symmetries can stand as evidence for counterlegal assertions seems problematic. Certainly, the practice of physics does not provide unequivocal evidence of the truth of such counterlegals and, again, it is certainly not obvious that we need to posit something that fixes the truth value of such claims. Given that, such counterlegals do not constitute a strong objection to taking laws as fundamental.

Furthermore, is it the case that laws and symmetries must be related in such a way that we can hold one fixed and straightforwardly vary the other, leading to these kinds of counterlegal assertions? We recall (from Chapter 6, for example) the two most well-known views of the relationship between symmetries and laws: the first takes the former to constrain the latter; the second takes the former to be by-products of the latter. On the first view, symmetries, as constraints on laws, belong to a nomically stable set that excludes the relevant dynamical laws, force laws, etc. It is in such terms that we can understand the claim that the symmetries would have held had the other laws been different.

Notice, however, that a kind of modal 'gap' has been opened up, such that we can effectively hold the symmetries fixed, and then entertain the (meta-?) possibility of the laws being different. Concern about the counterlegals considered here might lead one to suggest that opening up such a gap is problematic, or, more strongly, that on the basis of the view of physics practice sketched previously, it should not be opened up in the first place. Can we make any sense of this notion of a 'constraint' without such a gap? In particular, can we make any sense of such a notion, or of symmetries acting as 'meta-laws' from the structuralist perspective?

We might begin by thinking of symmetries and laws as distinct 'aspects' or 'features' of the structure of the world. This allows us to distinguish them to the extent that we can now think of the relationship between them. First of all, even in those cases where the relationship is such that the relevant symmetry is associated with the conservation of a quantity the interrelationship between whose instances is described by the relevant law, the fact that the symmetry can be taken to express a regularity at the (meta-) level of the laws themselves does not in itself imply any kind of priority of the former over the latter. At this stage, at least, we might appeal to Cassirer's notion of the symmetries acting as 'highest rules and principles' which 'hold together' the web of relations represented by the laws. However, one should not read too much into the 'holding together': it is not as if the symmetries 'govern' the laws, in the way that the laws 'govern' the relevant phenomena. Just as, from the structuralist perspective, the laws do not govern the phenomena, or the objects that are taken to fall under them, but rather the latter are dependent on the former, so the symmetries do not govern the laws, but rather there is a further kind of dependence in so far as the symmetries represent that which is invariant about the web of relations. Indeed, by doing so, they contribute to the dependence between the structure and the (putative) objects, by effectively grounding what Chakravartty calls the 'sociability' of the properties associated with these objects (or better, in terms of which these objects are effectively constructed).

We further recall that Cassirer locates the symmetries of physics among his 'statements of third order' or statements of principle that express how the laws themselves are interrelated. These various kinds of statement—of measurement, of laws, and of principles—all mutually condition and support one another and their mutual and 'well-rounded' interconnection means that although they can be logically distinguished they cannot be ascribed an independent existence in the sense that we could have one such particular set without the other. Thus we should be careful not to be in thrall to just the kind of spatial metaphor that Cassirer urged we should reject, in which the structure is viewed as a pyramid, with the symmetry principles at the top, the laws in the middle, and the results of measurement at the bottom. This would suggest that one or other layer could be removed, as it were, without affecting the others, in just the kind of counterlegal move envisaged earlier and which produces the 'gap' between laws and symmetries. As Cassirer insisted, this would be untenable since the truth of all such statements at whatever level is due to their interconnectedness. In our terms, there can be no gap; at least not while preserving the structure of the world.[82]

In a sense, then, the structuralist stance blurs the distinction between symmetries as by-products and as constraints. If the latter is understood in terms of some further modal strength that symmetries are supposed to have, then this is ungrounded in the practice of physics. And if the symmetries are taken to constrain the laws, in the manner indicated, then they do so by representing the interconnections between the latter but in such a way that, as Cassirer indicated, we should not conceptually imagine the symmetries as existing distinct from the laws; in that sense, they are like by-products. The counterlegals thus lose their force from the perspective afforded by OSR.

[82] Of course, it might be responded that while it is acknowledged that by removing the relevant laws while keeping the symmetries fixed, we are no longer talking about the structure of 'our' world, nevertheless such a move enables us to get a handle on the relationship between symmetries and laws, as expressed by the greater necessity of the former. In response, I can only repeat the worries about the counterlegal claims that this move depends on and say that just as we do not need the necessity of laws as understood via the kind of possible worlds exercise that the dispositionalist appeals to in order to explain why they are not violated, for example, so we do not need the broader necessity that might be associated with symmetry principles in order to understand their role.

10.11 Conclusion

From this perspective, then, laws and symmetry principles are simply features of the underlying structure (or, better, law statements, etc., are descriptions of significant features of that structure). The properties of purported objects have the ontological features they do because of the law-like (and other) relations they enter into and the purported objects themselves, qua substantive metaphysical entities (or at the very least qua the kinds of entities that might be said to have an 'individuality profile'), are reconceptualized as nothing more than nodes or metaphorical 'crossing points' in this network of relations (and hence can be eliminated). The 'governing' metaphor that is used to articulate the relationship between the laws and these purported objects and their properties is then replaced by the relation of metaphysical dependence:[83] the purported objects are dependent on the structures (and here the role of symmetries in presenting that dependence is fundamental) and the properties are dependent on the laws themselves. Indeed, if we take it that it is laws that are 'read off' our theories, rather than objects per se, then it is difficult to see where this governing role comes into the picture.

Furthermore, these laws and symmetries encode the relevant possibilities and in so far as they are stable or robust under certain changes to initial conditions can be regarded as necessary, in a certain sense. In this way the structure of the world can be said to be modally informed.

This concludes my exposition of the metaphysics of structure. In the spirit of the 'Viking' Approach I hope to have indicated how the advocate of OSR can appropriate various metaphysical concepts, distinctions, and strategies to flesh out this position. The next two chapters will be concerned with its extension (in some form or other) to, first, quantum field theory, and, secondly, biology (or, better, certain features of biology). With regard to the former I shall consider a fundamental obstacle to this extension and suggest how it might be overcome by drawing upon the metaphysical features presented in this chapter, to do with how we should understand possibility from the structuralist perspective. With regard to the latter, I shall indicate how some notion of structure might be delineated in the biological domain, again drawing on points from this and previous chapters and note that there are problems associated with an object-oriented stance here as well.

[83] Here again we may hark back to history: Eddington likewise rejected the governing conception of laws in the context of General Relativity, arguing that it introduced a kind of dualism that sat at odds with his structuralism (Ryckman 2005: 7.4.2). Instead, he advocated a view of Einstein's equations as 'definitions of the way in which certain states of the world (described in terms of indefinables) impress themselves on our perceptions' (Eddington 1923). If 'states of the world' are understood as features of the world-structure, the view I advocate here can perhaps be seen as a close relative of Eddington's.

11

Structure, Modality, and Unitary Inequivalence[1]

11.1 Introduction

As I have emphasized, one of the principal motivations for OSR concerns the treatment of objects in quantum mechanics. And certain aspects of the elaboration of this position have drawn on features of that theory. In this regard, not least, it represents an advance upon ESR and object-oriented realism, but the question naturally arises whether OSR can be extended to quantum field theory (QFT). One might expect a straightforward and positive answer, given Redhead's early advocacy of a structural realist stance towards fields (1995). However, there is another serious obstacle that OSR must overcome, one that is based not on problematic metaphysics but on certain fundamental features of the relevant physics, namely the existence of unitarily inequivalent representations.

In this chapter I shall explore two possible ways round this obstacle. The first involves adopting Wallace's 'naïve Lagrangian' interpretation of QFT and dismissing the generation of inequivalent representations as either a mathematical artefact or as non-pathological. The second takes up Ruetsche's 'Swiss Army Knife' approach and understands the relevant structure as spanning a range of possibilities, drawing on the discussion in Chapter 10. Both options present interesting implications for structural realism and I shall also consider related issues to do with the underdetermination of theories, the significance of spontaneous symmetry breaking, and how we should understand superselection rules in the context of quantum statistics. Finally, I shall suggest a way in which these options might be combined.

11.2 Being a Realist about QFT

So, where does the realist stand when it comes to QFT? The answer obviously depends on what she takes to be the elements of her ontology. 'Object-oriented' realism faces well-known problems in spelling these out. They cannot be particles, if

[1] This chapter is based on a version of French 2012a; I'd like to thank David Baker, Doreen Fraser, and David Wallace for helpful comments.

by 'particle' we mean entities that are *countable*, even if we allow that they do not have to be localizable (Fraser 2008).[2] If being countable is regarded as an essential feature of individuals then a metaphysics of particles as individual objects is ruled out.

Furthermore, according to Haag's theorem, QFT-incorporating interactions cannot be represented in the same Hilbert space as the theory of free fields. Hence an interacting field cannot be represented in terms of the superposition of free particles (Fraser 2008). One possibility then might be to pursue Bain's 'asymptotic particle' approach (Bain 2000), in which rather than trying to extend the particle concept from the 'free' context to that of interactions—which is what Haag's theorem blocks—one begins with the interaction picture and constructs a (limiting) notion of particle on that basis. The obvious objection is that asymptotic freedom yields too 'thin' a notion to satisfy the realist: with the number occupation operator only being defined in the asymptotic limit, countability is again a problem.[3]

These arguments against a particle interpretation of QFT may not be decisive but they certainly push the realist to place her ontological commitments elsewhere. A natural alternative would be a field-based metaphysics. However, similar arguments can also be turned against fields (Baker 2009) and thus this option also becomes problematic.[4] As in the particle case, there may be a way of sustaining some kind of field interpretation[5] but, again, the realist might feel it best to look for a further alternative ontology.

11.3 Field-Theoretic Structuralism

Here again it might be useful to recall some of the 'lost history' outlined in Chapter 4.

So, Cassirer, again, argued that the metaphysics of the 'material point' qua individual objects cannot be sustained once we make the transition to field theory (1936/1956: 178). In its place he offers a structuralist conception:

> The field is not a 'thing'; it is a system of effects (*Wirkungen*), and from this system no individual element can be isolated and retained as permanent, as being 'identical with itself'

[2] Colosi and Rovelli draw a distinction between 'global' particles, defined in terms of n-particle Fock states, and 'local' particles, which are eigenstates of local field operators. The latter, but not the former, are countable and correspond to what is detected in finite-sized particle detectors. In the limit in which these detectors are appropriately large, the two converge in a weak topology (2008).

[3] One could perhaps defend something akin to an object-oriented stance by adopting the formal framework of quasi-set theory (see French and Krause 2006: ch. 9). This provides a quasi-cardinal assigned to aggregates of quanta, without there being a (classically) correlative ordinal—so in one sense we lose countability but we could still regard the aggregate in an object-like manner.

[4] See also Halvorson and Müger (2006) on the difficulties associated with defining field quantities at space-time points.

[5] Despite the problems, field operators can still be defined, allowing the possibility of an interpretation of interacting states as yielding probabilities for the manifestation of 'field-like quantities' (Baker 2009: 606).

through the course of time. The individual electron no longer has any substantiality in the sense that it *per se est et per se concipitur*; it 'exists' only in its relation to the field, as a 'singular location' in it. (1936/1956: 178)

However, this characterization of the field as 'a system of effects' raises the 'how can we have structures without objects?' question in a new guise: how can we have an effect without a something which is doing the effecting? Again we face the issue of coming up with a thoroughgoing structuralism which avoids that which is ontologically non-structural. The problem is that in contemporary accounts, fields are *local* in the sense that field quantities are attributed to space-time points (or, taking into account quantum effects, space-time regions). Again, a form of metaphysical underdetermination arises here with the physics supporting both the view of fields as substances whose properties are instantiated at space-time points (or regions) and the view of fields as nothing but properties of those space-time points (or regions).

The former option reintroduces a non-structural substantiality, whereas the latter shifts the focus to space-time. Understanding *this* in substantival terms again brings a non-structural element back into play, but the standard relationist interpretation threatens to generate a circularity. One possible way out of the dilemma, of course, is to explore a structuralist understanding of space-time itself:[6]

To say that spacetime exists just means that the physical world exemplifies, or instantiates, a web of spatiotemporal relations that are described mathematically. (Dorato 2000: 7)

Again, however, the issue of the relata raises its ugly head,

to the extent that real relations, as it is plausible, presuppose the existence of *relata, then spatiotemporal relations presuppose physical systems and events.* (2000: 7; author's emphasis)

Now, we may avoid the *supervenience* of such relations on relata by adopting a form of 'bundle' theory as sketched in Chapter 7, but a further problem arises in that the identity of the space-time points is grounded in the relational structure provided by the gravitational field (2000: 3; Cao 1997). This, of course, throws the issue back to the ontology of fields.[7]

[6] See again Auyang (1995), who proposes a view of space-time according to which it is absolute, in the sense that it is presupposed by the 'concept' of individual things, but not substantival, structural but not relational, in that the relations involved are only 'implicit' (1995: 138). It is the space-time structure that keeps 'events' numerically distinct. The events are entities in an interacting field system (1995: 129) which are identified by a parameter, x, of the relevant base space in a fibre bundle formulation, and divided into kinds via the appropriate group (1995: 130–2; note that here again the kinds are entirely structural). Events are thus individuated structurally within 'the whole' and the conceptual structure of the world as a field is represented by a fibre bundle (1995: 133). Returning to the nature of space-time, Auyang is clear that neither the space-time structure, nor the event structure should be given ontological priority: '[t]he event structure and the spatio-temporal structure of the objective world emerge together' (1995: 135).

[7] A version of the Redhead argument arises here: 'I don't know how one can attribute existence as a *set of relations* in an observable or unobservable domain without also requiring that these relations be exemplified by non-abstract *relata*, namely the field itself, to be regarded as a new type of substance, radically different from the traditional, Aristotelian *ousia*' (Dorato 2000: 3; his emphasis). That Dorato

Of course, the field concept is used to generate the field equations of QFT and these can be understood as describing the structural aspects of 'these hypothetical entities', from which can be 'extracted' the concept of particle which is the 'observable manifestation' of the same hypothetical entity (Cao 2003). But then, what is this 'hypothetical entity', over and above the structural aspects? OSR has an answer, of course—the field is the structure, the whole structure and nothing but the structure.

11.4 The Generation of Inequivalent Representations[8]

However, the structural realist now has to face a fundamental problem: how to accommodate the unitarily inequivalent representations that arise in QFT, described as '[p]erhaps the single most important problem in the foundations of QFT' (Baker 2009: 592). If we assume that unitary equivalence is a necessary condition for the physical equivalence of Hilbert space quantizations, then the existence of these inequivalent representations implies that there are many physically inequivalent quantizations of QFT all competing for the claim that they represent the 'correct' quantization (Ruetsche 2003: 1332; 2011).

This raises a problem when it comes to adopting any kind of realist stance towards QFT, since the realist will be faced with a set of non-isomorphic choices. Which of these will she take as the 'correct' (in some sense) representation? She cannot simply acknowledge that they are not equivalent and leave it at that. On the other hand, if she picks one, she must give grounds for doing so but, as we shall see, in so far as such grounds can be given, they run the risk of dismissing as physically non-significant features that we would not want to so dismiss (Ruetsche 2003).

As Howard observes (2011), this problem seems particularly damaging for the structural realist, for the following reason. In his original presentation of OSR, Ladyman (1998) gave as an example of the virtues of this position the equivalence between matrix and wave mechanics, with both understood in terms of functionals on Hilbert space. Features of the latter are then taken to represent relevant aspects of the structure of the world. In this case, where we have systems with finite degrees of freedom, the Stone–von Neumann theorem ensures equivalence of representations (for extensive discussion see Ruetsche 2011). In the case of systems with infinite

inclines towards a form of epistemic structural realism here is clear from his insistence that, although we often identify physical entities via their relations, 'epistemic strategies for identifications should not be exchanged for ontological claims' (2000: 3). The latter claims obtain their warrant from the sorts of lab-based practices that support entity realism and Dorato argues that structural realism needs entity realism to be plausible (2000: 4). However, these practices crucially involve or depend on *causal* relations and given this, there is nothing in the practices themselves that particularly tells against structural realism and in favour of entity realism. Indeed, as I have said, in so far as causal relations are taken to be a fundamental feature of the structure of the world, the ontic structural realist can simply take such practices and insist that they reveal the dissolution of physical objects into structures, including causal ones (cf. Chakravartty 1998).

[8] For details see Haag 1992; Halvorson and Mueger 2006: §7; Ruetsche 2003 and 2011.

degrees of freedom, such as those covered by QFT, the theorem fails. Thus, in the absence of isomorphism in the case of these representations of QFT, the structural realist appears to be in difficulty. As Howard emphasizes, 'it's not just that there *happen* to be a variety of alternative ontological pictures among which theory is impotent to choose. No, in this case there exist, *of necessity*, a variety of structurally inequivalent representations of physical reality' (2011: 231). The problem then is that, on the one hand, the structural realist cannot respond in the same way as she did with the wave and matrix mechanics representations; but on the other, it is not clear that any grounds for choosing one inequivalent representation over another can be justified in structuralist or indeed any realistically acceptable terms. How might the structural realist respond to this dilemma?

Before I begin exploring possible responses to this question, let me briefly recall the origin of the problem, which lies with the 'algebraic' programme in the foundations of QFT.

According to this programme, the theory should be fundamentally described mathematically in terms of a net of (observable) algebras, where an algebra of (bounded) operators on Hilbert space is associated with open regions of space-time, and the algebra is generated by 'smeared out' fields with test functions having their support in the relevant region (Haag 1992: 104).[9] The role of the fields on this conception is then only to provide a 'co-ordinatization' of the net (Haag 1992: 104).

According to the 'algebraic chauvinist' (Ruetsche 2003: 1334), all the physical content of the theory is encoded in that net, with the representations of the algebra seen as having diminished or, more strongly, *no* ontological significance. Furthermore, by the Gelfand–Neimark–Segal theorem, all the Hilbert spaces we need are 'hidden inside' the algebra (Halvorson and Müger 2006: 38). However, this may appear not to yield all the features of QFT that we want, such as the connection between spin and statistics (Halvorson and Müger 2006). Hence we need to shift to another form of chauvinism—'Hilbert space chauvinism' (Ruetsche 2003: 1330)—that identifies physically relevant observables with the set of bounded operators on a particular Hilbert space. The problem then is how to justify that choice of Hilbert space in the face of the existence of inequivalent representations.[10]

[9] Alternatively, as Baker has suggested (and as we have just noted), one could claim that the expected value of an operator in the local algebra assigned to an open region of space-time is a physical property of that region. Of course that raises the issue of how we should interpret such regions and space-time in general from a realist perspective. A structuralist interpretation of space-time is again an obvious option here (see, for example, Stachel 2002; Rickles and French 2006).

[10] Kronz (2004) suggests that the algebraic quantum field theorist views this feature as an embarrassment and contrasts this attitude with that adopted in quantum statistical mechanics where these inequivalent representations are seen as having physical significance. However, as we shall see, the stance of those working in the foundations of algebraic QFT, at least, may have shifted towards the latter attitude, as the role of such representations in understanding superselecting rules and spontaneous symmetry breaking is emphasized.

How should the structural realist respond to this problem? One option is to reject the algebraic programme entirely and insist that the appropriate characterization of QFT should draw on the practices of physicists.

11.5 Option 1: Adopt 'Lagrangian' QFT

The Standard Model, from which OSR draws much of its force, is articulated in what can be called Lagrangian QFT (see Wallace 2006 and 2011). Here one begins with a classical field, expresses it in Lagrangian form, and then quantizes it. However, when interactions are incorporated, integration over arbitrarily short-length scales is required, yielding the infamous infinities of QFT. The standard move is to take the cue from condensed matter physics and introduce a cut-off length beyond which the theory is no longer taken to apply. This resolves the problem and the inequivalent representations can then be dismissed, depending on their type, as either not arising within the mathematics, or physically real, but non-pathological (Wallace 2006).

Since these inequivalent representations are only a feature of systems with infinite degrees of freedom, there are two ways in which they can arise: when we go to short distances and high energies, or when we go to long distances. In the former case the inequivalences occur because of the existence of degrees of freedom at arbitrarily short-length scales, but given the cut-off at these length scales, they simply no longer arise.

In the 'long-distance' case, the inequivalences are generated through the imposition of different boundary conditions imposed on the relevant wave-functionals at infinity (see Wallace 2006: 57–8 for examples). However, an analogy can be drawn between the inaccessibility of the representations and that of the long-distance structure of the universe given measurements confined to a local spatial region. In other words, the inequivalent representations encode inaccessible information but since 'we are always analysing a theory in a finite region—and idealizing the system beyond that region in whatever manner is convenient— . . . different choices of representation should not affect our conclusions' (2006: 59).[11]

So, if the structural realist were to adopt this understanding of QFT, she could respond to Howard's challenge by simply dismissing inequivalent representations in the way just indicated. Still, given that Algebraic AQFT might seem to be the 'natural' programme for the structural realist to adopt, it is worth exploring it a little further.

[11] In his more recent work Wallace (2011) makes it clear that algebraic methods may be drawn upon to tackle these representations but maintains that this is not tantamount to acceding to the requirements of the AQFT programme.

11.6 Response: AQFT, Inequivalence, and Underdetermination

Advocates of Algebraic QFT (AQFT) insist that introducing a cut-off length inserts an arbitrary element into the theory and that the search must continue for a theory that is well defined at *all* length scales. Here, then, we have competing programmes, or even a form of underdetermination holding between 'Lagrangian' and Algebraic QFT (Fraser 2009; 2011). Recalling our discussion from Chapter 2, can the underdetermination be broken in favour of the latter?[12]

A positive answer can be constructed on the understanding that Haag's theorem identifies an inconsistency in the foundations of QFT:

> Let F be the statement that the system described is free. By Haag's theorem, {T} ⇒ F. But, the interaction picture was introduced for the purpose of treating interacting systems; thus, by assumption, the system described by the interaction picture is not free. This sets up a *reductio ad absurdum*: {T}&~F ⇒ F&~F. Thus, Haag's theorem informs us that the source of the problem with the interaction picture is that it is inconsistent. Furthermore, Haag's theorem establishes that this is an entirely generic problem; the theorem does not hinge on any assumptions about the specific form taken by the interaction. (Fraser 2009: 547)

One way to respond to this inconsistency is to introduce a 'cut-off' and reduce the number of degrees of freedom from infinite to finite, thereby blocking the applicability of Haag's theorem. The AQFT approach, on the other hand, seeks to modify or reject one of the core assumptions of the theory in the hope of obtaining a consistent set of axioms. Fraser characterizes these as 'pragmatic' and 'principle' approaches respectively. This is not an uncommon situation in the history of science and various pragmatic moves have been adopted for dealing with inconsistencies (see da Costa and French 2003: ch. 5). However, in the case of QFT there may be a compelling reason to demand a consistent formulation, since QFT is by definition the theory that unifies quantum mechanics and Special Relativity:

> the project of formulating quantum field theory cannot be considered successful until either a consistent theory that incorporates both relativistic and quantum principles has been obtained or a convincing argument has been made that such a theory is not possible. (Fraser 2009: 550)

Of course, one can adopt a heuristic stance towards such inconsistencies, and regard theories containing them as stepping stones towards consistent successors, as in the cases of Bohr's theory of the atom, or the old quantum theory in general (Vickers

[12] However, Wallace notes that in her (2009) Fraser only considers the case of $\lambda\phi^4{}_2$, a scalar field theory in two space-time dimensions, with a ϕ^4 interaction term (2011). Here, he argues, we do have genuine underdetermination but as soon as we move to QED, QCD, or the Standard Model, the underdetermination is broken by the empirical success of Lagrangian QFT. Fraser (2011) considers this evidence of success in the context of a form of 'No Miracles Argument' and argues that given the application of renormalization group methods that underpin the underdetermination, the inference to the approximate truth of the theory is weak.

2013). Thus one could view the inconsistency here pragmatically and defend a form of the 'cut-off' interpretation, say, but only as a preliminary stage towards a 'deeper' consistent theory.

Indeed, it is not as if the AQFT programme has conclusively demonstrated that a consistent QFT is possible, since there is as yet no acceptable model of the relevant axioms that accommodates interactions. Of course, it must be acknowledged that AQFT is a programme that has yet to be completed (Fraser 2009: 557).[13] However, this diminishes the force of the attempt to break the underdetermination, since both AQFT and the Lagrangian form now seem to adopt the same stance with regard to the inconsistency: accept it, or rather the associated theory, as a staging post towards a consistent successor (where this may be different for the AQFT and 'cut-off' approaches).

There is a further difference between the Lagrangian and AQFT approaches in that the latter generates inequivalent representations and the former does not; hence they differ in content:

> the cutoff variant does not have even approximately the same content as algebraic QFT because the cutoff variant has a finite number of degrees of freedom and therefore does not admit unitarily inequivalent representations; in contrast, algebraic QFT has an infinite number of degrees of freedom and therefore admits unitarily inequivalent representations. (2009: 560)

Now one option would be to simply eliminate these representations and hence the difference in content but, as we'll see, they are put to use in accounting for spontaneous symmetry breaking, for example (Earman 2004; Fraser 2009), as well as other features including the Unruh effect/Rindler quanta and superselection rules.[14] In these cases, Wallace acknowledges that algebraic methods may be illuminating but insists that the latter are not what are at issue in his debate with Fraser; rather it is that of real cut-offs versus no cut-offs (2011).

Furthermore, dismissing the representations as not pathological in the long-distance case is not a compelling move, since the issue of whether the relevant degrees of freedom need to be taken into account is precisely the point of contention. Still, one could accept that long-distance divergences must be tackled using algebraic methods, but insist that the maintenance of the short-distance cut-off means that introducing such methods does not render the Lagrangian formulation equivalent to

[13] Relatedly Fraser argues (2011) that the empirical success of conventional QFT cannot form the basis of a 'No Miracles Argument' (NMA) (see note 12) because, in part, the lack of models of AQFT means that the class of candidate theories to bring within the scope of the NMA is being 'illegitimately' restricted. Again she maintains that were such models to be constructed we would have a superior account to conventional QFT and under these circumstances we should refrain from making NMA-type inferences. Of course Wallace might respond that were a whole range of things to be undertaken we would have all kinds of different accounts but all we have to work with, for NMA purposes, is our current empirically successful theory, namely conventional QFT.

[14] I'd like to thank David Baker (personal correspondence) for emphasizing this.

AQFT (Wallace 2011). Such moves will presumably be regarded as part of a diverse range of such devices to which the advocate of Lagrangian QFT can help herself.

The upshot then is that it is not clear that the supposed underdetermination can be straightforwardly broken, one way or the other. So, the structural realist could stick with the Lagrangian programme, and articulate a field-theoretic informed notion of structure (French and Ladyman 2003). Or she could go with AQFT, take the net of algebras as presenting the quantum field-theoretic structure of the world and face the problem of inequivalent representations head on.

11.7 Option 2: Use the Swiss Army Knife

We recall that tackling this problem is not merely a matter of recovering, in some sense, the 'appearances'. In the case of 'ordinary' quantum mechanics with finite degrees of freedom, the structural realist can insist that the relevant structure, or features of the structure of the world, can be represented via Hilbert space and that it doesn't matter whether one then chooses the Heisenberg or Schrödinger representations since these are unitarily equivalent. In the case of QFT, such indifference is not an option. The choice is then as follows: one can pick out just one Hilbert space representation as appropriately representing the way the world is. However, this rules out the option of using states associated with alternative representations, and may unnecessarily restrict the set of states taken to be possible. Alternatively, one might invest all physical significance with the underlying algebraic structure but this rules out certain observables as unphysical and may yield too thin and minimal a basis for even a structural realist interpretation.

The dilemma can be understood as holding between 'algebraic imperialism' and 'Hilbert space chauvinism' (Ruetsche 2003; 2011). One could adopt the former, and take all the physical content of QFT as invested in the net of algebras. But this would select only a subset of the bounded operators defined on the relevant Hilbert space representation and all the rest—that can be deemed 'parochial' to the representation (Ruetsche 2003)—would have to be dismissed as 'unphysical accretions'. However, since these would include most of the projection operators, including those in the spectrum of the total number operator, this would mean 'investing with physical significance fewer observables than either scientific practice or our favored approaches to interpreting quantum theories can bear' (2003: 1330). On the other hand, going the chauvinist route and picking just one Hilbert space representation (which would raise the issue again of the grounds for doing so) runs the risk of investing with physical significance fewer states than our practices can bear, since it would rule out using those states associated with an alternative representation.

The significance of this issue is illustrated by the case of quantum statistical mechanics where the accommodation of phase transitions in terms of the existence of multiple distinct equilibrium states requires going to the thermodynamic limit of an infinite number of systems (i.e. the limit as the number N of micro-systems and the volume V they occupy goes to infinity, while their density remains finite) and characterizing these distinct equilibrium states in terms of unitarily inequivalent representations (2003: 1334–9). The analogue to the Hilbert space chauvinist would have to insist that only one equilibrium temperature is physically possible and all others are ruled out as impossible. Furthermore, this is incompatible with the explanation of phase transitions according to which different phases—corresponding to different representations—coexist at such a transition.[15] On the other hand, investing all the physical content in the algebra runs up against the objection that since the concrete representations correspond to the phase and temperature of a system at equilibrium, these too are contentful. This would not be a problem if the algebraic imperialist could understand the differences between such representations in purely algebraic terms, but that is not possible since temperature cannot be captured in this way (Ruetsche 2003: 1339).

Now the obvious question is whether such inequivalent representations do similar work in the QFT context. Here I shall briefly look at two cases where they appear to do so: spontaneous symmetry breaking and superselection principles.

11.8 Case 1: Symmetry Breaking and Structuralism

In the first case, inequivalent representations help explicate spontaneous symmetry breaking in the context of the Standard Model, where we encounter field theories with degenerate vacua, where the vacuum states differ from one another everywhere in space (see Wallace 2006, for other examples). The relevant global continuous symmetry is spontaneously broken so that each unitarily inequivalent representation has its own vacuum state. As a result, 'a full understanding of spontaneous symmetry breaking in QFT cannot be gained by beavering away within any one representation of the CCR... but must take into account structural features of QFT that cut across different representations' (Earman 2004: 183). Thus, as in the case of the thermodynamic example, it is not enough to insist that the representations are empirically distinguishable and hence present no problem for the realist (of any stripe). In particular, it is precisely such structural features that the structural realist would be too keen to invest with ontological significance. However, as we have seen, *in so far as these features are represented by the underlying algebra*, they may not capture everything of physical interest. Of course, one may seek to represent them in other—structurally acceptable—ways, but these will need to be spelled out and they are not immediately obvious.

[15] Thus, it is not the case that one can simply insist that the representations are empirically distinguishable and that we can allow the world to choose, as it were. What we want, as realists, is to be able to include within our representational scope, all the relevant inequivalent representations so that—in the case of this example—we can accommodate the phase transitions via multiple equilibrium states. That motivates adopting algebraic chauvinism, but taken as is, that precludes specific representations from being understood as having content.

As noted previously, symmetry breaking in general terms has been presented as another challenge to the ontic structural realist. The problem is that if a symmetry is broken, then it cannot be invested with ontological significance, since the laws of the appropriate domain do not manifest this symmetry, on the symmetries as by-products view, or are not constrained by it, according to the symmetries as constraints approach. However, instead of thinking of the symmetry as somehow 'lost', the situation is better understood as one where the relevant phenomena is characterized by a symmetry that is 'lower' than the unbroken symmetry (see, for example, Castellani 2003; Brading and Castellani 2008). And expressed in group-theoretic terms, this means that the group characterizing the latter is 'broken' into one of its sub-groups. Consider the simple example of a ping-pong ball subjected to an external force and which subsequently buckles (Stewart and Golubitsky 1992: 51). Here the spherical symmetry represented by O(3), the orthogonal group in three dimensions, is broken to yield the circular symmetry represented by O(2), where the symmetries of the latter are contained in the former. In the case of the Standard Model, as we noted, the fundamental symmetry of $SU(3)\times SU(2)\times U(1)$ spontaneously breaks down to $SU(3)\times U(1)$ at the electroweak energy. Thus, again as already noted, symmetry breaking can be described in terms of relations between transformation groups, something that OSR can easily accommodate. Indeed, the objection here is a curious one to make, since the notion of symmetry being spontaneously broken is generally regarded as providing a way to allow symmetries to apply to, in some sense, asymmetric phenomena.

Now, spontaneous symmetry breaking (SSB) does not occur with finite systems, since the relevant degenerate states can superpose uniquely to give a lowest energy state. In the infinite volume limit these states are all orthogonal to one another and separated by a superselection rule (Brading and Castellani 2008). It is these rules that connect the different representation classes associated with the unitarily inequivalent representations.

However, accounting for the (asymmetric) phenomena in terms of the breaking of some more fundamental symmetry obviously involves an inferential move that itself requires justification. This might be found in a form of Curie's principle: the symmetries of the causes must be found in the effects; or, equivalently, the asymmetries of the effects must be found in the causes. If this is extended to include SSB, it can be regarded as equivalent to a methodological principle according to which the asymmetry of the phenomena must come from the breaking (explicit or spontaneous) of the symmetry of the fundamental laws. It is this that underpins the inferential move but the question now is whether it can be justified (for a critical consideration of this issue see Morrison 2003).

However, such an extension is problematic since SSB itself appears to undermine Curie's principle because a symmetry is broken 'spontaneously', that is without the presence of any asymmetric cause (Brading and Castellani 2008). However, the symmetry of the 'cause' is not actually lost, since it is conserved in the entire ensemble of the relevant solutions (Brading and Castellani 2008). Thus consider a linear vertical stick with a compression force applied to the top and directed along its axis:

> The physical description is obviously invariant for all rotations around this axis. As long as the applied force is mild enough, the stick does not bend and the equilibrium configuration (the lowest energy configuration) is invariant under this symmetry. When the force reaches a critical value, the symmetric equilibrium configuration becomes unstable and an infinite number of equivalent lowest energy stable states appear, which are no longer rotationally symmetric but are related to each other by a rotation. The actual breaking of the symmetry may then easily occur by effect of a (however small) external asymmetric cause, and the stick bends until it reaches one of the infinite possible stable asymmetric equilibrium configurations. (Brading and Castellani 2008)

These configurations are all related via the relevant symmetry transformations and hence 'there is a degeneracy (infinite or finite depending on whether the symmetry is continuous or discrete) of distinct asymmetric solutions of identical (lowest) energy, the whole set of which maintains the symmetry of the theory' (Brading and Castellani 2008; see also the discussion in Stewart and Golubitsky 1992: ch. 3). Now I'll return to this point briefly later but it's worth noting that it nicely meshes with Earman's comments on the role of inequivalent representations in explicating SSB within QFT: one has to consider the whole ensemble—in this case, of representations—in order to understand what is happening.[16]

Furthermore, in the context of the broadly structural understanding of the relationship between laws plus the associated symmetries and the phenomena that I have explored here, one can argue that the broader methodological principle should be seen as a further component of that understanding that allows us to retain symmetries as part of our fundamental structuralist ontology while also accounting for the blatantly asymmetric phenomena.[17] With regard to its philosophical justification, this should perhaps piggyback on its physical counterpart: in so far as SSB is justified within physics—in the usual terms—we can accept the methodological principle.

The point to bear in mind is that if we are to understand the way in which inequivalent representations do some work in the context of SSB, we need to embrace, in some sense, all these representations, in just the way that although the stick eventually falls one way rather than another, we need to embrace all the *possible* ways it might fall if we are to understand what is going on in these terms. It is in this sense that the 'ensemble' of inequivalent representations does work within QFT.

[16] According to Earman, Curie's principle is vacuous in QFT if vacuum representations are demanded since the antecedent condition of an initially symmetric or semi-symmetric state is never fulfilled (2004).

[17] cf. Weyl, who wrote that symmetry is the norm 'from which one deviates under the influence of forces of a non-formal nature' (1952: 13).

11.9 Case 2: Superselection Sectors and Statistics

As we have seen, in non-relativistic quantum mechanics, quantum statistics (and in particular the distinction between Bose–Einstein, Fermi–Dirac, and paraparticle kinds) is in effect an 'add-on' to the theory, arising from the assumption of Permutation Invariance (see French and Rickles 2003). We recall that the action of the permutation group is such as to divide up the Hilbert space into sub-spaces representing symmetry sectors corresponding to the possible types of permutation symmetry possessed by the particles whose state vectors lie in that sub-space. AQFT recovers these features but without assuming any add-on to the theory, by virtue of the fact that inequivalent representations of the net of algebras yield superselection rules. And it is in its analysis of these rules that 'the algebraic approach most clearly displays its beauty, utility, and foundational importance' (Halvorson and Müger 2006: 55; see also Baker 2013[18]).

The idea is straightforward: superposition of states cannot be unrestricted (see the discussion in Haag 1992: 108). Thus, superposition of states with integral and half-integral spin (or, equivalently, corresponding to Bose–Einstein or Fermi–Dirac statistics respectively), or of states with different charge, for example, are not statistically pure states and hence do not exhibit quantum interference. Such states belong in different sub-spaces or superselection sectors of the overall Hilbert space. These distinctions—between particles with different statistics and those with different charges, or quantum numbers in general—are encoded in the structure of the net of observable algebras (Haag 1992: 149). Since they correspond to the natural kind structure of the world (or, at least, the world of elementary particles), and the structural realist has long maintained that this structure can be incorporated within her framework (see French 2006, for example; and our discussion in Chapter 10), it would seem that this consequence of the existence of inequivalent representations might also be accommodated.

Omitting a great deal of technical detail (see Haag 1992: ch. 4; Halvorson and Müger 2006: §11.4), the upshot is that permutation symmetry is treated as a kind of gauge symmetry and the explanation of quantum statistics arises from the structure of the category of representations of the observable algebra (Halvorson and Müger 2006: 126).[19] In particular, the superselection approach allows us to make sense of

[18] The relationship between Permutation Invariance and superselection rules has long been well known of course and indeed, is intimately tied up with the metaphysical underdetermination—between particles as individuals and as non-individuals—discussed in Chapter 2 (see French 1985; French and Krause 2006: 151). Baker insists that this underdetermination cannot be exported into the context of QFT (however, for caveats see French and Krause 2006: ch. 9), but this should not unduly bother the structural realist who (a) will point to the alternative underdetermination that holds between the views of fields as substances, of a sort, and as properties (of space-time points or regions) and (b) is more than happy to go along with the shift in focus to superselection rules, as we saw in the previous chapter.

[19] Halvorson and Müger are dismissive of the explanation given by French and Rickles in terms of Permutation Invariance on the grounds that this involves an 'overly simplistic formalism' (2003: 125). This

non-permutation invariant states and quantities within the framework of Permutation Invariance (2006). The standard view of such states is that they represent so much 'surplus structure'. As I have noted in Chapter 2, insisting that theories that generate such surplus structure should be rejected in favour of those that do not, or, at least, that generate less (Teller and Redhead 2001), is problematic as a methodological principle, since such structure can prove to be heuristically useful, as—again—the case of quarks and parastatistics demonstrates. On the other hand, if such structure is retained and taken to correspond to possibilities that are contingently not realized, the obvious question is why they do not correspond to the actual world. Adopting the 'principle of plenitude' suggests that there should be particles corresponding to every symmetry type, and the question then is why we do not see them (Halvorson and Müger 2006: 128). Indeed, the problem is even more acute: since any system that has a symmetry described by the permutation group has a symmetry described by the braid group, and since the latter has infinitely many irreducible representations, the principle of plenitude would imply that there are more particles 'than we could ever possibly describe' (2006: 133).

Now, what is the structural realist to make of all this? First of all, as I have already suggested, she can adapt what has already been said about permutation symmetry in the context of structural realism and claim that what inequivalent representations are doing here is capturing a fundamental feature of the natural kind structure of the world, namely the distinction between bosons and fermions, or more generally, parabosons and parafermions. Indeed, the analysis here can be understood as further advancing the structuralist stance since its alternative explanation of quantum statistics means that we don't have to impose Permutation Invariance on the theory as a further symmetry condition; rather, as sketched previously, the statistics and the associated kind classification arise naturally from within the (AQFT) formulation of the theory itself. Thus, the structure of the world has one less fundamental feature. Far from presenting a problem for the structural realist then, inequivalent representations in this context help her cause![20]

There is still the issue of what to do about all the representations that don't seem to correspond to any kinds of particles we observe in the actual world. On the one hand, as already noted, eliminating this surplus structure would be not only ad hoc in this context but would also throw away a heuristically useful resource. On the other,

is perhaps a little unfair, given that the context of these considerations was primarily non-relativistic quantum mechanics and when QFT was discussed, the 'conventional' formulation was typically adopted. Still, it is undeniable that AQFT offers a useful insight on this issue, particularly as far as the structuralist is concerned.

[20] Of course, the (non-structural) 'object-oriented' realist can also avail herself of this analysis but the manner in which the classification arises within the AQFT formulation may not help her cause quite so much, if she adheres to a non-structuralist account of the natural kinds involved. The point is, if one does have such a structuralist account, the approach outlined here helps advance it because one does not have to *impose* Permutation Invariance.

retaining them opens the door to the principle of plenitude and the kinds of concerns outlined earlier. Now, the structural realist does not have to accept that principle and she could plausibly maintain, of course, that not all mathematical structures correspond to reality. Again drawing in the relevant history, we recall Weyl's point:

> The various primitive sub-spaces are, so to speak, worlds which are fully isolated from one another. But such a situation is repugnant to Nature, who wishes to relate everything with everything. She has accordingly avoided this distressing situation by annihilating all these possible worlds except one—or better, she has never allowed them to come into existence! The one which she has spared is that one which is represented by anti-symmetric tensors, and this is the content of Pauli's exclusion principle. (1968: 288)[21]

The problem is that if the structural realist wishes to avail herself of the work done by inequivalent representations in showing how quantum statistics drops out of the AQFT formulation, she can hardly then back away from what this work yields. In particular, it might be argued, she can't then pick and choose which features of that structure she is going to take as real or actual and which as merely surplus. Now, as I've argued, there are good reasons for thinking of this structure as modally informed and one can respond to this issue of inequivalent representations by appealing to this modally informed sense of structure and understanding it as covering or encoding a range of possibilities in the way I've indicated in Chapter 10. Let's see how this works.

11.10 Back to Inequivalent Representations

Returning to the debate between advocates of 'Lagrangian' QFT and AQFT, we recall the concerns about investing the relevant infinities associated with inequivalent representations with physical significance (Wallace 2006). In particular, one might insist that going to the limit, where the infinities appear, is a significant idealization and one should be wary of interpreting the elements of such a limit in a realistic manner. The problem is that these limits yield precisely those features of the idealization that allow it to explain the phenomena and do representational work (Ruetsche 2003: 1342). One could argue that in the QFT case it is not so clear that the relevant idealized features do similar work: given the spatio-temporal limits we are constrained by, one could insist that we can in fact do everything we need to do and explain everything we need to explain in terms of the Lagrangian formulation.[22] Nevertheless, concerns remain. One could insist that when it comes to cut-off variants of QFT, it is the assumption of finite degrees of freedom that is the idealization, since we know (unlike the case of quantum statistical mechanics) that QFT systems should be taken to have an infinite number of degrees of freedom

[21] See also Weyl 1931: 238 and 347.

[22] This would also constrain the limits of our realism, although perhaps only in the same 'in principle' way that dismissing worries about events beyond the event horizon does.

(unless space-time is taken, on non-ad-hoc grounds, to be discrete and finite; Fraser 2009).[23] And unlike the case of quantum statistical mechanics, there is an alternative to the cut-off approach.

The dilemma can be resolved by adopting what has been called the 'Swiss Army Knife' approach (Ruetsche 2003: 1339–41). We recall the issue of how we should read off our commitments from the theory. But even before we specify the physical content of the theory, before we decide whether that content should be understood in terms of objects or structures, we must distinguish that which is physically possible from the (physically) impossible (2003). According to the Hilbert space chauvinist, that sorting of possibilities is achieved by the Hilbert space structure of observables; according to her algebraic counterpart it is via the abstract algebraic structure. Ruetsche's suggestion is that we simply refuse to sort and specify the content in 'one fell swoop'; rather we should take such a specification as appropriately tiered, with a corresponding gradation in the possibilities allowed by the theory. Thus at the first tier, corresponding to the broadest set of possibilities, we have the space of algebraic states on the appropriate abstract algebra. At the next tier, physical contingencies are taken into account by distinguishing the narrower set of possibilities corresponding to the empirical situation through appeal to the relevant features of that situation. Other algebraic states are then to be thought of as more or less remote possibilities, rather than dismissed as physically impossible.

We can also think of this tiered specification of content in terms of the universal representation of an algebra, which is the direct sum over the set of algebraic states of the relevant representations. This would then yield the theory's broadest set of physical observables, and,

> [a]t this stage of content specification, this vast host of physical observables is just sitting there, like blades folded up in a Swiss army knife. The next (coalescence) stage appeals to contingent features of the physical situation to focus on a small set of representations, which are summands in the universal representation. Observables parochial to those representations are extracted for application to the situation at hand. Thus coalescence is something like opening the Swiss army knife to the appropriate blade or blades, once you've figured out what you're supposed to do with it. (2003: 1341)

Can such an approach serve the realist cause? In the case of the object-oriented realist it would seem it would push her towards a form of 'representation-specific realism', according to which she could only make realist claims about specific objects within the domain corresponding to the given representation. Beyond such a domain no such claims are warranted. However, such a realist would have to give some account

[23] Earman also suggests that reflection on the nature of QFT might be used to support the idea that it is the finite case that is the idealization, fostered by thinking classically in terms of physical systems 'as consisting of hunks of spatially localized matter' (2004: 192).

of the relationship between these domains and the underlying universal algebra. Focusing on the latter takes us to structural realism, of course.

Again we recall that this fundamental structure can be understood as *inherently* modal in the sense of encoding the full range of allowable physical possibilities. Adopting a broadly model-theoretic stance and thinking of the models that the given theory makes available, we can take the modality as *represented* by the 'shared structure' spanning the models of the theory, in the sense that this structure encodes the possibilities that obtain when we move from the high-level generality of Newton's laws, say, to two-body models, for example (Brading 2011). Here too, as in the earlier analysis, we have a tiered hierarchy, and taking the modality of this shared structure as representing the modal features of the structure of the world, we get a corresponding gradation in the possibilities allowed by the theory.

Of course, this structure cannot be entirely, or merely, modal. It must also incorporate determinate features, as I have argued, such as, in the case of symmetries, specific representations corresponding to the way the world is. Again, this seems to accommodate Ruetsche's understanding of how physical contingencies are to be taken into account by appealing to the appropriate features of the actual physical situation. Thus, to return to the example of particle statistics, the way our world appears to be (and this is a fallible claim of course) corresponds to the Bose–Einstein and Fermi–Dirac representations. Thinking of Ruetsche's metaphor, actuality then emerges, depending on the representational blade that is pulled out of this structural knife, as it were. Thinking in modal terms, the abstract algebra would have to be viewed as a structure that effectively encodes these modal features.[24]

Furthermore, this meshes well with the understanding of SSB touched on previously. The latter gives us a further reason to incorporate all the relevant inequivalent representations (corresponding to the different vacuum states) into the structuralist picture: thus if we recall the example of the vertical stick under compression,[25] we need to consider the 'possibility space' covered by all the post-collapse orientations of the stick in our explanation.[26] From the SSB perspective (from which the inequivalent representations are deemed to 'do work') the representations represent the post-break situation, and the symmetry—which the structuralist will want to focus on—is preserved across all of these possibilities, or in Ruetsche's engaging terms, for all the blades of the knife.[27] It is for this reason

[24] If one wished one could adopt the possible-worlds analysis of modality, insist that the actual world 'contains' no inherent modality, and understand the relevant structure as spanning physically possible worlds, rather than being confined to the actual one. (Thus we would have a kind of 'Trans-World Structuralism'.)

[25] Also used by Nambu in his Nobel prize presentation on SSB.

[26] Another example would be that of ferromagnetism (see, for example, Morrison 2003: 354) where the symmetry is not lost but is effectively hidden, as it still exists over all possible directions (cf. Castellani 2003: 325).

[27] As Fujita notes, in a related context (the Goldstone theorem), the mathematics is straightforward but the physics is difficult because 'one has to examine all the possible conditions in nature when the symmetry

that spontaneous symmetry breaking has been characterized as 'symmetry spreading' (Stewart and Golubitsky 1992).[28]

Let us return to the point that 'a full understanding of spontaneous symmetry breaking in QFT cannot be gained by beavering away within any one representation of the CCR... but must take into account structural features of QFT that cut across different representations' (Earman 2004: 183). Such a 'full understanding' requires us to consider all the inequivalent representations, thus undermining the stance of the Hilbert space chauvinist.[29] But now the structuralist could raise the following dilemma against her critic: if the QFT situation is like that of the stick, in the sense that SSB is invoked to account for the asymmetric phenomena, then the role of the inequivalent representations is not so fundamental; or at least, not so threatening to the structuralist. Here we are invoking SSB in order to retain a symmetry-based ontology. If, on the other hand, we accept that we cannot dismiss the inequivalent representations, since they are doing some work for us, then we should take the quote from Earman to heart, but then it is the structural features that cut across different representations that are important, and again the structuralist can only murmur her approval.

Of course, on either alternative further issues arise. In the first case, one might feel that the goalposts have shifted somewhat. The original criticism was that the structural realist cannot identify the relevant representations as isomorphic as in the case of the wave and matrix mechanics and so is faced with the dilemma set by Ruetsche. The response here acknowledges that the relevant structure (as characterized by the net of algebras) gives too thin an ontology, as it were, but redirects attention to the way that SSB allows us to preserve the ontological primacy of symmetry. However, it is not really a case of goalposts having been moved, but rather that of different aspects of structure and structural realism being emphasized. In the case of the wave and matrix formulations of quantum mechanics, Ladyman emphasized the underlying Hilbert space structure as the structuralist's ontological locus. But in that context the structure and Permutation Invariance (as one of the fundamental symmetries) are not tied so closely together as in the case of QFT, or, perhaps, are tied together in a different way. In the case of QFT, Howard is obviously right that we cannot appeal to isomorphism, but we can identify an underlying structure. And the inequivalence is tied up with the asymmetry of the actual phenomena. In order to accommodate that, we appeal to SSB and the structuralist retains the ontological emphasis on

is broken spontaneously' (2007: 50). What I am suggesting is that in order to accommodate inequivalent representations in QFT, the structuralist must likewise incorporate the examination of 'all possible conditions in nature' into her structuralist ontology.

[28] Of course, as Earman notes, what distinguishes SSB in QFT is that a symmetry of the laws of motion is not unitarily implementable, a feature that implies but is not implied by the failure of the vacuum state to exhibit the symmetry (2004).

[29] One could think of the different degenerate vacuum states as belonging to the same Hilbert space, but as lying in different 'superselection sectors' (Earman 2004). This might be characterized as a form of Hilbert space 'enthusiasm' that does not imply chauvinism; thanks to one of the readers for pointing this out.

symmetry.[30] Indeed, the symmetry connects the unitarily inequivalent representations each with its own vacuum state. As Earman notes, 'this is the precise sense in which spontaneous symmetry breaking in QFT involves degeneracy of the vacuum' (2004).

Turning finally to Permutation Invariance and quantum statistics, we can easily see how an appropriately modal notion of structure can provide the relevant metaphysical framework here. As in the case of SSB, a full understanding and explanation of quantum statistics must take into account the structural features of QFT that cut across different representations. And what this account shows is that, 'it is the structure of the category of representations that provides the really interesting theoretical content of QFT' (Halvorson and Müger 2006: 57). Following the schema given here then allows the structural realist to take *this* structure seriously: invoking the knife metaphor again, we can think of the bosonic and fermionic cases as corresponding to two of the blades that are deployed in the actual world. Others, such as those corresponding to the parastatistics cases, are retained as heuristic resources; indeed, as we have noted, they were in fact deployed in the early history of quantum chromodynamics and may be again. Eliminating these latter cases as features of the structure we are interested in would not only remove a valuable resource, but would lose the kinds of important intra-structural connections briefly indicated previously.

Nevertheless, algebraic chauvinism is an inappropriate attitude, since the representations also feature as part of the content of the theory (Halvorson and Müger 2006: 118). It is precisely because of this contribution to the structuralist account of the natural kind classification of the world that the existence of inequivalent representations is again not an issue. Indeed, this is only seen as a problem because the representations are unhelpfully understood to be 'competitors', in the sense of offering competing descriptions of the phenomena (Halvorson and Müger 2006: 121–2). This is a further feature that I suspect lies behind the challenge to the structural realist: if the representations are not isomorphic and therefore equivalent, they must be competitors but then (the argument goes) any grounds for selecting one over another will introduce a non-structural element and sully the structuralist picture. However (leaving aside this second point, since one might suppose one could present structuralistically acceptable grounds for selection or argue that such elements do not in fact introduce a blemish into the structuralist picture), this

[30] Of course, I am not suggesting that either the non-structural realist or indeed the empiricist cannot accommodate the role of symmetries in explanations, etc. The point (again) is that given that, as I have indicated, structural realism emphasizes the ontological status of both laws and symmetries as aspects of structure, the account presented here offers a way around Howard's problem. Of course, in this context the ontological weight given to each might shift through further physics research—and not just on the basis of (perhaps overblown) claims that (gauge) symmetries help pin down the dynamics but also because of the role of SSB in cosmology: 'as the universe expands and cools down, it may undergo one or more SSB phase transitions from states of higher symmetries to lower ones, which change the governing laws of physics' (Nambu 2008).

understanding is unhelpful. Compare the situation with disjoint group representations, for example (2006): there we are not inclined to see these as merely competitors precisely because of the relations that hold between them. Likewise, what the algebraic analysis reveals are the additional relations on the category of physical representations in the superselection case. And again, an understanding of these representations as merely competitors is not entirely appropriate. Consider the statistics case again: of course, if a particle is in the Bose–Einstein sector it cannot be in the Fermi–Dirac; nor, as I have already said, can it leave the former sector and move to the latter (because the Hamiltonian is an observable and thus commutes with the permutation operator). In *that* sense—a sense that I have explored in Chapter 10—they are competitors. But of course, to understand the actual world we need both, and the relations between them (as resulting from the action of the permutation group in the non-relativistic case), so in *this* sense—which is what the structural realist will seize on—they do not compete.

Nevertheless, using the importance of the representations to motivate a shift to a form of Hilbert space chauvinism here would be far too simplistic a response (Halvorson and Müger 2006: 119). Indeed, if we were to focus ontological attention on just one representation we would not only be ignoring the really interesting structure—namely the relations between representations—but we would not be able to define Bose and Fermi fields, among other things.[31]

11.11 Conclusion

There may be more options for the structural realist but the considerations given here illustrate ways in which she can respond to the challenge posed by inequivalent representations.

As I have emphasized, like most realists, she urges that one take the 'best' theory we currently have available and invest significance in the relevant structures it presents. Now, one criterion for being the best obviously has to do with empirical success and this the Lagrangian form of QFT has in spades (Wallace 2011). Furthermore, as we have seen, it may be able to assuage concerns about inequivalent representations. On this basis, the structuralist might be inclined to suggest that the structure of the world is (approximately) as given by Lagrangian QFT with cut-offs. She is not compelled to take the algebraic route and of course her emphasis on symmetry as captured group-theoretically can still be satisfied within this approach as manifested by the Standard Model.[32]

[31] Halvorson and Müger's 'Representational Realism' focuses on this structure that arises from the inter-representational relations and its explanatory role, and takes it as comprising the content of the theory, along with the net of algebras and the dynamics, as captured by the representation of the translation group (2006). One can understand this as a form of structural realism, along the lines suggested here.

[32] The role of the Lagrangian here would also offer a corrective to North's espousal of Hamiltonian formulations as representing the structure of the world as discussed in Chapter 2.

Of course, the defender of the AQFT approach will insist that the Lagrangian formulation cannot be 'the best', precisely because it incorporates critical idealizations and lacks foundational coherence. Furthermore, the emphasis on fundamental algebraic structure is alluring to the ontic structural realist who, as Howard notes, has precisely focused on similar underlying structure in the case of quantum mechanics. Appropriating Ruetsche's attempted dissipation of the problem of inequivalent representations then supports the structural realist in her adoption of a profoundly modal conception of this structure. However, such a conception seems entirely appropriate in the context of understanding the role of these representations for explaining both SSB and quantum statistics. Again we recall that the fundamental problem with regard to the former is that the dynamics does not determine the representation and hence SSB must be appealed to. The focus then is on the symmetry that connects the inequivalent representations but then the problem appears to have dissolved; or at least, it is now the 'problem' of accounting for asymmetric phenomena on the basis of the fundamental symmetry and dynamics.[33] With regard to the statistics, the algebraic analysis can actually help the structuralist cause by showing how these need not be understood as arising from the imposition of Permutation Invariance, but as being encoded within the algebraic formulation of the theory. Furthermore, this analysis reveals the importance of the structure of these representations and the structural realist can incorporate this structure into her worldview, noting the issues it raises as to whether the representations at issue should be regarded as competitors in the sense that the challenge seems to require, and thereby draw the latter's sting.

Indeed, returning again to the context of SSB, the structural realist might articulate the following third option: to fully understand SSB we must take into account structural features of QFT that cut across different representations. These features are captured by the algebra which can be understood, from the perspective of the Swiss Army Knife approach, as representing a modal form of structure stretching across possible worlds, as it were. However, tracking the break in symmetry and shifting down through the energy regimes, we can adopt, as realists, the best theory we currently have available, namely the Standard Model and the associated Lagrangian form of QFT with cut-offs. And as structural realists we can then invest with ontological significance the relevant structures this theory presents us with, including the associated symmetries such as $SU(3) \times SU(2) \times U(1)$. This investment is, as always, fallible and provisional, in particular since we have good reason to believe that the theory will be superseded (by something like loop quantum gravity or string theory).[34] Far from presenting a problem for structural realism, then, inequivalent representations indicate how that position can be further developed and strengthened.

[33] There remains the further issue of justifying SSB but this is not particular to structural realism.

[34] This suggestion obviously requires further elaboration. As Doreen Fraser has noted (personal communication) there is a difference in kind between the representations offered by Lagrangian QFT and AQFT and the unitarily inequivalent representations that feature in SSB. Due attention may then need to be paid to the issues regarding how we understand the scope of these theories.

12

Shifting to Structures in Biology and Beyond[1]

12.1 Introduction

As the previous chapters exemplify, both the elaboration of and debate over structural realism in general and OSR in particular have typically been articulated in the context of theories of physics. They are motivated by, first of all, the presence within such theories of the appropriate mathematics that allows the straightforward presentation of the relevant structures; and secondly, the implications of such theories for the individuality and identity of putative objects at what might be called the 'micro-level'. My aim in this chapter is to explore the possibility of developing similar views in the chemical and biological domains[2] (see also Ladyman, Ross, et al. 2007). An obvious concern is that within these contexts we may not be able to find the kinds of highly mathematized structures that structural realism can point to in physics. I shall indicate how a focus on models might help allay such concerns. Furthermore, it turns out that issues of object identity and individuality arise here as well. Thus, when it comes to biology, Dupré insists that there exists a 'General Problem of Biological Individuality' that relates to the issue of how one divides 'massively integrated and interconnected' systems into discrete components. His solution is to advocate a form of 'Promiscuous Realism' with regard to biological kinds. Instead I shall urge serious consideration of those aspects of the work of Dupré and others that lean towards a structuralist interpretation. By doing so I hope to suggest possible ways in which a structuralist stance might be elaborated within biology. Let me begin, however, with chemistry and a consideration of molecular structure.

[1] Much of this chapter is based on French 2011e and 2012b. I am particularly grateful to Angelo Cei, Phyllis Illari, Holger Lyre, and Marcel Weber for discussion and helpful remarks.

[2] This is contra Lyre (2012), for example, who argues that there is no need to 'structuralize' other sciences beyond physics.

12.2 Reductionism and the Asymmetry of Molecular Structure

Now, an easy option would be to insist that chemistry reduces to physics and thus any associated issues are more appropriately tackled in terms of the fundamental structures of the latter. However, it is now commonplace to acknowledge that reduction in the traditional sense of recovering (in some sense) the laws of the reduced theory (the relevant theory of chemistry, for example) from those of the reducing, or more fundamental theory (quantum mechanics) is a dead end. At the representational level, of the relevant philosophy of science, attempts to capture such moves within the framework of standard approaches have not met with much success. However, at the object level of chemistry itself, many would nevertheless accept that *ontological* reduction, in some form or other, is still viable. Consider the example of methane: certainly one can argue quite forcefully that there is nothing, ontologically speaking, to the atoms of carbon and hydrogen that is not grounded in quantum physics. What about the bonds between the atoms? Here too one can make a good case for ontological reduction since the ontological nature of chemical bonding can also be understood in quantum mechanical terms. Indeed, from the structuralist perspective, a reductionist stance can draw considerable support from the role of permutation symmetry in explaining the formation of bonds and chemical valency in general.[3]

But what about the *structure* of the molecules? This also has to be introduced as an ontological component in order to account for the difference between, say, butane and isobutane. Here problems arise and ironically for the structuralist, it is the symmetry inherent in the quantum mechanical reductive base that creates an obstacle (for a useful overview of these issues see Hendry 2011). Thus Wooley (1998) has argued that, factoring in only the relevant nuclear and electronic states, isomers like butane and isobutane should share the same Schrödinger equation but then that leaves their different structures unaccounted for. More generally, arbitrary solutions to Schrödinger equations should be spherically symmetrical but molecules clearly are not. Now, one can introduce certain idealizations and shift from Schrödinger's equations to structures with less symmetry—by, for example, holding the nuclei fixed and considering only electronic motion—and account for some of the relevant phenomena by effectively assuming a certain molecular structure. In this way one can account for the phenomena associated with enantiomers, for example, where one has isomers with different chirality, yielding different optical polarization rotation angles (Hendry 2011: 304). But, of course, the structure is assumed, not obtained from 'exact' quantum mechanics. In general, these asymmetries in molecular structures 'are essential to all kinds of explanation at the molecular level' (Hendry 2011: 304) and the conclusion has been drawn that such structures are simply not

[3] We recall Heitler's cry of 'Now we can eat Chemistry with a spoon!' following his work with London on the explanation of covalent bonds through Permutation Invariance.

'there', ontologically, in 'exact' quantum mechanics, which incorporates the very symmetries the structuralist sets such store by.

One can set reactions to this conclusion along a spectrum: at one extreme one might argue that molecular structure should be regarded as existing at a distinct ontological level in some sense and treated accordingly; at the other, one might be tempted to insist that if the structure is not 'there' in quantum mechanics, it is not 'there' in the world, and should be excluded from our ontological purview. In the spirit of 'Big Tent' structuralism, let me briefly consider how the structuralist might accommodate these responses as well as a more moderate position, before indicating, again but in this new context, how eliminativism might not be such a worry.

If molecular structure is treated as ontologically distinct from 'exact' or 'symmetrized' quantum mechanical structure, the structuralist's conception of structure will have to be different at this level than it is at the quantum mechanical. Certainly, the role of symmetry will be greatly reduced, at the very least, and whatever form of OSR one adopts, one could hardly call it group-theoretic. One might try to articulate an appropriate notion of 'chemical' structure in this context, but such a notion is obviously going to have to abstract away from the various particularities if it is going to cover different molecular structures, such as those of butane and isobutane. One might resist being a realist about that structure, on pain of facing the kinds of concerns we looked at in Chapter 8. On the other hand, if one insists that one's realism is directed towards the particular molecular structures themselves (that is, the physico-chemical structures in the world and not just the kinds 'butane structure', 'isobutane structure', etc., that again are abstracted away from these structures), then one is going to have to accept a plurality of structures—one for each molecule. Given that, it is hard to see what advantages this form of OSR would bring, as one might just as well stick with an ontology of molecules as objects.[4]

Alternatively one could adopt a looser understanding of 'structure' and, consequently, a broader conception of OSR. In effect this is what Ladyman does in his interesting attempt to give a structuralist reading of the shift from phlogiston to oxygen (Ladyman 2011; see also Carrier 2004). Here he argues—convincingly, I think—that phlogiston theory correctly described the causal or nomological structure of the world, at least to some extent, but that attempts to accommodate the success of theory in terms of ESR fail because

> The core theoretical structure that is correct in phlogiston theory is not the unknowable entity that we know relationally as what is released on combustion but rather the relational structure expressed by the theory of Redox reactions. (Ladyman 2011: 100; where 'Redox [REDuction-Oxidation] reactions' are those that involve a change in oxidation state, such as the oxidation of metals or the burning of hydrocarbons)[5]

[4] Of course, one might still resist taking the constituent atoms of these objects as themselves objects!

[5] Furthermore, if the advocate of ESR chooses the Ramsey sentence as her mode of representation, then one obtains a sentence asserting the existence of something that is released on combustion—but of course there is no such thing (Ladyman 2011).

Following Ladyman, Ross, et. al (2007), what phlogiston theory latched on to were the 'real patterns in nature'. However, an understanding of structure in terms of patterns might be too loose for some folk, so it is worth exploring some alternatives.[6]

Moving now to the other extreme, one could argue that if molecular structure cannot be recovered from quantum structure, then, given that the latter should be regarded as fundamental, so much the worse for molecular structure. This would lead to the adoption of some form of eliminativism about such structure. However, as Hendry notes (2011: 305–6), chemists invoke such structures in a huge range of chemical explanations that account for an equally vast array of phenomena. It is one thing to demand a radical revision of such explanations as a consequence of one's eliminativism, but quite another to actually come up with the revisions required. As Hendry says, until the advocate of this view can actually show how one can explain the phenomena concerned on the basis of 'exact' quantum mechanics—and thereby overcome the obstacles to accounting for molecular structure—such demands for revision seem idle (2011: 306).

However, there are intermediate positions that offer a more plausible basis for accommodating chemistry and that can be nicely enhanced through one of the metaphysical moves discussed in Chapter 7. So, one might follow Primas (1983; Hendry 2011) who takes molecular structures to be artefactual in the sense of arising in the context of our model building and then being read into the world, as it were. This world is correctly described by 'exact' quantum mechanics and thus does not contain or manifest molecular structure. That chemists find such structures to be useful in their explanations is then no surprise, given the role of the aforementioned models in these explanations. Typically, of course, such model building may be complex and involve diverse data models and representations of the phenomena, as well as theoretical features, and in particular incorporates those patterns of phenomena that precisely lead us to ascribe a particular structure to the molecule concerned. The truth of the relevant statements about this structure must then be taken to be grounded in the patterns of phenomena from which the structure is ultimately derived, rather than anything unobservable. This might generate worries about the explanatory power of molecular structure that chemists extol: in effect, the advocate of Primas' view would have to say 'It's no wonder your explanations appear so good, as the molecular structures they use are effectively derived from the patterns of phenomena you are seeking to explain!' That would not be a particularly comfortable position to adopt.

[6] Perhaps one could say that the phrase 'patterns in nature' has a kind of umbrella role, covering group-theoretic structures in the case of physics, and a greater plurality of molecular structures in the case of chemistry.

Alternatively, as Hendry notes (2011: 305), there is evidence from molecular beam experiments that 'exact' quantum mechanics can be applied to isolated molecules in certain contexts. If molecular structure is not given by quantum mechanics, then in these contexts the molecules cannot be said to have that structure. This suggests that '[a] determinate molecular structure is... something a quantum-mechanical system of nuclei and electrons may or may not have, depending on its interactions with its environment' (2011: 305). Molecular structure might thus be regarded as contextual, in this sense. Indeed Ramsey (2000) has advocated a form of 'contextual' realism on this basis, according to which molecular structure is to be understood as a relational, non-intrinsic feature of molecules that is dependent on their environment. But as Hendry notes, acknowledging that a molecule's state, in general, is dependent upon its interactions with the environment is surely compatible with any 'sane version of realism' (2011). And of course, the structural realist (less sane than most, perhaps) will happily accept this relational aspect of molecular structure. Furthermore, if 'the environment' is taken, as it should of course, to be part of the world, and therefore amenable to structural analysis, then again, and putting it broadly, the contextual exhibition of molecular structure is merely the manifestation of certain features of the structure of the world, under certain circumstances (that is, the interaction with other features).

The issue remains, however, as to how one is going to accommodate the role of molecular structure in chemists' explanations if one is going to be an eliminativist about it, in at least the sense of taking it to be contextual or otherwise non-fundamental. Here we can deploy our Viking Approach again. Thus, chemists' talk about molecular structure and, by extension, the role of such structure as an explanatory resource can be accommodated by Cameron's truthmaker theory. In short, what makes such talk—and the corresponding explanations—true is not molecular structure as an ontological feature of the world per se, but rather the quantum mechanical system plus environment, where that combination can be reconceived in fundamentalist structural terms. Thus, as in the case of tables and physical objects, molecular structure can be eliminated from our fundamental ontology without giving up the corresponding scientific 'talk'.

Stepping outside the Big Tent for a moment (for a quick fag and a chat with the clowns) this seems to me to offer a viable way of developing a form of OSR appropriate for chemistry that obviously meshes nicely with the eliminativist approach taken in Chapter 7: just as there are no objects at the macro-level, so they can be eliminated at the intermediate levels inhabited by molecules. Nevertheless, one might wish to step away from the reductionism debate, yet still be able to articulate something akin to OSR which incorporates at least a family resemblance to the focus on laws and symmetries that I've explored in the context of physics. I shall now sketch such a form in the biological domain where, traditionally, it has been argued that no such notion of law can apply.

12.3 Shifting to Structuralism in Biology[7]

Let me begin by recalling the twin motivations for structural realism in responding to theory change and to the ontological implications of our theories with regard to putative objects. I shall consider the latter shortly, where I shall argue that although we do not have the same kind of metaphysical underdetermination regarding identity and individuality that can be articulated in physics, we can still motivate a biologically informed version of OSR. With regard to the former, the response of the structural realist is to uncover the structural 'commonalities' between the relevant theories and urge the realist to place her ontological emphasis on these. Now, it might seem that the kind of broad correspondence underlying such claims of commonality could also be claimed to exist in the biological domain; think, for example, of the claim that chromosome inheritance theory reproduces Mendel's laws of inheritance (where it is granted that the inherited factors are not quite as Mendel conceived them; this being analogous to changes in our understanding of the underlying nature of light). Nevertheless, of course, in biology we face the obvious problem of a comparative paucity of mathematized equations or laws by means of which we can identify and access the relevant structures.[8] Even in those cases where we can identify relevant law statements, concerns have been expressed.

So, take the so-called 'Ancestral Law of Inheritance', extracted from Galton's 'stirp' theory, where ancestral contributions are given by the formula: $(0.5) + (0.5)^2 + (0.5)^3 + \ldots$ According to one well-known commentator, '[t]oday Galton's Ancestral Law of Inheritance still stands as a mathematical representation of the average distribution of continuously varying characters in a population of freely outbreeding individuals not subject to selection' (Olby 1966: 81–2). Stanford, however, insists that, 'contemporary genetics does not recognize the fractional relationships expressed in Galton's Ancestral Law as describing any fundamental or even particularly significant aspect of the mathematical structure of inheritance' (Stanford 2006: 182). The law is expressed in terms of generational contributions but, Stanford argues, there is nothing in contemporary genetics corresponding to this fractional distribution. Of course, one might argue that put so baldly, this is a poor comparison. Even in the example taken from physics, it is not the case that Fresnel's equations are written on the face of Maxwell's theory; likewise, the equations of Newtonian mechanics have to be recovered from Special Relativity under certain constraints and one might suggest that a similar operation needs to be effected here. Still, that leaves very few opportunities for the structuralist to get a grip on the relevant common elements between

[7] Just to give a little more of the 'lost history': in urging the functional unity of science, Cassirer also adopted a broadly structuralist approach to biology (see Krois 2004: 1–19). Biology, he insisted, had to be understood as the study of systems in which the relationships between elements produce a complex whole and structural changes are studied morphologically, rather than causally (Cassirer 1950).

[8] Although this is not true of all biological fields—population genetics and theoretical ecology are the exceptions (my thanks to Alirio Rosales for pointing this out to me).

theories. Hence appeal might be made to a more general notion of structure in this case. Stanford rejects this as too vague but the force of such a complaint may be diminished if we insist that the relevant structures we should be realists about are revealed through appropriate models, statespaces, etc.[9]

Still, the lack of laws represents a fundamental problem, particularly when it comes to the characterization of structure that I have outlined in the previous chapter. Let us see how we might overcome this obstacle through a focus on models and abandoning the object-oriented articulation of law-like necessity.

12.4 Laws and the Lack Thereof

Again, let's recall how, as realists, we should read off our ontological commitments from theories: we begin with the laws and (crucially, in physics at least) the symmetries of the theory, and regard these as representing the way the world is. The relevant properties are then identified in terms of the role they play in these laws. And... we stop there and do not make the further move of taking these properties to be possessed by objects. This is the structuralist way of looking at things and we take the laws (and symmetries) as representing the structure of the world.

Instead of governance—of laws over objects—the relevant relation is one of dependence, in the sense that properties depend on laws, since their identity is given by their nomic role. Furthermore, how we conceive of the necessity of laws must also be understood differently, as we saw. That feature of laws by which we can distinguish them from accidental generalizations is now understood in terms of the modal 'resilience' of laws, in the sense of remaining in force despite changes of circumstances. On the view I've set out here, this resilience is an inherent feature of laws, as elements of the structure of the world. And it is this resilience that gives laws their explanatory power—explaining why in every case, like charges repel, for example. The explanation of this regularity—the reason why it obtains, and why it is, in a sense, unavoidable (cf. Lange 2007: 472–3)—lies with the laws and their inherently modal nature by which they have this resilience.

This then opens up some metaphysical space in which to consider laws in biology, or, rather, the supposed lack of them. Interestingly, this feature rests on a characterization of laws as necessary. Consider, for example, Beatty's well-known 'Evolutionary Contingency Thesis':

> All generalizations about the living world: (a) are just mathematical, physical, or chemical generalizations (or deductive consequences of mathematical, physical, or chemical generalizations plus initial conditions) or (b) are distinctively biological, in which case they describe contingent outcomes of evolution. (Beatty 1995: 46–7)

[9] Stanford has a general response to all such moves: they fail to meet the demand for historically reliable and prospectively applicable criteria for realist belief. However, this seems too demanding as it's hard to see what could satisfy the second conjunct in particular.

If (a) is true, then biological laws 'reduce' (in whatever sense) to physical ones and there are no biological laws per se (this obviously presupposes some form of reductionism and needs further argument to the effect that if *a* reduces to *b*, then *a* can be eliminated, in the sense outlined previously; see Esfeld and Sachse 2011). If (b) is true, then the relevant generalizations are 'merely' contingent and thus cannot be necessary. (b) is certainly supported by the current conceptions of mutation and natural selection which imply that all biological regularities must be evolutionarily contingent (I shall return to this point shortly). On that basis, they cannot express any natural necessity and hence cannot be laws, at least not on the standard understanding of the latter (Beatty 1995: 52). It then follows that if either (a) or (b) is true, there are no biological laws.

One option would be to accept Beatty's thesis but insist that even though contingent, the relevant biological generalizations are still not 'mere' accidents in the way that, say, the claim that I have 67 pence in my pocket is. Thus one might argue that biological generalizations are fundamentally evolutionary, in the sense that under the effects of natural selection they themselves will evolve. In this sense, they cannot be said to hold in all possible worlds and thus cannot be deemed 'necessary'. If lawhood is tied to necessity, then such generalizations cannot be regarded as laws. However, given their role in biological theory, they cannot be dismissed as mere accidents like the claim about the contents of my pocket. They have more modal resilience than that. Perhaps then they could be taken to be laws in an inherently modal sense, where this is weaker than in the case of physical laws but still stronger and more resilient than mere accidents. Moreover, they are evolutionarily contingent in Beatty's sense. Putting these features together in the structuralist framework yields a form of 'contingent structuralism' in the sense that, unlike the case of physical structures where the structural realist typically maintains that scientific progress will lead us to *the* ultimate and fundamental structure of the world, biological structures would be temporally specific, changing in their fundamental nature under the impact of evolution. Again, I shall return to this suggestion shortly.[10]

Alternatively, one might challenge the standard understanding of laws assumed by Beatty, as Mitchell (2003) does. She argues that this standard view assumes that natural necessity must be modelled on, or is taken to be isomorphic to, logical necessity (2003: 132). But the crucial roles of laws—that they enable us to explain, predict, intervene, and so on—can be captured without such an assumption. Indeed, what characterizes laws as they feature in practice on her view is a degree of 'stability', in terms of which we can construct a kind of continuum (2003:138): at one end are those regularities the conditions of which are stable across space and time; at the other, are the accidental generalizations and somewhere in between are where most scientific laws are to be found. And even though biological generalizations might be

[10] I was led to think along such lines by a talk given by Alexander Rosenberg in which he sought to apply Beatty's view to regularities in the social sciences; see (Rosenberg 2012).

located further towards the 'accidental' end of the continuum than the physical ones, this does not justify their dismissal as 'non-laws'. Within such a framework Morgan has suggested that the Caspar–Klug formula for virus structure can be considered a biological law, contrary to Beatty's claim (Morgan 2010; for a further defence of the existence of such laws see also Elgin 2006). Likewise, Dorato has argued that biological laws differ from physical only in degree of stability and universality (2012). Such claims clearly mesh nicely with, and can be pressed into the service of, OSR, with 'resilience' equated with 'stability' and biological regularities regarded as features of the (evolutionarily contingent) biological structure of the world. It is this latter aspect that accounts for their (relative) resilience/stability and the way that aspect of their nature can explain why certain biological facts obtain.

Of course, one might still complain that nevertheless there are fewer such laws in the biological domain than in physics, say, but this hardly seems strong grounds for blocking the development of structuralism. Indeed, one can respond to Beatty's arguments by looking at the kinds of models and 'structures' in general that biology presents. Let us consider, then, how the role and nature of models in biology might be of service to the structural realist.

12.5 Models and Structures in Biology

Let me begin by re-emphasizing the distinction between the 'object' level, where biological models and theories themselves live, as it were, and the meta-level, where we can place the representational devices of the philosopher. A particularly nice summary with regard to the former is given by Odenbaugh (2009; see also his 2008), who begins by characterizing biological models very generally as 'idealised representations of empirical systems', a characterization that, of course, could equally apply to models in physics. There are two features of this characterization that I would like to emphasize (again, following Odenbaugh). The first has to do with the role of idealization. Thus, for example, infinitely large populations might be assumed in models of natural selection; certain statistical summaries are deployed in order to yield simple and tractable equations at the macro-level that represent situations of considerable complexity at the micro-level, as in the case of the Lotka–Volterra Equations; geometrically simple representations of butterfly wings are utilized in models of wing pattern formation (see Murray 2003).

In all these cases there appears to be little if anything to distinguish biological idealizations from those we come across in physics and chemistry—not surprisingly perhaps given both the mathematical nature of the models in these examples, and the role of analogies (with, for example, ideal gas models in physics) in their construction. And as in the case of idealizations in physics, the same kinds of moves are made in relating these idealizations with, to put it bluntly, 'reality' (see Rowbottom 2009). Here the structural realist needs to adopt a slightly more nuanced stance than the 'standard' realist: the latter will typically dismiss idealizations as clearly false, and likewise the models in which they are embedded;[11] the former, however, may argue that certain idealizations and abstractions represent fundamental structural features of the world. So, the idea here is that, again, certain 'high-level' laws and features of theories such as Schrödinger's equation in quantum physics, or the Hamiltonian representation of classical mechanics (discussed in Chapter 2), may be regarded as true and as representing the 'ultimate' structures of the world. Similar claims might then be made from the perspective of a structuralist view of biology.

The second feature has to do with the nature of the representation involved and here the usual claim is that biological models exhibit much more diversity than their physical counterparts. It is important to be a little cautious with regard to the statements of biologists themselves. As in the case of physicists, biologists may switch between the terms 'theory' and 'model' to describe the same element of scientific practice (for want of a better description) and as with statisticians or behavioural scientists, for example, may even take a model to be simply a set of equations or quantitative assumptions (Odenbaugh 2009: 2). Suppes noticed this tendency many years ago (Suppes 1960), and insisted that the *meaning* of 'model' across the sciences could be appropriately articulated via the model-theoretic approach covered in Chapter 5, while the *use* of the concept may differ considerably between domains (see again da Costa and French 2003).

Thus, as well as mathematical models, to which I shall return shortly, biologists not only use non-quantitative models, similar in kind to those one finds in physics, say, but also physical models, such as the classic tinplate and wire (these days, of course, plastic) Crick and Watson model of DNA as well as, and most notably, *model organisms*, such as fruit flies, flour beetles, and so on (Griesemer 1990). Both these latter kinds of models have been put forward as counterexamples to the model-theoretic approach but they can in fact be brought within its scope, as long as we keep in mind what we are doing as philosophers of science: namely, representing at the 'meta-level' of the philosophy of science, the relevant elements of practice that exist at the 'object-level' of science itself (da Costa and French 2003; French and Ladyman 1999; see also Rowbottom 2009).

So, as I suggested in Chapter 5, what we have at the object level (here, that of biological practice) are a whole range of representational devices and elements, representing, of course, systems and processes in the 'world', and at the meta-level of the philosophy of science we have various (meta-level) means of representing those devices. Obviously, the constraints on those representational relationships will be different but with the materiality or otherwise of the devices themselves set to one side, there appears to be no obstacle to the use of set theory to capture their *representational* function. And of course, the materiality of these elements may

[11] But again I must emphasize that there are ways in which such models can be regarded as pragmatically and approximately, or partially, true; see da Costa and French 2003.

feature in discussions of the ontological status of theories and models themselves (see the discussion in Contessa 2010).

Indeed, the apparent lack of the kinds of laws one finds in physics, for example, has led many philosophers of biology to embrace the model-theoretic approach, particularly those versions, such as Giere's, that downplay the role of laws in general (Giere 1999). On this latter view the focus is on a characterization of the relevant model together with a hypothesis to the effect that the model is similar to the system in relevant respects and degrees. So, for example, we might have an equation that 'describes a mathematical structure which may be claimed to be similar to the spatial dynamics or persistence times of metapopulations of checkerspot butterflies in Santa Clara, California or the Glanville fritillary butterflies on the Åland Islands in the Baltic Sea' (Odenbaugh 2009: 3).

These mathematical structures may also be represented in terms of something akin to the phase spaces or, more generally, statespaces of physics. Thus the Lotka–Volterra models of predator–prey interactions can be explicated as statespaces or 'phase portraits' (2009: 10). It is precisely the ubiquity and significance of such spaces in physics that led proponents of the model-theoretic approach such as van Fraassen to adopt them in their meta-level characterization of models. But the point is that, even without the kinds of 'laws' that one finds in physics, these models appear to have the sorts of features that the structuralist can get her teeth into.

Still, the diversity of biological models and the purported 'patchwork' nature of the coverage of the biological domain may be seen as a further obstacle (Odenbaugh 2009: 4). Thus Mitchell has written,

> If science is representing and exploring the structure of the world, it is reasonable to ask why there is such a diversity of representations and explanations in some domains. (Mitchell 2003: 2)

Thus we often find various features of a particular system represented in multiple ways by different models. In particular, we may have one model that is highly focused and quantitative, that allows for precise predictions, and another that is more general and qualitative, capturing broader features, with both together mapping the 'contours of biological theory' (2003). So, in the case of metapopulation theory, for example, which models the interactions between spatially separated populations of the same species, we find diverse models being employed—some that do not assume that the rate of colonization is constant or that extinction is constant, others that combine colonization and competition, and so on (2003).

Again, this is not something that is peculiar to biology. Hacking long ago noted that physicists work with multiple models of the electron, for example, rather than 'the' theory (Hacking 1983). More recently Wilson has constructed a very wide-ranging framework encompassing theories, models, and linguistic concepts in general in terms of stitched together 'façades' (Wilson 2006). Are such patchworks a problem for the structural realist? Again invoking the Big Tent, one could, presumably, be a 'disunificationist' and accept that we cannot arrive at a completely unified structuralist representation of the world but that each of the patches or façades represents some piece of the underlying structure. Of course, an obvious question is whether these patchworks at the level of representation are matched by ontological 'dappling'. Odenbaugh notes that the combination of the characterization of biological systems in terms of large numbers of weakly interacting variables and our cognitive resources means we have no choice but to use multiple models (2009: 4), but this is a pragmatic issue. Even if it is computationally impossible to tackle the relevant equations, forcing us to use idealized models, this does not necessarily mean that the underlying structure of the world is fundamentally façade-like. That's not to say that it may not be multifaceted, in just the way that structures in physics are and the structuralist's answer to Mitchell's question may be a simple 'that's the way the structure of the world is!'

As well as the diversity of biological models, their relative independence from high-level theory may also be brought into play as a block on adopting a structural realist stance. I have already touched on this in Chapter 5 and a similar response holds here: we should distinguish a weak and strong form of the independence claim. So, the weak form states that the relationship between a theory and model is such that it allows for a degree of independence, at least with regard to the cognitive attitude taken towards the model. This allows models to mediate between theories and phenomena models and thereby possess a certain degree of autonomy in terms of acting as the focus of scientific developments (Morrison 1999). A stronger form holds that in mediating between theories and the world, models contain some form of 'surplus structure' or extra cognitive resources such that they cannot be straightforwardly deduced or otherwise obtained from the relevant theory (Cartwright, Shomar, and Suarez 1996). However, as in the case of 'horizontal' diversity noted earlier, there appears to be nothing to prevent an advocate of the model-theoretic approach from directly representing such independent models set-theoretically and accepting that they cannot be appropriately related to higher-level theory (da Costa and French 2003). Likewise, there do not appear to be any obstacles to adopting a structuralist stance towards them.

Of course, as Odenbaugh notes, if we do not have any high-level theory, one might wonder what it is that biological models mediate between (2008: 22). His suggestion is that as long as we have an appropriate hierarchy to draw on, we can still maintain that there is mediation of a kind. I think that is right, but without a sole highest-level theory, as it were, the mediation will have to be taken as relative to whatever is above that particular model in the hierarchy and that, of course, may vary depending on the particular context. This may weaken the notion of mediation to the point where one begins to wonder about its usefulness here; perhaps in the biological domain all one can say is that there are interrelated hierarchies of models and that any particular model 'mediates' between those above and below it in the hierarchy (except those 'at the top'!). Certainly, with such a weakened notion it's hard to see how a robust understanding of 'autonomy' could be maintained.

As an example of the stronger claim, whereby further features are introduced that are not obtained from theory, consider the Lotka–Volterra predator–prey model again, where in order to develop models applicable to the relevant situation we have to incorporate empirical assumptions, such as 'predator satiation' (Odenbaugh 2008: 23–6). But now the independence claim seems to amount to little more than the point that to get workable models from an idealized and often quite abstract representation such as the Lotka–Volterra equations, we need to add reasonable auxiliary assumptions based on a lower-level understanding (e.g. that predators can't glutton themselves to the point of consuming an infinite amount of prey). Again, I can see no reason why this kind of independence would present an obstacle to the structuralist programme.

Biological models may also be 'internally' diverse, as in the case of 'hybrid' models; these combine mathematical and physical elements in ways that are claimed to be interestingly different from the case in physics. Thus consider gene regulation mechanisms in the context of synthetic biology (Loettgers 2007), where models are seen as 'engineered genetic networks' that can be used to both investigate certain regulatory mechanisms and engineer biological components. In particular, these models offer a way of 'getting a more complete understanding of the structure of genetic networks and how the structure relates to specific functions' (2007: 135). Within this approach, scientists may use mathematical models, synthetic models, and model organisms, and variants of all three. This diversity can be in part explained by the different modelling traditions of different communities (geneticists and theoretical biologists). More importantly, these 'synthetic' models combine features of mathematical models and model organisms. Consider the 'Repressilator', for example, in which a cyclic negative feedback loop containing three genes is introduced into a bacterium: the loop results in temporal oscillations in protein concentrations which can then be made visible via a gene for making a fluorescent protein (Loettgers 2007: 140–1). Here we have biological components used in a model the performance of which is explored under the constraints of particular biological systems, and the construction of this model was guided by mathematical modelling, using network designs based on feedback loops. Now of course, this is not a truly hybrid model in the sense of somehow combining both biological and mathematical elements—that would obviously be ontologically provocative![12] What we have here is a material model, the construction of which involved reflection on certain forms of mathematical modelling. In this sense, it does not appear so different from Crick and Watson's wire and tinplate model, based as it was on mathematical considerations as well as well-known empirical data. Certainly the use of mathematics here seems unproblematic and although it provides another useful example of the diversity of

[12] cf. Morrison's claim—touched on in Chapter 5—that spin represents a hybrid property in a similar sense (Morrison 2007; see French 2015).

biological models, I again see no fundamental difficulty in accommodating such models within either the model-theoretic approach or a structuralist framework.

Finally, there is the further dimension touched on earlier, namely that biological structures are evolutionarily contingent. If we consider the invariant biological regularities represented by models, then a striking feature is the variety and heterogeneity of the limitations on these invariances. The point is nicely put as follows:

> once environments come to include creatures and their effects on one another, the life-times of regularities about creatures' adapted traits fall from the scale of billions of years (archebacteria—whose environment has not changed for 3 billion years) to multiple geological epochs (oxygen-respirators) to hundreds of millions of years (vertebrates) to weeks and months in the case of others (the AIDS-virus). (Rosenberg 2012: 12)

Consider the following example (which exemplifies the typical form of laws considered by philosophers): 'All genes are composed of DNA'. Over a long period, this regularity remained invariant but by virtue of being subject to no exceptions, 'its operation provided an environment that would allow for the selection for any new biological system that could take advantage of the fact that all genes are composed of DNA' (2012: 12 n. 11). Of course, such a system eventually evolved, namely RNA viruses, which parasitize the machinery of DNA replication (as an example, consider the HIV virus). Thus a kind of 'arms race' of evolutionary competition generated a shift from 'All genes are made of DNA' to 'All genes are made of nucleic acids (either RNA or DNA)', with further shifts possible in the future.

How can we explain this variety of limitations on invariances? Clearly, we need to appeal to laws and in the biological domain the laws required are those of natural selection (Rosenberg 2012). It is in this manner that we can explain the differences in both the limits and success of models. So, for example, in the case of the Lotka–Volterra model the invariance is broader than that exhibited by Nicholson–Bailey models of bacterial parasites and hosts. More importantly, there are no spatio-temporally unrestricted regularities in biology, as there are, supposedly, in physics. Thus, either the principles of Darwinian evolution themselves must be nomological generalizations akin to those in physics (in the sense of being invariant under relevant changes), or, if it turns out they are only locally invariant, then there must be more fundamental invariances, that together with specific conditions establish the limits of this invariance (Rosenberg 2012). In other words, if biological structures are conceived of as spatio-temporally limited and evolving structures, this needs to be understood as holding within a more encompassing or fundamental structure. Then, there are two options: if biology is not reduced to chemistry and physics, then this more encompassing structure will be that of the principles of natural selection, understood as globally invariant nomological generalizations as in physics; or, if reductionism holds, then this more fundamental structure will be physical structure. So, either we have a *sui generis* OSR for the biological domain, or, ultimately, biological structure is reduced to the kinds of structures I have already mapped in previous chapters.

Nevertheless, if we take the first option and attempt to articulate a biological form of OSR, drawing on the kinds of models outlined previously, we do not typically find the other feature of physical structures in biology, namely symmetries. However, one can identify similar 'high-level' features of biological structures. There is, of course, Price's Equation (for discussion see Okasha 2006: §1.2 and, in a different context, Rowbottom 2010), sometimes presented as representing 'The Algebra of Evolution', and which one could take as characterizing a certain fundamental—if, perhaps, abstract—and 'high-level' feature of biological structure. Put simply, this states that,

$$\Delta z = \mathrm{Cov}\,(w, z) + \mathrm{E}w(\Delta z)$$

where Δz is the change in average value of character from one generation to the next; Cov (w,z) represents the covariance between fitness w and character (action of selection) and $\mathrm{E}w(\Delta z)$ represents the fitness weighted average of transmission bias (difference between offspring and parents). Thus the equation separates the change in average value of character into two components, one due to the action of selection, and the other due to the difference between offspring and parents. There is a sense in which this offers a kind of 'meta-model' that represents the structure of selection in general (for a useful overview, see Gardner 2008; also Okasha 2006: §1.2 and Jones 2008). Okasha writes that it reveals the 'common logic underlying all selection processes, at all scales and at all hierarchical levels' (Okasha 2011: 245; see also 2006).[13] Indeed, as Gardner notes, it can be viewed as reflecting an even more general feature of reality:

> The importance of the Price equation lies in its scope of application. Although it has been introduced using biological terminology, the equation applies to any group of entities that undergoes a transformation. (Gardner 2008: 199)[14]

Although obviously not a symmetry principle, this covariance equation is independent of objects, rests on no contingent biological assumptions, and can be understood as representing the modal, relational structure of the evolutionary process (see Rosales preprint). Just as the laws and symmetries of physics 'encode' the relevant possibilities, so Price's equation encodes how the average values of certain characters changes between generations in a given biological population.

Let me now turn to the motivation for dropping the object-oriented stance in philosophy of biology, namely the 'Problem of Biological Individuality' and the

[13] Waters, on the other hand, takes the above formulation to represent simply a partial decomposition of evolutionary causes, as he sees the Price equation as just one tool in the toolbox that biologists have available (Waters 2011).

[14] As Gardner goes on to note, Price himself emphasized that his equation could be used to describe the selection of radio stations with the turning of a dial just as easily as it could to describe biological evolution.

heterogeneity of biological objects in general. As we shall see, there are relevant similarities with the case in physics, but also significant differences.

12.6 Identity and Objecthood in Biology

There are, I would claim, at least four issues that could be invoked in support of adopting a structuralist stance here:

i. gene identity
ii. gene pluralism vs the hierarchical approach
iii. metagenomics and the general problem of biological individuality
iv. the heterogeneity of biological objects

12.7 Gene Identity

It has been claimed that the notion of 'gene' has undergone such a radical transformation during the history of genetics that there are simply no straightforward identity conditions that it could be said to satisfy throughout the course of that history. This has been pointed to in support of an anti-realist stance towards the term; namely that there is no object in the world to which it refers. In this respect the term might be usefully compared to that of 'atom' or even 'electron', which have undergone similar transformations. In the latter cases, as we have seen, a structuralist stance allows us to retain a realist interpretation in the face of the Pessimistic Meta-Induction.

As Fox-Keller famously put it,

> [This] little word [gene], so innocently conceived in the early days of this century, has had to bear a load that was veritably Herculean. One single entity was taken to be the guarantor of intergenerational stability, the factor responsible for individual traits, and, at the same time, the agent directing the organism's development. Indeed, one might say that no load seemed too great... as long, that is, as the gene was seen as a quasi-mythical entity. But by the middle part of the century, the gene had come to be recognized as a real physical molecule... [and]... that load has become steadily easier to discern. (Fox-Keller 2000: 144–5)

Thus, the term is required to fulfil a number of roles and it simply cannot do so; or at least not in a way that allows us to pin down a consistent set of identity conditions. Similarly Burian insisted that, although '[t]here is a fact of the matter about the structure of DNA,... there is no single fact of the matter about what the gene is' (Burian 2005: 142).[15]

How should the (standard) realist respond to this history of shifting roles? A dilemma arises: if the term is taken to refer to some 'essence' or hidden nature

[15] A useful overview can also be found in Burian and Zallen 2009.

of the gene, then referential continuity has been secured at the cost of no little mystery (or, more bluntly, it has simply been asserted); if it is taken to refer to the entities as they feature in scientific practice, then we are stuck with ontological discontinuity and Ladyman's complaint that what we have is an ersatz form of realism.[16]

At the core of this dilemma lies the problem of setting down clear identity conditions for the gene qua object. One option is to articulate a form of 'contextual' identity, so that the gene could be regarded as a 'thin' object, as outlined in Chapter 3. Thus, we might adopt a functional approach in which a 'gene' is understood to be identical to a particular sequence of DNA typically assembled in a particular way in certain typical contexts and which typically produces specific protein(s). In other words, we should 'Identify a gene as a specific DNA sequence, S, that plays a typical functional role, R, of typically producing protein P in context C in an organism O' (Kraemer 2008).[17] On this account, the 'gene' as a biological entity would be reconceptualized structurally in terms of a (multi-aspected) nexus of biological relationships, and individuated in a 'thin' manner, via those relationships, or, to put it another way, perhaps, functionally identified via the role(s) it takes on. The worry, of course, is that the complexity of the multiple role-shifts may be such that even such a contextual approach offers too little to get an ontological grip on.

Thus Fox-Keller, again, noted that since Burian's observation quoted earlier,

> things have only gotten worse... The complications brought by the new data are vast... taken together, they threaten to throw the very concept of 'the gene'—either as a unit of structure or as a unit of function—into blatant disarray.... Techniques and data from sequence analysis have led to the identification not only of split genes but also of repeated genes, overlapping genes, cryptic DNA, antisense transcription, nested genes, and multiple promoters (allowing transcription to be initiated at alternative sites and according to variable criteria). All of these variations immeasurably confound the task of defining the gene as a structural unit.... Similarly, discovery of the extensive editorial process to which the primary transcript is subject, of regulatory mechanisms operating on the level of protein synthesis, and others operating even on the level of protein function confound our efforts to give a clear-cut functional definition of the gene. (Fox-Keller 2002: 66–7)

One might conclude, then, that *no* appropriate identity conditions can be given in the case of the gene and that not even a 'thin' notion of object can be retained. Of course, as in the case of the elementary particles of physics, this would not prevent biologists and others from using the term in scientific discourse; here again we can avail ourselves of such devices as Cameronian truthmaker theory to explicate that use

[16] Alternatively, one could take the reference of the term as simply not fixed or 'open' in some sense (Burian 2005). Here we might recall our discussion in Chapter 5 and the suggestion that, in the context of the Suppesian 'dual perspective' framework, the structural realist might avail herself of certain, more flexible, accounts of reference, where these might include some form of open-ended theory.

[17] For more on a comparison between functionalism and structuralism, see McCabe 2006 who suggests a structural realist approach to the mind.

and ground the truth of statements made about genes without having to accept genes-as-objects as our truthmakers. The point, once again, is that when we come to (philosophically) reflect on what the term 'gene' refers to, the concerns outlined here suggest that it is not to an object per se, but rather a node in an interrelated set of biological structures.

Criticism of what has been called the 'gene-centred' stance in foundations of biology has also emerged from 'Developmental Systems Theory' (DST), the aim of which is to study the interactions between the various factors that influence biological development whilst avoiding the usual dichotomies between genes and the environment, nature and nurture, etc. (Oyama et al. 2001). According to DST genes should not be regarded as the fundamental biological individual, or as some kind of 'master molecule' that encodes the relevant traits; rather they should be understood as a developmental resource for the construction of biological systems. The focus is then shifted away from the intrinsic properties of the gene-as-fundamental-biological-object, to the relevant context (such as the environment), the contingent features of which will also contribute to development (see Wilson 2007). Furthermore, DST suggests a form of explanatory symmetry or 'parity' between genetic and non-genetic causal factors (although such claims may be overdone; see Godfrey-Smith 2000). Such shifts may of course be viewed sympathetically by structuralists, particularly if, in this case, the context concerned can be characterized in broadly structural or relational terms. And taking objects to be mere nodes in structure may be used to underpin a form of explanatory symmetry according to which objects are not causally privileged. Thus, as we have seen, one of the differences between OSR and 'object-oriented' realism concerns the typical metaphysics associated with the latter that insists on the role of objects as the 'seat' of causal powers. Allowing structure to be causally informed allows for the kinds of parity associated with DST.

The idea of the 'gene' as a resource or tool also features in practice-oriented accounts that understand genetics as organized via an integration of explanatory reasoning (associated with a theory) and investigative strategies rooted in the relevant practices (Waters 2008). Here the emphasis is on 'bottom-up' manipulability whereby '[g]enes are used as levers to manipulate and investigate a wide variety of biological processes' (Waters 2008: 260). There is an obvious comparison with Hacking's entity realism and so it might suggest a return to some form of object-oriented stance. But of course, Hacking's insights can be accommodated within a form of structural realism, as Chakravartty has shown (1998). Thus, we recall from Chapter 9, the distinction between detection properties and auxiliary properties, where it is the former that is associated with manipulability. But then, of course, paraphrasing Hacking's famous slogan, even if we were to agree that '[i]f you can lever them, they're real', it does not follow from an understanding of genes as levers that they must be understood as objects, in any metaphysically robust sense. Since the leverage is effected via certain properties—just as the supposed manipulability of electrons is in Hacking's own example—relying on this feature for one's ontology does not get you all that far—certainly not to entities-as-objects. A further step is needed. For Chakravartty this step is underpinned by the argument that objects are required as the 'seat' of causal properties but I suggested in Chapter 8 that there is no need to take that further step and that we can, in effect, 'unseat' these properties.

Furthermore, if we do away with the associated dispositional analysis, as outlined in Chapter 9, we can retain their relational feature and thus the structuralist elements of Chakravartty's picture, without the object-oriented aspects. Applying this to the biological domain, we come to see genes as phenomenological entities—just like elementary particles in that respect—that in a certain sense can be said to be manipulated (in precisely the sense that we can effect certain changes via the relevant properties) but not as metaphysically fundamental objects (at the biological level). And as also in the case of the particles of physics, we can still talk of causal powers in the sense of difference making but now understand these powers as invested in the relevant biological structures.

Let me just make it clear that my intention is not to harness biological structuralism to any one of the approaches discussed here but again, within the spirit of the 'Big Tent', to indicate how certain work in the foundations of biology might find a sympathetic context within a broadly structuralist framework. These suggestions are further supported by reflection on the second issue, where something akin to, but not identical to, a form of metaphysical underdetermination may be discerned.

12.8 Gene Pluralism vs the Hierarchical Approach

The units and levels of selection debate is one of the most prominent and significant in the philosophy of biology and here I shall only present a crude sketch that brings out the relevant points for my purposes.

There are multiple concerns and questions that have arisen in this debate (for a useful overview, see Lloyd 2005) but as Okasha (2006) has noted, the core issue has to do with the circumstances under which entities at different levels of the biological hierarchy are subject to selection, where these entities include genes, chromosomes, individual organisms, and so on. The debate centres on two images of selection: one that involves a hierarchy of entities and their traits' environment, and another that focuses on genes with properties that enable copying. These images underpin alternative representations of the relevant processes, between which—it has been claimed—formal and empirical equivalence can be established. The so-called 'multi-level selection' approach argues that selection operates simultaneously at different levels, so that advocating genic selection, for example, as opposed to selection of individuals, is to commit a conceptual error. Furthermore, as Okasha has also emphasized, the aforementioned hierarchy is itself the product of evolution, with multilevel selection contributing to the explanation of the move from one level to the next.

A significant theme that has emerged in this debate concerns the issue of realism versus pluralism (Okasha 2006). The former holds that there is always a fact of the matter as to which level selection is operating at, whereas the latter maintains that there may be no such 'fact'. This pluralist approach is captured in a now famous declaration:

> Once the possibility of many, equally adequate, representations of evolutionary processes has been recognized, philosophers and biologists can turn their attention to more serious projects than that of quibbling about the real unit of selection. (Kitcher, Sterelny, and Waters 1990)

The core claim of this approach, then, is that the distinction between levels of selection is conventional and, furthermore, that the fundamental error underlying the debate is the positing of entities ('targets of selection') that do not exist (see, for example, Waters 2006).

This has been dismissed as a 'weak' pluralism, to be contrasted with a strong form in which the different representations of selection are *jointly* relevant and required to describe the phenomena (Lloyd 2005). In terms of the contrast here between the representation of selection that involves a hierarchy of entities and their traits' environment and the other that focuses on genes with copiable properties, gene-based representation has been argued to be dependent on the hierarchical image and hence cannot be relevant in the manner required. Thus,

> the genic account does not give us a theory independent of individuating causal interactions at various levels of the biological hierarchy, nor does it solve or dissolve the problem of how to individuate those very interactions. (Lloyd 2005)

Indeed, not only are genic models not independent from hierarchical representations, they are arguably derivative from them:

> 'genic' level causes are derivative from and dependent on higher level causes. Their genic level models depend for their empirical, causal, and explanatory adequacy on entire mathematical structures taken from the hierarchical models and refashioned. (Lloyd 2005)

The pluralists in turn have responded by insisting that hierarchical models don't 'own' the relevant structures and hence they can be drawn upon by genic representations.

Now, it is not clear that the characterization of this debate in terms of pluralism, whether weak or strong, is entirely appropriate. Rather, what we seem to have here is a form of underdetermination, involving empirically equivalent but interpretationally distinct models. Interestingly, in this context, the role of metaphysical elements in the genic representation has been emphasized:

> Given that the genic model construction and metaphysical conclusions are inextricably bound together in the arguments as the pluralists have formulated them, they are not free to slice off metaphysical questions as they wish. (Lloyd 2005)

Of course, a comparison with the metaphysical underdetermination that I presented in Chapter 2 is not straightforward. It is certainly not the case here that we have two models both positing objects, but one taking them to be individuals, the other non-individuals in some sense. Instead, we have one 'horn' that itself points to the positing of entities as problematic. Nevertheless, if we take the distinction between alternatives to be fundamentally metaphysical, we can perhaps see this debate as opening the door to a structural understanding of the unit of selection. Certainly, such a stance would allow us to move away from metaphysical quibbling over the *entities* underlying selection.

Let me now turn to further and more general considerations in which concerns about individuality arise.

12.9 The General Problem of Biological Individuality

In discussions of biological ontology, the following implicit assumptions have been identified:

a) 'life' is organized in terms of the 'pivotal unit' of the individual organism
b) such organisms constitute biological entities in a hierarchical manner

These assumptions have come under pressure from the 'metagenomic' stance which represents a shift in focus away from individual genomes to 'large amounts' of DNA 'collected from microbial communities in their natural environments' (Dupré and O'Malley 2007: 836). Correspondingly, a shift in philosophical attention has been urged from individual organismal lineages to the 'overall evolutionary process in which diverse and diversifying metagenomics underlie the differentiation of interactions within evolving and diverging ecosystems' (2007: 838).

Crucially, this shift threatens the standard understanding of individuality in biology:

> To the extent that such individual autonomy requires just an individual life or life history, then it surely applies much more broadly than is generally intended by biological theorists. Countless non-cellular entities have individual life-histories, which they achieve through contributing to the lives and life-histories of the larger entities in which they collaborate, and this collaboration constitutes their claim to life. But—and this is our central point—no more and no less could be said of the claims to individual life histories of paradigmatic organisms such as animals or plants; unless, that is, we think of these as the collaborative focus of communities of entities from many different reproductive lineages. (Dupré and O'Malley 2009: 15)

The 'deep and extensive' collaborations between biological entities mapped by metagenomics blurs the distinction between putative individual organisms and the larger groupings of which these entities are parts (2009: 12). Hence, '[i]ndividual organisms, from this viewpoint, are an abstraction from a much more fundamental entity' (Dupré and O'Malley 2007: 842). Biological objects are no more than 'temporarily stable nexuses in the flow of upward and downward causal interaction' (2009: 842). In particular, 'a gene is part of the genome that is a target for external (that is, cellular) manipulation of genome behaviour and, at the same time, carries resources through which the genome can influence processes in the cell more broadly' (2009: 842).

Again, one response to this is to adopt a form of pluralism that suggests that 'there are countless legitimate, objectively grounded ways of classifying objects in the world' (Dupré 1993: 18). Extended from kinds to objects, this 'Promiscuous Realism' would hold that there are countless, objectively grounded ways of delineating such biological objects. As applied to kinds, and in the context of the debate whether species count as such or as individuals, Promiscuous Realism has come under critical fire (see, for example, Wilson 1996) and an object-oriented realist might well baulk at giving up the claim that there is a 'unique and privileged set of categories' in the world (Dupre 1996: 443).

However, a structuralist perspective on the metagenomic 'facts' here would undercut this debate by eliminating the central notion of object. From this perspective *there are no biological objects (as metaphysically robust entities).* All there is are biological structures, interrelated in various ways and causally informed. Putative objects, such as genes, individual organisms, and so forth, can then be seen as emergent entities, or as dependent upon the appropriate structures, where the notions of emergence and dependence here will be both informed by the relevant biology and framed in terms of an appropriate metaphysics, along the lines I have tried to indicate throughout this book. Thus there is no need for Promiscuous Realism since we can adapt a (dynamical) form of structuralism which will allow us to be realist about the relevant biological structures, without being ontologically pluralist about the entities.

Here we might draw a useful comparison with the accommodation of causation within physical structures presented in Chapter 8: it is not necessary for such an accommodation to always be in terms of causal loci instantiated in terms of specific entities; rather, causality can be seen as arising relationally (in this case, from the relevant interactions) and holistically, in the sense that it is 'located' across the relevant structure as a whole. Furthermore, in the biological domain at least, one can sidestep the contentious issue of positing 'relations without relata' by allowing the relata in this context to be non-biological (i.e. chemical or physical) structures. Of course, this would be tantamount to the adoption of Beatty's position (a), or Rosenberg's second option, according to which all generalizations about the living world are ultimately just mathematical, physical, or chemical generalizations (or deductive consequences thereof plus initial conditions). From the biological perspective there is no reintroduction of *biological* objects to serve as such relata (although of course the problem remains when it comes to physical structures). Of course, if one were to insist on the standard forms of anti-reductionism, the concern cannot be assuaged in this way.

Let me say a little more about causation in this context.

12.10 Causation in Biology

In the context of physics there is always the fall-back option of adopting the Russellian line that here there is little scope for any robust notion of causation in the first place (Ladyman, Ross, et al. 2007). When it comes to biology, however, such a fall-back move may not be straightforwardly available. Thus, Okasha has suggested that distinctive issues arise here that have no parallel in the physical sciences (2009). He argues that the kinds of Darwinian explanations that one finds in this context must be regarded as causal, but at the population level, rather than singular. In so far as natural selection is 'blind to the future' and genetic mutation is undirected, these explanations certainly can be taken to have pushed teleology out of biology (2009: 719–20). When it comes to genetics, matters are more nuanced. Here the distinction between singular and population-level causality is crucial as heritability analyses pertain only to the latter. In particular, such analyses 'tell us nothing about individuals' (2009: 722). Furthermore, and recalling our brief considerations previously, the idea of the gene as the sole causal locus has been undermined by the implicit relativity to background conditions (2009: 721). Further challenges to the notion of the gene as the seat of causal power have also been posed by proponents of the DST approach who advocate a form of 'causal democracy'. Nevertheless, genes might still play the more dominant causal role, although this is something that should be determined by further research (2009: 724). And of course, as we have seen, even if that is granted, the structuralist can apply well-known pressure to the concept of the 'gene' and argue that even if this does play the dominant role in biological causation, it should not be understood in object-oriented terms.

The point I'd like to emphasize is that talk of causal powers and associated causal loci per se does not represent a major obstacle to the structuralist. Even if one were entirely comfortable with such talk, one could follow the metagenomic line and insist that these causal powers are derived from the interactions of individual components and are controlled and coordinated by the causal capacities of the 'metaorganism' (Dupré and O'Malley 2009). This sort of account seems entirely amenable to a structuralist metaphysics. Alternatively, as we saw in Chapter 8, one could acknowledge that causation is a kind of 'cluster' concept, under whose umbrella we find features such as the transmission of conserved quantities, temporal asymmetry, manipulability, being associated with certain kinds of counterfactuals, and so on. Even at the level of the 'everyday' this cluster may start to pull apart under the force of counterexamples. And certainly in scientific domains only certain of these features apply at best. Thus in the context of physics we can only say, at best, that a very 'thin' notion of causation holds, understood in terms of the relevant dependencies. We may talk, loosely, of one charge 'causing' the acceleration of another, but what does all the work in understanding this relationship is the relevant law and from the perspective of OSR, it is this that is metaphysically basic and in terms of which the property of charge must be understood. It is the law—in the classical context, Coulomb's—that encodes the relevant dependencies that hold between the instantiations of the property and that, at the phenomenological level, we loosely refer to as causal.

But once we move out of that domain, the possibility arises of 'thickening' our concept of causation in various ways. We might, for example, insist that for there to be causation there must be, in addition to those conditions corresponding to what are designated the 'cause' and the 'effect', a process connecting these conditions, where this actual process shares those features with the process that would have unfolded under ideal, 'stripped down' circumstances in which nothing else was happening and hence there could be no interference (Hall 2011: 115). Here one might draw upon mechanism-based accounts of causation and explanation (see, for example, Machamer et al. 2000; for criticism, see Psillos 2011).

In particular, if such accounts were to drop or downplay any commitment to an object-oriented stance, possible connections can be established with various forms of structuralism. In general, characterizations of mechanisms can be broken down into two features: one that says something about what the component parts of the mechanism are, and another that says something about the activities of these parts (McKay-Illari and Williamson 2013). This suggests a dual ontology with activities as well as entities—of which the parts of mechanisms are composed—in the fundamental base. Here consideration of putative asymmetries between activities and entities (McKay-Illari and Williamson 2013) mirrors to a considerable degree consideration of, again putative, asymmetries between objects and relations within the structuralist context. Indeed, a useful comparison can be drawn between the insistence that activities are not reducible to entities—so that one needs both in one's ontology—and certain forms of 'moderate' structural realism that set objects and relations ontologically on a par (Esfeld and Lam 2008):

> Activities are real causes, they give us the modal structure of the bundles of mechanism schemas that are biological theories. And biological entities do indeed depend on biological structure. So we have both the basic realist claim, that is also recognizably structural due to a characteristic dependence claim. (McKay-Illari forthcoming: 12)

One can go even further and identify a deeper structure, namely that corresponding to the functional causal roles that are experimentally established in developing mechanism schemas (McKay-Illari and Williamson 2013). On this view, both entities and activities alike should be regarded as the 'locators' of the patterns that Ladyman, Ross, et al. focus on (2007). The crucial difference between biology and physics is that in the former, unlike the latter, 'these patterns are local, patchwork and diverse, which is why we need many many locators to track them' (2007: 16). Both entities and activities can be regarded as locally specific locators for the production of the phenomena that act as explananda yielding a 'deep priority of structure', corresponding to that which persists through theory change, and a full-blown biological form of OSR.

12.11 The Heterogeneity of Biological Entities

This structuralist perspective can be further reinforced by consideration of the heterogeneity and general 'fluidity' of biological objects (see Clarke 2010 and forthcoming; Godfrey-Smith 2011). Going back to assumptions (a) and (b) in section 12.9, they underpin the decomposition of biological organisms into individuals that are commonly taken to have the following fundamental characteristics:

possessing three-dimensional spatial boundaries;
bearing properties;
acting as a causal agent (see Wilson 2007).

In addition, biological individuals are generally taken to be

countable;[18] and
genetically homogenous.[19]

However, there are well-known confounding cases that raise problems for one or more of these characteristics.[20] So, consider the case of the so-called 'humungous fungus', or *Armillaria ostoyae* which, in one case, covers an area of 9.65 square km. Previously thought to grow in distinct clusters, denoting individual fungi, researchers established through the genetic identity of these clusters that they were in fact manifestations of one contiguous organism that, as one commentator noted, challenges 'traditional notions of what constitutes an individual organism' (USDA Forest Service 2003). Or take the example of the Pando trees in Utah, covering an area of 0.4 square km, all determined—again by virtue of having identical genetic markers—to be a clonal colony of a single 'Quaking Aspen'. In both cases, obvious problems to do with counting arise (how many 'trees' are there?) and at the very least force a liberal notion of biological individual to be adopted.

More acute problems for this notion arise with examples of symbiotes, such as that of a coral reef, which consists not just of the polyp plus calcite deposits but also zooanthellae algae that are required for photosynthesis. Or consider the Hawaiian bobtail squid, whose bioluminescence (evolved, presumably, as a defence mechanism against predators who hunt by observing shadows and decreases in overhead lighting levels) is due to bacteria that the squid ingests at night and which are then vented at the break of day, when the squid is hidden and inactive. The presence of the bacteria confers an evolutionary advantage on the squid and thus renders the squid the individual that it is, from the evolutionary perspective, but they are, of course, not genetically the same as the squid, nor do they remain spatially contiguous with it.

[18] Also emphasized by Godfrey-Smith (2011).

[19] An assumption that forms part of what Dupré calls 'genomic essentialism'.

[20] Some of these examples are taken from the papers and discussion at the symposium on 'Heterogeneous Individuals', at the *PSA 2010*, Montreal.

Again, one can try to construct a unitary account of biological individuals that can cover these cases, or, alternatively, abandon any such attempt and insist that there is no one such framework of biological individuality.

Thus, one option is to abandon the genetic homogeneity assumption of biological individuality by shifting to a 'policing'-based account. Pradeu offers an immunological approach to individuation which, he claims, moves away from the self/non-self distinction and is based on strong molecular discontinuity in antigenic patterns. A biological organism is then understood as a set of interconnected heterogeneous constituents, interacting with immune receptors (Pradeu 2012). This is an interesting line to take but concerns have been raised over its extension to plants, for example, where genetic heterogeneity may not be appropriately policed (Clarke 2010 and 2013) or to viruses (O'Malley 2014). And in general its reliance on an account of immune responses that are taken to be similar across a range of organisms leads to the criticism that it ignores the differences in how the relevant immune systems work (O'Malley 2014).

Alternatively, one might adopt a 'tripartite account', according to which an organism is (a) a living agent; (b) belongs to a reproductive lineage, some of whose members have the potential to possess an intergenerational life cycle; and (c) has minimal functional autonomy (Wilson 2007). Underlying this view is the assumption that organisms and the lineages they form have stable spatial and temporal boundaries but recent commentators have suggested that if we pay attention to the microbial world as well as the macroscopic examples we are used to discussing, then rather than a 'tree' of life composed of such lineages, we have a 'web or network of life' in which the idea of stable and well-defined lineages begins to break down. Again, the example of symbiosis and indeed its pervasiveness suggests that lineages/individuals are fluid and ephemeral (see, for example, Bouchard 2010).

Perhaps then one might be tempted by a pluralistic approach that distinguishes two kinds of biological individuals (Godfrey-Smith 2011): Darwinian individuals, which are members of a collection in which there is variation, heredity, and differences in reproductive success; and organisms, which are systems comprised of diverse parts which work together to maintain the system's structure. Some Darwinian individuals are organisms and for both kinds, there are clear and less clear, or more marginal, cases. Darwinian individuals that are not organisms include viruses and genes, and organisms that are not Darwinian individuals include symbiotic collectives (as in the case of the bobtail squid). And just as certain metabolic collaborations become Darwinian individuals, so certain of the latter 'reach out' to other individuals to form new organisms, leading to an interesting to-and-fro on the border between organisms and Darwinian individuals.

Now, the category of organisms that are not Darwinian individuals may be larger than many people appreciate (and, indeed, includes people!) and, as noted earlier, this can suggest a shift from individual organismal lineages to the 'overall evolutionary process in which diverse and diversifying metagenomics underlie the differentiation of interactions within evolving and diverging ecosystems' (Dupré and O'Malley 2007: 838). We recall the claim that the notion of the autonomous individual breaks down across biological domains, and so rather than thinking of biological objects in this way we should regard them as the product of multiple collaborations. This suggests a more radical form of pluralism that we might call 'Promiscuous Individualism': there are countless, objectively grounded ways of individuating or, more generally, delineating biological objects. However, we also recall the worry about the extent to which we can legitimately call this a form of realism: if an object-oriented stance is assumed—as it typically is—so that biological theories are taken to represent or refer to biological objects, then pluralism will lead at best to a form of contextual reference or at worst to a kind of indeterminacy that may be incompatible with realism as typically understood.

Alternatively, we may eschew both unitary and pluralistic options, while retaining the insights that power the latter and adopt the structuralist stance. From this perspective, there are no biological objects (as metaphysically substantive entities); all there is, are biological structures, interrelated in various ways and causally informed. Putative objects—whether Darwinian individuals or organisms—should be seen as dependent upon the appropriate structures ('nodes') and from the realist perspective, eliminable, or, at best, regarded as secondary in ontological priority. This then accommodates the 'fluidity' and 'ephemerality' of biological entities, as evidenced in the example of symbiotes, and also the 'to-and-fro' across the boundary between Darwinian individuals and organisms. Again, from this perspective, biological individuals come to be seen as nothing more than abstractions from the more fundamental biological structure (cf. Dupré and O'Malley 2007), or as 'temporarily stable nexuses in the flow of upward and downward causal interaction' (2007: 842). This still allows for there to be appropriate 'units of selection', but such units are not to be conceptualized in object-oriented terms. In particular, we can accommodate the view that 'a gene is part of the genome that is a target for external (that is, cellular) manipulation of genome behaviour and, at the same time, carries resources through which the genome can influence processes in the cell more broadly' (2007: 842).

There are, of course, numerous issues to be tackled within this framework. Does the view of a biological object as a 'temporarily stable nexus' imply the elimination of objects (as elements in our metaphysics of biology—I am not suggesting the elimination of genes or organisms as phenomenologically grasped) or can we hold a 'thin' notion of object, in the sense of one whose individuality is grounded in structural terms? Is the temporary stability of such objects sufficient for fitness to be associated with it? And can we articulate appropriate units of selection in such terms? I shall leave such issues for another occasion.

12.12 Conclusion

It is a contingent fact of the recent history of the philosophy of science that structuralism in general, and the more well-known forms of structural realism in particular, have been developed using examples from physics. This has shaped these accounts in various ways but it would be a mistake to think that because of that, forms of structuralism could not be articulated within other contexts. When it comes to the biological, the apparent obstacle of the lack of laws crumbles away under the appreciation that even in physics the standard connection between lawhood and necessity is not well grounded. Adopting an understanding of laws in terms of their modal resilience allows one to accept certain biological regularities as law-like and there are models a-plenty to form the basis for a structuralist framework. Furthermore, the central claim of this chapter is that there are good reasons for shifting one's ontological focus away from biological objects and towards something that is more fluid and contextual and, ultimately, structurally grounded. Causality can then be 'de-seated' and possible connections open up with activity-based accounts of biological processes. Certainly I would argue that the realist need not be promiscuous in this context, but can, and should, be a 'staid' structuralist instead.

As I have said throughout this chapter there are still numerous issues to deal with in order to arrive at a viable form of biological structuralism. And even setting the more purely biological issues aside, we still have to articulate a suitable representation of structure in this domain together with an appropriate metaphysics of biological structure. Further work is also required to develop the motivations for structural realism in this context. Nevertheless, I hope I have indicated both how certain apparent obstacles—such as the purported lack of mathematical equations—can be overcome, and how certain issues in the foundations of biology can be drawn upon as motivations for such a position. In this way, I further hope to have laid some of the groundwork for a form of biological structural realism and more generally indicated how structuralism may be conceived as a broad framework for biological ontology.

12.13 Further Developments

Chemistry and biology are not the only fields beyond physics where a structuralist stance might gain some purchase. Thus, a form of OSR appropriate for economics has been suggested (Ross 2008). The key 'bridge' between OSR-as-articulated-in-the-context-of-physics and OSR-in-the-context-of-economics lies in the claim that just as the former takes physical objects to be merely heuristic, 'book-keeping' devices, so economics from a structural point of view should regard economic agents (i.e. people) as they figure in economic theory and social science in a similar manner. Thus what is represented by economic models should not be regarded as individuals, their properties, or even proxies thereof, but rather 'aggregate properties' of idealized markets and agents. At best there is a 'thin' notion of agent that identifies such with 'the gravitational centres of consistent preference fields'. And the central claim, common to both epistemic and ontic structural realism, of course, is that in so far as one can say that progress has been made in economics this has consisted in 'deepening our knowledge of abstract structures'. What is cumulative through such progress are the relevant patterns, in particular of optimization and maximization.

And likewise Floridi draws on Dennett's view that macro-objects should be regarded as patterns in the development of his 'Informational Structural Realism' (ISR) (Floridi 2008). Here in particular he gives the example of 'Object-Oriented Programming' (OOP) in which discrete informational objects are constituted by data structures and computational procedures and systems are formed from collections of such objects. This concept of an 'informational object' can then be usefully applied to the relata of OSR, and ISR can be seen as a flexible methodology for making precise the Dennettian view. The scalability of a structural ontology is then effected by computational approaches, such as the methodology of OOP, with portability underpinned by group theory. From the perspective of ISR, 'the ultimate nature of reality is informational' (Floridi 2008: 241), providing a 'full-blooded ontology of objects as structural entities' (Floridi 2008: 241). Cashing out such an ontology whilst avoiding the criticism that this amounts to the hypostatization of an abstract noun, in this case, 'information' (Timpson 2008), is yet another challenge that structural realism must face. Nevertheless, I would argue that we have the tools to meet such challenges—particularly if we adopt the Viking Approach!—and this view has the flexibility and, as Floridi notes, the portability to apply across a range of domains in science (for further consideration of ISR, see Bueno 2010).

Of course, there are still numerous issues to tackle. But by paying attention to the relevant scientific details and applying the 'Viking Approach' to the range of metaphysical options, I am confident that structuralism in general and OSR in particular can be developed and extended as a broad framework for scientific ontology in general. Philosophy has spent too long pursuing objects. A realism fit for modern science will only be achieved once we abandon that pursuit and instead pay attention to the interrelated structure of reality.

Bibliography

Ainsworth, P.M. (2009) 'Newman's Objection', *British Journal for the Philosophy of Science* 60 (1): 135–71.

Ainsworth, P.M. (2010) 'What is Ontic Structural Realism?', *Studies in History and Philosophy of Science Part B* 41 (1): 50–7.

Anjum, R. and Mumford, S. (2011) 'Causal Dispositionalism', in A. Bird, B. Ellis, and H. Sankey (eds), *Properties, Powers and Structure*. New York: Routledge: 101–8.

Apostel, L. (1961) 'Towards the Formal Study of Models in the Non-Formal Sciences', in H. Freudenthal (ed.), *The Concept and the Role of the Model in Mathematics and Natural and Social Sciences*. Dordrecht: Reidel: 1–37.

Arenhart, J. (2012) 'Finite Cardinals in Quasi-Set theory', *Studia Logica* 100: 437–52.

Armstrong, D.M. (1961) *Perception and the Physical World*. London: Routledge and Kegan Paul.

Armstrong, D.M. (1978) *A Theory of Universals (Volume II of Universals and Scientific Realism)*. Cambridge: Cambridge University Press.

Armstrong, D.M. (1986) 'In Defence of Structural Universals', *Australasian Journal of Philosophy* 64 (1): 85–8.

Armstrong, D.M. (1997) *A World of States of Affairs*. Cambridge: Cambridge University Press.

Armstrong, D.M. (2004) *Truth and Truthmakers*. Cambridge: Cambridge University Press.

Ash, M.G. (1998) *Gestalt Psychology in German Culture, 1890–1967: Holism and the Quest for Objectivity*. Cambridge: Cambridge University Press.

Auyang, S. (1995) *How is Quantum Field Theory Possible?* Oxford: Oxford University Press.

Averill, E.W. (1990) 'Are Physical Properties Dispositions?', *Philosophy of Science* 57: 118–32.

Bacon, J. (1997) 'Tropes', *The Stanford Encyclopedia of Philosophy*, Edward N. Zalta (ed.), <http://plato.stanford.edu/archives/win2011/entries/tropes/>.

Baez, J. and Weis, M. (undated) 'Is Energy Conserved in General Relativity?', <http://www.phys.ncku.edu.tw/mirrors/physicsfaq/Relativity/GR/energy_gr.html>.

Bain, J. (2000) 'Against Particle/Field Duality: Asymptotic Particle States and Interpolating Fields in Interacting QFT (or: who's afraid of Haag's theorem?)', *Erkenntnis* 53: 375–406.

Bain, J. (2013) 'Category-Theoretic Structure and Radical Ontic Structural Realism', *Synthese*, 190: 1621–35.

Bain, J. (preprint) 'Toward Structural Realism', <http://ls.poly.edu/~jbain/papers/SR.pdf>.

Bain, J. and Norton, J.D. (2001) 'What Should Philosophers of Science Learn from the History of the Electron?', in J.Z. Buchwald and A. Warwick (eds), *Histories of the Electron. The Birth of Microphysics*. Cambridge, MA: MIT Press: 451–66.

Baker, D. (2009) 'Against Field Interpretations of Quantum Field Theory', *British Journal for the Philosophy of Science* 60: 585–609.

Baker, D. (2011) 'Broken Symmetry and Spacetime', *Philosophy of Science* 78: 128–48.

Baker, D. (2013) 'Identity, Superselection Theory and the Statistical Properties of Quantum Fields', *Philosophy of Science* 80: 262–85.

Ballentine, L. (1998) *Quantum Mechanics: A Modern Development*. Singapore: World Scientific Publishing Company.

Bangu, S. (2008) 'Reifying Mathematics? Prediction and Symmetry Classification', *Studies in History and Philosophy of Science Part B* 39 (2): 239–58.

Bangu, S. (2012) *The Applicability of Mathematics in Science: Indispensability and Ontology.* London: Palgrave-Macmillan.

Bartels, A. (1999) 'Objects or Events?: Towards an Ontology for Quantum Field Theory', *Philosophy of Science* 66 (3): 170–84.

Batterman, R. (2010) 'On the Explanatory Role of Mathematics in Empirical Science', *British Journal for the Philosophy of Science* 61 (1): 1–25.

Bauer, W.A. (2011) 'An Argument for the Extrinsic Grounding of Mass', *Erkenntnis,* 74: 81–99.

Beatty J. (1995) 'The Evolutionary Contingency Thesis', in G. Wolters and J.G. Lennox (eds), *Concepts, Theories, and Rationality in the Biological Sciences.* Pittsburgh: University of Pittsburgh Press: 45–81.

Beebee, H., Hitchcock, Ch., and Menzies, P. (eds) (2009) *The Oxford Handbook of Causation.* New York: Oxford University Press.

Bell, J.L., and Korté, H. (2011) 'Hermann Weyl', *The Stanford Encyclopedia of Philosophy,* Edward N. Zalta (ed.), <http://plato.stanford.edu/archives/spr2011/entries/weyl/>.

Belot, G. (2006) 'The Representation of Time and Change in Mechanic', in J. Butterfield and J. Earman (eds), *Philosophy of Physics (Handbook of the Philosophy of Science).* Dordrecht: North-Holland: 133–228.

Belot, G. (2009) 'Healey, Method and Metaphysics', *Pacific APA.*

Belot, G. (2011) *Geometric Possibility.* Oxford: Oxford University Press.

Benacerraf, P. (1965) 'What Numbers Could Not Be', *Philosophical Review* 74: 47–73; reprinted in P. Benacerraf and H. Putnam, *Philosophy of Mathematics: Selected Readings,* 2nd edn. Cambridge: Cambridge University Press, 1983: 272–94.

Berenstain, N., and Ladyman, J. (2012) 'Ontic Structural Realism and Modality', *The Western Ontario Series in Philosophy of Science* 77: 149–68.

Berry, M.V., and Robbins, J.M. (1997) 'Indistinguishability for Quantum Particles: Spin, Statistics and the Geometric Phase', *Proceedings of the Royal Society* 453: 1771–90.

Bhushan, N., and Rosenfeld, S. (eds) (2000) *Of Minds and Molecules: New Philosophical Perspectives on Molecules.* Oxford: Oxford University Press.

Bigaj, T., and Ladyman, J. (2010) 'The Principle of Identity of Indiscernibles and Quantum Mechanics', *Philosophy of Science* 77: 117–36.

Bigelow, J., Ellis, B., and Lierse, C. (1992) 'The World as One of a Kind: Natural Necessity and Laws of Nature', *British Journal for the Philosophy of Science* 43: 371–88.

Bird, A. (2006) 'Looking for Laws', symposium review by B. Ellis, A. Bird, and S. Psillos, with a reply by Stephen Mumford, *Metascience* 15: 437–69.

Bird, A. (2007) *Nature's Metaphysics: Laws and Properties.* Oxford: Oxford University Press.

Bitbol, M., Kerszberg, P., and Petitot, J. (eds) (2009) *Constituting Objectivity, Transcendental Perspectives on Modern Physics.* The Western Ontario Series in Philosophy of Science. Dordrecht: Springer.

Black, R. (2000) 'Against Quidditism', *Australasian Journal of Philosophy* 78: 87–104.

Blackburn, Simon (1999) 'Morals and Modals', in J. Kim and E. Sosa (eds), *Metaphysics: An Anthology.* Oxford: Blackwell: 634–48.

Bogen, J., and Woodward, J. (1988) 'Saving the Phenomena', *Philosophical Review* 97 (3): 303–52.

Bokulich, A. (2008) *Reexamining the Quantum-Classical Relation: Beyond Reductionism and Pluralism.* Cambridge: Cambridge University Press.

Bokulich, A. (2010) 'Bohr's Correspondence Principle', *The Stanford Encyclopedia of Philosophy*, Edward N. Zalta (ed.), <http://plato.stanford.edu/archives/win2010/entries/bohr-correspondence/>.

Bokulich, A. and Bokulich, P. (eds) (2011) *Scientific Structuralism*. Boston Studies in the Philosophy of Science. Dordrecht: Springer.

Bonolis, L. (2004) 'From the Rise of the Group Concept to the Stormy Onset of Group Theory in the New Quantum Mechanics. A Saga of the Invariant Characterization of Physical Objects, Events and Theories', *Rivista Del Nuovo Cimento* 27: 4–5.

Born, M. (1926) 'Zur Quantenmechanik der Stoßvorgänge', *Zeitschrift für Physik* 37 (12): 863–7.

Born, M. (1956) *Physics in my Generation*. London: Pergamon Press.

Born, M. (1962) *Physics and Politics*. Edinburgh and London: Oliver and Boyd.

Born, M. (1971) *The Born-Einstein Letters: The Correspondence between Albert Einstein and Max and Hedwig Born, 1916–1955*, trans. by Irene Born. New York: Walker.

Born, M. (1978) *My Life: Recollections of a Nobel Laureate*, London: Taylor & Francis.

Bouchard, F. (2010) 'Symbiosis, Lateral Function Transfer and the (Many) Saplings of Life', *Biology and Philosophy* 24 (4): 623–41.

Bowler, P., and Pickstone, J. (eds) (2009) *The Cambridge History of Science*, volume 6. Cambridge: Cambridge University Press.

Boyd, R.N. (1973) 'Underdetermination and a Causal Theory of Evidence', *Nous* 7: 1: 1–12.

Braddon-Mitchell, D., and Nola, R. (eds) (2009) *Conceptual Analysis and Philosophical Naturalism*. Cambridge, MA: MIT Press.

Brading, K. (2011) 'Structuralist Approaches to Physics: Objects, Models and Modality', in A. Bokulich and P. Bokulich (eds), *Scientific Structuralism*. Boston Studies in the Philosophy of Science. Dordrecht: Springer: 43–65.

Brading, K. (forthcoming) 'On Composite Systems: Newton, Descartes and the Law-Constitutive Approach'.

Brading, K. and Brown, H. R. (2003) 'Symmetries and Noether's Theorems', in K. Brading and E. Castellani (eds) *Symmetries in Physics: Philosophical Reflections*. Cambridge: Cambridge University Press: 89–109.

Brading, K. and Brown, H.R. (2004) 'Are Gauge Symmetry Transformations Observable?', *British Journal for the Philosophy of Science* 55 (4): 645–65.

Brading, K. and Castellani, E. (2008) 'Symmetry and Symmetry Breaking', *The Stanford Encyclopedia of Philosophy*, Edward N. Zalta (ed.), <http://plato.stanford.edu/archives/fall2008/entries/symmetry-breaking/>.

Brading, K. and Castellani, E. (eds) (2010) *Symmetries in Physics: Philosophical Reflections*. Cambridge: Cambridge University Press.

Brading, K. and Landry, E. (2006) 'Scientific Structuralism: Presentation and Representation', *Philosophy of Science* 73: 571–81.

Brading, K. and Landry, E. (2007) 'Shared Structure and Scientific Structuralism', *Philosophy of Science (Proceedings)* 73: 571–81.

Brading, K. and Skiles, A. (2012) 'Underdetermination as a Path to Ontic Structural Realism', in E. Landry and D. Rickles (eds), *Structural Realism: Structure, Objects, and Causality*. The Western Ontario Series in Philosophy of Science 77. Dordrecht: Springer: 99–116.

Braithwaite, R.B. (1929) 'Professor Eddington's Gifford Lectures', *Mind* 38 (152): 409–35.

Broad, C.D. (1940) 'Sir Arthur Eddington's *Philosophy of Physical Science*', *Philosophy* 15: 301–12.

Brown, H., and Holland, P. (2004) 'Dynamical versus Variational Symmetries: Understanding Noether's first theorem', *Molecular Physics* 102: 1133–9; also in *Studies in History and Philosophy of Modern Physics* 38 (2007): 457–81.

Brown, H.R. (2005) *Physical Relativity: Space-time Structure from a Dynamical Perspective.* Oxford: Oxford University Press.

Bueno, O., (1997) 'Empirical Adequacy: A Partial Structures Approach', *Studies in History and Philosophy of Science Part A* 28 (4): 585–610.

Bueno, O., (2000) 'Empiricism, Scientific Change and Mathematical Change', *Studies in History and Philosophy of Science* 31: 269–96.

Bueno, O., (2006) 'The Methodological Character of Symmetry Principles', *Abstracta* 3: 3–28.

Bueno, O., (2010) 'Structuralism and Information', *Metaphilosophy* 41 (3): 365–79.

Bueno, O., (forthcoming) 'Indispensability and Configuration Space Realism'.

Bueno, O., and French, S. (1999) 'Infestation or Pest Control: The Introduction of Group Theory into Quantum Mechanics', *Manuscrito* 22: 37–86.

Bueno, O., and French, S. (2011) 'How Theories Represent', *British Journal for the Philosophy of Science* 62 (4): 857–94.

Bueno, O., and French, S. (2012) 'Can Mathematics Explain Physical Phenomena?', *British Journal for the Philosophy of Science* 63 (1): 85–113.

Bueno, O., and French, S. (forthcoming) *From Weyl to von Neumann: An Analysis of the Application of Mathematics to Quantum Mechanics.*

Bueno, O., French, S., and Ladyman, J. (2002) 'On Representing the Relationship between the Mathematical and the Empirical', *Philosophy of Science* 69: 452–73.

Bueno, O., French, S., and Ladyman, J. (2012) 'Models and Structures: Phenomenological and Partial', *Studies in History and Philosophy of Science* Part B 43 (1): 43–6.

Burian, R. (2005) *The Epistemology of Development, Evolution and Genetics.* Cambridge: Cambridge University Press.

Burian R.M., and Zallen, D.T. (2009) 'Genes', in P. Bowler and J. Pickstone (eds), *The Cambridge History of Science*, volume 6. Cambridge: Cambridge University Press: 432–50.

Busch, J. (2003) 'What Structures Could Not Be', *International Studies in the Philosophy of Science* 17: 211–25.

Busch, J. (2008) 'No New Miracles, Same Old Tricks', *Theoria* 74: 102–14.

Butterfield, J., and Caulton, A. (2012) 'Symmetries and Paraparticles as a Motivation for Structuralism', *British Journal for the Philosophy of Science* 63 (2): 233–85.

Busch, J. and Earman, J. (eds) (2006) *Handbook of the Philosophy of Physics.* Dordrecht: Kluwer.

Callender, C. (2011) 'Philosophy of Science and Metaphysics', in S. French and J. Saatsi (eds), *The Continuum Companion to the Philosophy of Science.* London: Continuum: 33–54.

Busch, J. and Cohen, J. (2009) 'A Better Best System Account of Lawhood', *Philosophical Studies* 145 (1): 1–34.

Cameron, R. (2008) 'Truthmakers and Ontological Commitment: Or How to Deal with Complex Objects and Mathematical Ontology without Getting Into Trouble', *Philosophical Studies* 140: 1–18.

Cameron, R. (2010) 'On the Source of Necessity', in B. Hale and A. Hoffmann (eds), *Modality*. Oxford: Oxford University Press: 137–52.

Camino, F.E., Zhou, W., and Goldman, V.J. (2005) 'Realization of a Laughlin Quasiparticle Interferometer: Observation of Fractional Statistics', *Physical Review*, B 72, 075342.

Cao, T. (1997) *Conceptual Developments of Twentieth Century Field Theories*. Cambridge: Cambridge University Press.

Cao, T. (2003) 'Structural Realism and the Interpretation of Quantum Field Theory', *Synthese* 136: 3–24.

Cao, T. (2010) *From Current Algebra to the Genesis of QCD–A Case for Structural Realism*. Cambridge: Cambridge University Press.

Cao, T. (ed.) (1999) *Conceptual Foundations of Quantum Field Theory*. Cambridge: Cambridge University Press.

Carnap, R. (1963) 'Intellectual Autobiography', in P.A. Schilpp (ed.), *The Philosophy of Rudolf Carnap*. La Salle, IL: Open Court, 1999: 3–84.

Carrier, M. (2004) 'Experimental Success and the Revelation of Reality: The Miracle Argument for Scientific Realism', in M. Carrier et al. (eds), *Knowledge and the World: Challenges Beyond the Science Wars*. Heidelberg: Springer: 137–61.

Carson, C. (1996a) 'The Peculiar Notion of Exchange Forces I: Origins in Quantum Mechanics, 1926–1928', *Studies in History and Philosophy of Modern Physics* 27: 23–45.

Carson, C. (1996b) 'II: From Nuclear Forces to QED, 1929–1950', *Studies in History and Philosophy of Modern Physics* 27: 99–131.

Cartwright, N. (1983) *How the Laws of Physics Lie*. Oxford: Oxford University Press.

Cartwright, N. (1999) *The Dappled World: A Study of the Boundaries of Science*. Cambridge: Cambridge University Press.

Cartwright, N. Shomar, T., and Suárez, M. (1996) 'The Tool Box of Science (Tools for Building of Models with a Superconductivity Example', in W.E. Herfel et al. (eds), *Theories and Models in Scientific Processes*. Amsterdam: Editions Rodopi: 137–49.

Cassirer, E. (1907a) *Das Erkenntnisproblem in der Philosophie und Wissenschaft der neueren Zeit*. Berlin: Bruno Cassirer.

Cassirer, E. (1907b) 'Kant und die moderne Mathematik', *Kant-Studien* 12: 1–40.

Cassirer, E. (1913) 'Erkenntnistheorie nebst den Grenzfragen der Logik', *Jarbücher der Philosophie* I: 1–59.

Cassirer, E. (1936/1956) *Determinismus und Indeterminismus in der Modernen Physik*. Göteborg: Göteborgs Högskolas Årsskrift 42. Translated as *Determinism and Indeterminism in Modern Physics*. New Haven: Yale University Press, 1956.

Cassirer, E. (1938) 'The Concept of Group and the Theory of Perception', *Philosophy and Phenomenological Research* 5 (1944): 1–36.

Cassirer, E. (1944) 'The Concept of Group and the Theory of Perception', *Philosophy and Phenomenological Research* 5: 1–36.

Cassirer, E. (1950) *The Problem of Knowledge. Philosophy, Science, and History since Hegel*, trans. by W.H. Woglom and C.W. Hendel. New Haven: Yale University Press.

Cassirer, E. (1953) [1910] *Substance and Function*. (*Substanzbegriff und Funktionsbegriff*), trans. by William Curtis Swabey and Marie Collins Swabey. New York: Dover.

Cassirer, E. (1957) *Zur modernen Physik*. Darmstadt: Wissenshaftliche Buchgesellshaft.

Castellani, E. (1993) 'Quantum Mechanics, Objects and Objectivity', in C. Garola and A. Rossi (eds), *The Foundations of Quantum Mechanics—Historical Analysis and Open Questions*. Dordrecht: Kluwer, 105–14.

Castellani, E. (2003) 'Symmetry and Equivalence', in K. Brading and E. Castellani (eds), *Symmetries in Physics: Philosophical Reflections*. Cambridge: Cambridge University Press: 422–33.

Castellani, E. (ed.) (1998) *Interpreting Bodies: Classical and Quantum Objects in Modern Physics*. Princeton: Princeton University Press.

Cat, J. (2007) 'Switching Gestalts on Gestalt Psychology: On the Relation Between Science and Philosophy', *Perspectives on Science* 15 (2): 131–77.

Cei, A. (2005) 'Structural Distinctions. Entity, Structures and Change in Science', *Philosophy of Science* 72: 1385–96.

Cei, A. and French, S. (2006) 'Looking for Structure in all the Wrong Places: Ramsey Sentences, Multiple Realizability, and Structure', *Studies in History and Philosophy of Science* 37: 633–55.

Cei, A. and French, S. (2009) 'On the Transposition of the Substantial into the Functional: Bringing Cassirer's Philosophy of Quantum Mechanics into the 21st Century', in M. Bitbol, P. Kerszberg, and J. Petitot (eds), *Constituting Objectivity, Transcendental Perspectives on Modern Physics*. The Western Ontario Series in Philosophy of Science. Dordrecht: Springer: 95–115.

Cei, A., and French, S. (2014) 'Getting Away from Governance: Laws, Symmetries and Objects', *Méthode—Analytic Perspectives* 3 (DOI: http://dx.doi.org/10.13135/2281-0498%2F4).

Chakravartty, A. (1998) 'Semirealism', *Studies in History and Philosophy of Science Part A* 29 (3): 391–408.

Chakravartty, A. (2003a) 'Review of N. Cartwright, *The Dappled World: A Study in the Boundaries of Science*', *Philosophy and Phenomenological Research* 66: 244–7.

Chakravartty, A. (2003b) 'The Structuralist Conception of Objects', *Philosophy of Science* 70: 867–78.

Chakravartty, A. (2007) *A Metaphysics for Scientific Realism*. Cambridge: Cambridge University Press.

Chakravartty, A. (2012) 'Ontological Priority: The Conceptual Basis of Non-Eliminative, Ontic Structural Realism', in E. Landry and D. Rickles (eds), *Structure, Object, and Causality*. The Western Ontario Series in Philosophy of Science. Dordrecht: Springer.

Chakravartty, A. (2013) 'Realism in the Desert and in the Jungle: Reply to French, Ghins and Psillos', *Erkenntnis* 178: 39–58.

Chalmers, D., Manley, D., and Wasserman, R. (2009) *Metametaphysics: New Essays on the Foundations of Ontology*. Oxford: Oxford University Press.

Chayut, M. (2001) 'From the Periphery: The Genesis of Eugene P. Wigner's Application of Group Theory to Quantum Mechanics', *Foundations of Chemistry* 3 (1): 55–78.

Clarke, E. (2010) 'The Problem of Biological Individuality', *Biological Theory* 5: 312–25.

Clarke, E. (2013) 'The Multiple Realizability of Biological Individuals', *Journal of Philosophy* 110: 413–35.

Clifton, R., and Halvorson, H. (2001) 'Are Rindler quanta real?', *British Journal for the Philosophy of Science* 52: 417–70.

Coleman, J.A. (1997) 'Groups and Physics—Dogmatic Opinions of a Senior Citizen', *Notices AMS* 44 (1): 8–17.

Colodny, R. (ed.) (1970) *The Nature and Function of Scientific Theories*. University of Pittsburgh Series in the Philosophy of Science: Volume 4. Pittsburgh: University of Pittsburgh Press.

Colosi, D. and Rovelli, C. (2008) 'What is a particle?', <http://arxiv.org/abs/gr-qc/0409054>.

Colyvan, M. (unpublished) 'Causal Explanation and Ontological Commitment'.

Cometto, M. (2009) 'Preludes to Ontic Structural Realism: Eddington and Weyl; A Theoretic Route from General Relativity to Quantum Mechanics Through the Theory of Groups'. PhD Thesis, University of Rome.

Condon, E.U., and Shortley, G.H. (1935) *The Theory of Atomic Spectra*. Cambridge: Cambridge University Press.

Contessa, G. (2010) 'The Ontology of Scientific Models', *Synthese* 172 (2): 215–29.

Contessa, G. (2011) 'Scientific Models and Representation', in S. French and J. Saatsi (eds), *The Continuum Companion to the Philosophy of Science*. London: Continuum Press.

Cooper, S.B., Löwe, B., and Sorbi, A. (eds) (2007) *Computation and Logic in the Real World: Third Conference on Computability in Europe, CIE*.

Correia, F. (2008) 'Ontological Dependence', *Philosophy Compass* 3/5: 1013–32.

Corsi, G. et al. (eds) (1991) *Bridging the Gap: Philosophy, Mathematics, Physics*. Dordrecht: Kluwer Academic Press.

Crilly, T. (1999) 'The Emergence of Topological Dimension Theory', in I.M. James (ed.), *History of Topology*. Amsterdam: North-Holland: 1–25.

Crull. E. (preprint) 'Scientific Realism and the Case of Weak Interactions', *PhilSci Archive*.

Cruse, P., and Papineau, D. (2002) 'Scientific Realism without Reference', in M. Marsonet (ed.), *The Problem of Realism*. London and Aldershot: Ashgate: 174–89.

Curiel, E. (2014) 'Classical Mechanics Is Lagrangian; It Is Not Hamiltonian', *British Journal of Philosophy of Science* 65: 269–321.

Curiel, E. (preprint) 'On the Propriety of Physical Theories as a Basis for their Semantics', *PhilSci Archive*.

Cushing, J.T. (1985) 'Is There Just *One* Possible World? Contingency vs. the Bootstrap', *Studies in History and Philosophy of Science* 16: 31–48.

Da Costa, N., and French, S. (2003) *Science and Partial Truth*. New York: Oxford University Press.

Dalla Chiara, M.L., Doets, K., Mundici, D., and van Benthem, J. (eds) (1997) *Logic and Scientific Methods*. Dordrecht: Kluwer.

Dalla Chiara, M.L., and Toraldo di Francia, G. (1993) 'Individuals, Kinds and Names in Physics', in G. Corsi et al. (eds), *Bridging the Gap: Philosophy, Mathematics, Physics*. Dordrecht: Kluwer Academic Press: 261–83.

Darby, G. (2010) 'Quantum Mechanics and Metaphysical Indeterminacy', *The Australasian Journal of Philosophy* 88: 227–45.

Dasgupta, S. (2009) 'Individuals: An Essay in Revisionary Metaphysics', *Philosophical Studies* 145 (1): 35–67.

Debs, T., and Redhead, M. (2007) *Objectivity, Invariance, and Convention: Symmetry in Physical Science*. Cambridge, MA and London: Harvard University Press.

Demopoulos, W. (2011) 'Three Views of Theoretical Knowledge', *British Journal for the Philosophy of Science* 62 (1): 177–205.

Demopoulos, W. and Friedman, M. (1985) 'Bertrand Russell's *The Analysis of Matter*: Its Historical Context and Contemporary Interest', *Philosophy of Science* 52 (4): 621–39.

Di Francia, T.G. (1981) *The Investigation of the Physical World*. Cambridge: Cambridge University Press.

Di Francia, T.G. (1985) 'Connotation and Denotation in Microphysics', in P. Mittelstaed and E.W. Stachow (eds) (1988), *Recent Developments in Quantum Logics*. Mannheim: Bibliographishes Institut: 203–14.

Dicken, P. (2007) 'Constructive Empiricism and the Metaphysics of Modality', *British Journal for the Philosophy of Science* 58 (3): 605–12.

Dicken, P. (2008) 'Conditions May Apply', *Studies in History and Philosophy of Science* 39 (2): 290–3.

Dieks, D. (ed.) (2008) *Ontology of Spacetime*. Philosophy and Foundations of Physics Series, volume 2. Amsterdam: Elsevier.

Dirac, P.A.M. (1978) *The Principles of Quantum Mechanics*, 4th edn. Oxford: Oxford University Press.

Domenech, G., and Holik, F. (2007) 'A Discussion on Particle Number and Quantum Indistinguishability', *Foundations of Physics* 37: 855–78.

Domski, M. (2013), 'Mediating Between Past and Present: Descartes, Newton, and Contemporary Structural Realism' in M. Laerke, J. E. H. Smith, and E. Schliesser (eds), *Philosophy and Its History: New Essays on the Methods and Aims of Research in the History of Philosophy*. Oxford: Oxford University Press: 278–300.

Domski, M. (preprint) 'The Epistemological Foundations of Structural Realism: Poincaré and the Structure of Relations', paper given to the Research Workshop of the Division of History and Philosophy of Science, University of Leeds.

Doncel, M.G., Hermann, A., Michel, L., and Pais, A. (eds) (1987) *Symmetry in Physics (1600–1980)*. Barcelona: Bellaterra.

Dorato, M. (2000) 'Substantivalism, Relationism and Structural Spacetime Realism', *Foundations of Physics* 30: 1605–28.

Dorato, M. (2007) 'Dispositions, Relational Properties and the Quantum World', in M. Kistler and B. Gnassounou (eds), *Dispositions and Causal Powers*. Ashford: Ashgate: 249–70.

Dorato, M. (2012) 'Mathematical Biology and the Existence of Biological Laws', in D. Dieks et al. (eds), *Probability, Laws and Structures*. Dordrecht: Springer: 109–22.

Dorato, M. and Esfeld, M. (2010) 'GRW as an Ontology of Dispositions', *Studies in History and Philosophy of Modern Physics* 41: 41–9.

Dowe, P. (2000) *Physical Causation*. New York and Cambridge: Cambridge University Press.

Dowe, P. (2007) 'Causal Processes', *The Stanford Encyclopedia of Philosophy*, Edward N. Zalta (ed.), <http://plato.stanford.edu/archives/fall2008/entries/causation-process/>.

Dowe, P. (2009) 'Causal Process Theories', in H. Beebee, P. Menzies, and C. Hitchcock (eds), *The Oxford Handbook of Causation*. Oxford: Oxford University Press: 213–33.

Dowe, P. Gardner, S., and Oppy, G. (2007) 'Bayes Not Bust! Why Simplicity Is No Problem for Bayesians', *British Journal for the Philosophy of Science* 58 (4): 709–54.

Drake, K., Feinberg, M., Guild, D., and Turetsky, E. (2009) 'Representations of the Symmetry Group of Spacetime', <http://pages.cs.wisc.edu/~guild/symmetrycompsproject.pdf>.

Drewery, A. (2001) 'Dispositions and Ceteris Paribus Laws', *British Journal for the Philosophy of Science* 52: 723–33.

Drewery, A. (2005) 'Essentialism and the Necessity of the Laws of Nature', *Synthese* 144 (3): 381–96.

Dupré, J. (1993) *The Disorder of Things: Metaphysical Foundations of the Disunity of Science*. Cambridge, MA: Harvard University Press.

Dupré, J. (1996) 'Promiscuous Realism: Reply to Wilson', *The British Journal for the Philosophy of Science* 47: 441–4.

Dupré, J. and O'Malley, M. (2007) 'Metagenomics and Biological Ontology', *Studies in History and Philosophy of Science* 28: 834–46.

Dupré, J. and O'Malley, M. (2009) 'Varieties of Living Things: Life at the Intersection of Lineage and Metabolism', *Philosophy and Theory in Biology* 1: 1–25.

Earman, J. (1986) *A Primer on Determinism*, Dordrecht: Reidel.

Earman, J. (1989) *World Enough and Space-Time: Absolute Versus Relational Theories of Space and Time*. Cambridge, MA: MIT Bradford.

Earman, J. (2004) 'Curie's Principle and Spontaneous Symmetry Breaking', *International Studies in the Philosophy of Science* 18: 173–98.

Earman, J. Glymour, C., and Stachel, J. (eds) (1997) *Foundations of Space-Time Theories*. Minneapolis, MN: University of Minnesota Press.

Eckart, C. (1930) 'The Application of Group theory to the Quantum Dynamics of Monatomic Systems', *Reviews of Modern Physics* 2 (3): 305–80.

Eddington, A.S. (1920) 'The Philosophical Aspect of the Theory of Relativity', *Mind* 29: 415–22.

Eddington, A.S. (1920/1966) *Space, Time, and Gravitation: An Outline of the General Relativity Theory*. Cambridge: Cambridge University Press.

Eddington, A.S. (1923) *Mathematical Theory of Relativity*. Cambridge: Cambridge University Press.

Eddington, A.S. (1928) *The Nature of the Physical World*. Cambridge: Cambridge University Press.

Eddington, A.S. (1936) *Relativity Theory of Protons and Electrons*. Cambridge: Cambridge University Press.

Eddington, A.S. (1939) *The Philosophy of Physical Science*. New York: Macmillan.

Eddington, A.S. (1941) 'Discussion: Group Structure in Physical Science', *Mind* 50: 268–79.

Eddington, A.S. (1946) *Fundamental Theory*. Cambridge: Cambridge University Press.

Efird, D., and Stoneham, T. (2010) 'The Subtraction Argument for Free Mass', *Philosophy and Phenomenological Research* 80: 50–7.

Ehrenfels, C. (1980) 'Über Gestaltqualitäten', *Vierteljahrsschrift für wissenschaftliche Philosophie* 14: 249–92; reprinted in B. Smith (ed.), *Foundations of Gestalt Theory*. Munich: Philosophia Verlag, 1988: 82–117.

Elgin, M. (2006) 'There May Be Strict Empirical Laws in Biology, After All', *Biology and Philosophy* 21: 119–34.

Ellis, B. (1999) 'The Really Big Questions', *Metascience* 8: 63–73.

Ellis, B. (2005) 'Katzav on the Limitations of Dispositionalism', *Analysis* 65: 90–2.

Esfeld, M. (2003) 'Do Relations Require Underlying Intrinsic Properties? A Physical Argument for the Metaphysics of Relations', *Metaphysica* 5: 5–25.

Esfeld, M. (2004) 'Quantum Entanglement and a Metaphysics of Relations', *Studies in History and Philosophy of Modern Physics* 35B: 601–17.

Esfeld, M. (2009) 'The Modal Nature of Structures in Ontic Structural Realism', *International Studies in the Philosophy of Science* 23: 179–94.

Esfeld, M. (2012) 'Causal Realism', in D. Dieks et. al. (eds), *Probabilities, laws, and structures. The philosophy of science in a European perspective*. Dordrecht: Springer: 157–68.

Esfeld, M. and Dorato, M. (2010) 'GRW as an Ontology of Dispositions', *Studies in History and Philosophy of Modern Physics* 41: 41–9.

Esfeld, M. and Lam, V. (2008) 'Moderate Structural Realism about Space-time', *Synthese* 160: 27–46.

Esfeld, M. and Lam, V. (2009) 'Structures as the Objects of Fundamental Physics', in U. Feest and H.-J. Rheinberger (eds), *Epistemic Objects*. Berlin: Max Planck Institute for the History of Science, 374 preprint.

Esfeld, M. and Lam, V. (2011) 'Ontic Structural Realism as a Metaphysics of Objects, in A. Bokulich and P. Bokulich, *Scientific Structuralism*. Boston Studies in the Philosophy of Science. Dordrecht: Springer: 143–59.

Esfeld, M. and Sachse, C. (2011) *Conservative Reductionism*. Routledge Studies in the Philosophy of Science. London: Routledge.

Falkenberg, B. (2007) *Particle Metaphysics: A Critical Account of Subatomic Reality*. Dodrecht: Springer.

Fara, M. (2006) 'Dispositions', *The Stanford Encyclopedia of Philosophy*, Edward N. Zalta (ed.), <http://plato.stanford.edu/archives/spr2012/entries/dispositions/>.

Faraday, M. (1844) 'A Speculation on the Nature of Matter', *L.E.D. Philosophical Magazine* xxiv: 140–3.

Feest, U., and Rheinberger, H.-J. (eds) (2009) *Epistemic Objects*. Berlin: Max Planck Institute for the History of Science, preprint 374.

Feigl, H., Sellars, R.W., and Lehrer, K. (eds) (1972) *New Readings in Philosophical Analysis*. New York: Appleton-Century Crofts.

Fine, K. (2001) 'The Question of Realism', *Philosophers' Imprint* 1 (2):1–30.

Fine, K. (2011) 'An Abstract Characterization of the Determinate/Determinable Distinction', *Philosophical Perspectives* 25 (1): 161–87.

Fisher, M.E. (1999) 'Renormalization Group Theory: Its Basis and Formulation in Statistical Physics', in T. Cao (ed.), *Conceptual Foundations of Quantum Field Theory*. Cambridge: Cambridge University Press, 89–135

Floridi, L. (2005) 'Informational Realism', *ACS Conferences in Research and Practice in Information Technology* (CRPIT) (3): 7–12

Floridi, L. (2008) 'A Defence of Informational Structural Realism', *Synthese* 161: 219–53.

Floridi, L. and Sanders, J.W. (2004) 'On the Morality of Artificial Agents', *Minds and Machines* 14 (3): 349–79.

Folina, J. (2000) 'Review Article. Ontology, Logic, and Mathematics', *British Journal for the Philosophy of Science* 51 (2): 319–32.

Folina, J. (2010) 'Poincaré's Philosophy of Mathematics', *Internet Encyclopaedia of Philosophy*, <http://www.iep.utm.edu/poi-math/#H3>.

Fox-Keller, E. (2000) *The Century of the Gene*. Cambridge, MA: Harvard University Press.

Fraser, D. (2008) 'The Fate of "Particles" in Quantum Field Theories with Interactions', *Studies in History and Philosophy of Modern Physics* 39: 841–59.

Fraser, D. (2009) 'Quantum Field Theory: Underdetermination, Inconsistency, and Idealization', *Philosophy of Science* 76: 536–67.

Fraser, D. (2011) 'How to Take Particle Physics Seriously: A Further Defence of Axiomatic Quantum Field Theory', *Studies in History and Philosophy of Science Part B* 42 (2): 126–35.

French, S. (1985) 'Identity and Individuality in Classical and Quantum Physics'. PhD Thesis, University of London.

French, S. (1987) 'First Quantised Paraparticle Theory', *International Journal of Theoretical Physics* 26: 1141–63.

French, S. (1989) 'Identity and Individuality in Classical and Quantum Physics', *Australasian Journal of Philosophy* 67: 432–46.
French, S. (1995) 'The Esperable Uberty of Quantum Chromodynamics', *Studies in History and Philosophy of Modern Physics* 26: 87–105.
French, S. (1997) 'Partiality, Pursuit and Practice', in M.L. Dalla Chiara et al. (eds), *Structures and Norms in Science: Proceedings of the 10th International Congress on Logic, Methodology and Philosophy of Science*. Dordrecht: Reidel: 35–52.
French, S. (1998) 'On the Withering Away of Physical Objects', in E. Castellani (ed.), *Interpreting Bodies: Classical and Quantum Objects in Modern Physics*. Princeton: Princeton University Press: 93–113.
French, S. (1999) 'Models and Mathematics in Physics: The Role of Group Theory', in J. Butterfield and C. Pagonis (eds), *From Physics to Philosophy*. Cambridge: Cambridge University Press: 187–207.
French, S. (2000a) 'Identity and Individuality in Quantum Theory', *Stanford Encyclopedia of Philosophy*, Edward N. Zalta (ed.), <http://plato.stanford.edu/archives/sum2011/entries/qt-idind/>.
French, S. (2000b) 'The Reasonable Effectiveness of Mathematics: Partial Structures and the Application of Group Theory to Physics', *Synthese* 125 (1–2): 103–20.
French, S. (2000c) 'Putting a New Spin on Particle Identity', in R. Hilborn and G. Tino (eds), *Spin-Statistics Connection and Commutation Relations*. Melville, NY: American Institute of Physics: 305–18.
French, S. (2001) 'Getting Out of a Hole: Identity, Individuality and Structuralism in Space-Time Physics', *Philosophica* 67: 901–19.
French, S. (2002) 'A Phenomenological Approach to the Measurement Problem: Husserl and the Foundations of Quantum Mechanics', *Studies in History and Philosophy of Modern Physics* 22: 467–91.
French, S. (2003a) 'Scribbling on the Blank Sheet: Eddington's Structuralist Conception of Objects', *Studies in History and Philosophy of Modern Physics* 34: 227–59.
French, S. (2003b) 'A Model-Theoretic Account of Representation (or, I Don't Know Much About Art... but I Know It Involves Isomorphism)', *Philosophy of Science* 70 (5): 1472–83.
French, S. (2006) 'Structure as a Weapon of the Realist', *Proceedings of the Aristotelian Society* 106 (2): 167–85.
French, S. (2007) 'The Limits of Structuralism', British Society for the Philosophy of Science Presidential Address, <http://www.thebsps.org/society/bsps/events.html>.
French, S. (2010a) 'The Interdependence of Structures, Objects and Dependence', *Synthese* 175: 89–109.
French, S. (2010b) 'Keeping Quiet on the Ontology of Models', *Synthese* 172 (2): 231–249.
French, S. (2011a) 'Metaphysical Underdetermination: Why Worry?', *Synthese* 180: 205–21.
French, S. (2011b) 'Austere Realism: Contextual Semantics Meets Minimal Ontology—Terence Horgan and Matjaž Potrč', *Philosophical Quarterly* 61 (242): 201–2.
French, S. (2011c) 'Welcome to the Jumble', *Metascience* 20 (3): 543–8.
French, S. (2011d) 'The Law-Governed Universe—John T. Roberts', *Philosophical Quarterly* 61 (245): 872–3.
French, S. (2011e) 'Shifting to Structures in Physics and Biology: A Prophylactic for Promiscuous Realism', *Studies in History and Philosophy of Science Part C* 42 (2): 164–73.

French, S. (2012a) 'Unitary Inequivalence as a Problem for Structural Realism', *Studies in History and Philosophy of Modern Physics* 43: 121–36.

French, S. (2012b) 'The Resilience of Laws and the Ephemerality of Objects: Can a Form of Structuralism be Extended to Biology?', in D. Dieks et al. (eds), *Probability, Laws and Structures*. Dordrecht: Springer: 187–200.

French, S. (2013) 'Semi-realism, Sociability and Structure', *Erkenntnis* 178: 1–18.

French, S. (2015) 'Between Weasels and Hybrids: What Does the Applicability of Mathematics Tell us about Ontology?' in J-Y Beziau, D. Krause and J. Arenhart (eds.), *Conceptual Clarifications: Tributes to Patrick Suppes*, London: College Publications: 63–86.

French, S. and Kamminga, H. (eds) (1993) *Correspondence, Invariance and Heuristics: Essays in Honour of Heinz Post (Boston Studies in the Philosophy of Science* 148*)*. Dordrecht: Kluwer Academic Press.

French, S. and Krause, D. (2003) 'Quantum Vagueness', *Erkenntnis* 59 (1): 97–124.

French, S. and Krause, D. (2006) *Identity in Physics: A Historical, Philosophical, and Formal Analysis*. Oxford: Oxford University Press.

French, S. and Ladyman, J. (1997) 'Superconductivity and Structures: Revisiting the London Account', *Studies in History and Philosophy of Modern Physics* 28: 363–93.

French, S. and Ladyman, J. (1999) 'Reinflating the Semantic Approach', *International Studies in the Philosophy of Science* 13 (2): 103–21.

French, S. and Ladyman, J. (2003) 'Remodelling Structural Realism: Quantum Physics and the Metaphysics of Structure: A Reply to Cao', *Synthese* 136: 31–56.

French, S. and Ladyman, J. (2011) 'In Defence of Ontic Structural Realism', in A. Bokulich and P. Bokulich (eds), *Scientific Structuralism*. Dordrecht: Springer: 25–42.

French, S. and McKenzie, K. (2012) 'Thinking Outside the (Tool)Box: Towards a More Productive Engagement Between Metaphysics and Philosophy of Physics', *The European Journal of Analytic Philosophy* 8: 42–59.

French, S. and Redhead, M. (1988) 'Quantum Physics and the Identity of Indiscernibles', *British Journal for the Philosophy of Science* 39 (2): 233–46.

French, S. and Rickles, D. (2003) 'Understanding Permutation Symmetry', in K. Brading and E. Castellani (eds), *Symmetries in Physics: New Reflections*. Cambridge: Cambridge University Press: 212–38.

French, S. and Saatsi, J. (2006) 'Realism about Structure: The Semantic View and Non-linguistic Representations', *Philosophy of Science (Proceedings)* 73: 548–59.

French, S. and Saatsi, J. (eds) (2011) *The Continuum Companion to the Philosophy of Science*. London: Continuum Press.

French, S. and Vickers, P. (2011) 'Are There No Such Things as Theories?', *The British Journal for the Philosophy of Science* 62: 771–804.

Friedman, M. (1995) 'Poincaré's Conventionalism and the Logical Positivists', *Foundations of Science* 2: 299–314.

Friedman, M. (2000) *A Parting of the Ways: Carnap, Cassirer, and Heidegger*. Chicago: Open Court.

Friedman, M. (2004) 'Ernst Cassirer', *The Stanford Encyclopedia of Philosophy*, Edward N. Zalta (ed.), <http://plato.stanford.edu/archives/spr2011/entries/cassirer/>.

Frigg, R., and Votsis, I. (2011) 'Everything You Always Wanted to Know About Structural Realism but Were Afraid to Ask', *European Journal for Philosophy of Science* 1 (2): 227–76.

Fujita, T. (2007) *Symmetry and its Breaking in Quantum Field Theory*. New York: Nova Science Publishers.

Garber, D., and Longuenesse, B. (eds) (2008) *Kant and the Early Moderns*. Princeton: Princeton University Press.

Gardner, A. (2008) 'The Price Equation', *Current Biology* 18: R198–202.

Garola, C., and Rossi, A. (eds) (1993) *The Foundations of Quantum Mechanics—Historical Analysis and Open Questions*. Dordrecht: Kluwer.

Gavroglu, K. (1995) *Fritz London: A Scientific Biography*. Cambridge: Cambridge University Press.

Geach, P.T. (1979) *Truth, Love and Immortality*. London: Hutchinson.

Gell-Mann, M. (1987) 'Particle Theory from S-Matrix to Quarks', in M.G. Doncel, L.M. Hermann, and A. Pais (eds), *Symmetry in Physics (1600–1980)*. Barcelona: Bellaterra: 474–97.

Gethmann, C.F. (ed.) (2011) *Lebenswelt und Wissenschaft, Deutsches Jahrbuch Philosophie 2*. Hamburg: Meiner Verlag.

Ghins, M. (2007) 'Laws of Nature—Do We Need a Metaphysics?', *Principia* 11: 127–49.

Giere, R. (1988) *Explaining Science: A Cognitive Approach*. Chicago: University of Chicago Press.

Giere, R. (1999) *Science without Laws*. Chicago: University of Chicago Press.

Glynn, L. (2012) 'Review of *Getting Causes from Powers*', *Mind* 121: 1099–1106.

Godfrey-Smith, P. (2000) 'Explanatory Symmetries, Preformation, and Developmental Systems Theory', *Philosophy of Science* 67, Supplement. Proceedings of the 1998 Biennial Meetings of the Philosophy of Science Association. Part II: Symposia Papers (September, 2000): 322–31.

Godfrey-Smith, P. (2009) 'Causal Pluralism', in H. Beebee, C. Hitchcock, and P. Menzies (eds), *The Oxford Handbook of Causation*. New York: Oxford University Press: 326–37.

Godfrey-Smith, P. (2011) 'The Evolution of the Individual', Lakatos Award Lecture, LSE.

Gower, B.S. (2000) 'Cassirer, Schlick and "Structural" Realism: The Philosophy of the Exact Sciences in the Background to Early Logical Empiricism', *British Journal for the History of Philosophy* 8 (1): 71–106.

Gracia, J.J.E. (1988) *Individuality*. New York: SUNY Press.

Greene, B. (2011) *The Hidden Reality: Parallel Universes and the Deep Laws of the Cosmos*. London: Allen Lane, 2011.

Griesemer, J. (1990) 'Material Models in Biology', *PSA 1990* 2: 79–93.

Guay, A., and Hepburn, B. (2009) 'Symmetry and Its Formalisms: Mathematical Aspects', *Philosophy of Science* 76: 160–78.

Guyer, P., and Wood, A. (1998) 'Introduction', *Critique of Pure Reason*. Cambridge: Cambridge University Press: 1–73.

Haag, R. (1992) *Local Quantum Physics*. Dordrecht: Springer Verlag.

Hacking, I. (1975) 'The Identity of Indiscernibles', *Journal of Philosophy* 72: 249–56.

Hacking, I. (1983) *Representing and Intervening: Introductory Topics in the Philosophy of Natural Science*. Cambridge: Cambridge University Press.

Hale, B., and Hoffmann, A. (eds) (2010) *Modality*. Oxford: Oxford University Press.

Hall, A.R., and Hall, Boas M. (eds) (1962) *Unpublished Papers of Isaac Newton*. Cambridge: Cambridge University Press.

Hall, N. (2004) 'Rescued From the Rubbish Bin: Lewis on Causation', *Philosophy of Science* 71 (5): 1107–14.

Hall, N. (2011) 'Causation and the Sciences', in S. French and J. Saatsi (eds), *The Continuum Companion to the Sciences*. London: Continuum Press: 96–119.

Halvorson, H., and Clifton, R. (2002) 'No Place for Particles in Relativistic Quantum Theories?', *Philosophy of Science* 69: 1–28.

Halvorson, H., and Müger, M. (2006) 'Algebraic quantum field theory', in J. Butterfield and J. Earman (eds), *Handbook of the Philosophy of Physics*. Dordrecht: Kluwer: 731–922.

Hamermesh, M. (1962) *Group Theory and Its Application to Physical Problems*. Reading, MA: Addison-Wesley Pub. Co.

Hardin, C.L., and Rosenberg, A. (1982) 'In Defence of Convergent Realism', *Philosophy of Science* 49: 604–15.

Hawley, K. (2006a) 'Science as a guide to metaphysics', *Synthese* 149: 451–70.

Hawley, K. (2006b) 'Principles of Composition and Criteria of Identity', *Australasian Journal of Philosophy* 84: 481–93

Hawley, K. (2009) 'Identity and Indiscernibility', *Mind* 118: 101–19.

Hawley, K. (2010) 'Throwing the Baby Out with the Bathwater', critical notice of Ladyman, J., Ross, D., et al. (2007), *Every Thing Must Go: Metaphysics Naturalized*. Oxford: Oxford University Press, *Metascience* 19: 161–85.

Hawthorne, J. (2001) 'Causal Structuralism', *Philosophical Perspectives* 15: 361–78.

Healey, R. (2007) *Gauging What's Real: The Conceptual Foundations of Gauge Theories*. Oxford: Oxford University Press.

Healey, R. (2009) 'Causation in Quantum Mechanics', in H. Beebee, C. Hitchcock, and C. Menzies (eds), *The Oxford Handbook of Causation*. Oxford: Oxford University Press: 673–86.

Healey, R. (forthcoming) '*A Lego Universe? The Physical Construction of the World*', talk given at the workshop 'Part and Whole in Physics', Lorentz Center, Leiden 2010.

Heath, A.E. (1928) 'Contribution to the Symposium "Materialism in the Light of Scientific Thought"', *Proceedings of the Aristotelian Society*, Suppl. Vol. 8: 130–42.

Heisenberg, W., and Erwin Schrödinger, E. (1987) 'The Origins of the Principles of Uncertainty and Complementarity', *Foundations of Physics* 17 (5): 461–506.

Hendry, R.F. (2011) 'Philosophy of Chemistry', in S. French and J. Saatsi (eds), *The Continuum Companion to the Philosophy of Science*. London: Continuum Press: 293–313.

Henle, M. (ed.) (1971) *The Selected Papers of Wolfgang Köhler*. New York: Liveright.

Herfel, W.E. et al. (eds) (1996) *Theories and Models in Scientific Processes*. Amsterdam: Editions Rodopi.

Hesse, M. (1962) 'On What There is in Physics', *British Journal for the Philosophy of Science* 13 (51): 234–44.

Hesse, M. (1963) *Models and Analogies in Science*. London: Sheed and Ward.

Hesse, M. (1980) *Revolutions and Reconstructions in the Philosophy of Science*. Brighton: Harvester Press.

Hintikka, J. (1998) 'Ramsey Sentences and the Meaning of Quantifiers', *Philosophy of Science* 65: 289–305.

Hoefer, C. (2000) 'Energy Conservation in GTR', *Studies in History and Philosophy of Modern Physics* 31: 187–99.

Hoefer, C. (2009) 'Causation in Spacetime Theories', in H. Beebee, P. Menzies, and C. Hitchcock (eds), *The Oxford Handbook of Causation*. Oxford: Oxford University Press: 687–706.

Hopkins, M. (2009) 'Re: Unitary Representations of the Poincaré Group', <http://golem.ph.utexas.edu/category/2009/03/unitary_representations_of_the.html#c024879)>.

Horgan, T., and Potrc, M. (2008) *Austere Realism*. Cambridge, MA: MIT Press.

Howard, D. (2010) Talk given at the Workshop on Structural Realism, University of Notre Dame, November 2010.

Howard, D. (2011) 'Are Elementary Particles Individuals? A Critical Appreciation of Steven French and Décio Krause's *Identity in Physics: A Historical, Philosophical, and Formal Analysis*', *Metascience* 20 (2): 225–31.

Howard, D. (preprint) 'The Trouble with Metaphysics'.

Hoyhingen-Huene, P., and Sankey, H. (eds) (2001) *Incommensurability and Related Matters*. Dordrecht: Kluwer.

Huggett, N. (1999a) 'Atomic Metaphysics', *Journal of Philosophy* 96: 5–24.

Huggett, N. (1999b) 'On the Significance of the Permutation Symmetry', *British Journal for the Philosophy of Science* 50: 325–47.

Humphreys, P. (1997) 'Emergence, Not Supervenience', *Philosophy of Science* 64: 337–45.

Husserl, E. (1891) *Philosophie der Arithmetik*, trans. by D. Willard, *Philosophy of Arithmetic*. Dordrecht: Kluwer Academic Publishers, 2003.

Hutten, E.H. (1953–1954) 'The Rôle of Models in Physics', *British Journal for the Philosophy of Science* 4 (16): 284–301.

Ihmig, K.-N. (1999) 'Ernst Cassirer and the Structural Conception of Objects in Modern Science: The Importance of the "Erlanger Program"', *Science in Context* 12: 513–29.

Ihmig, K.-N. (2001) *Grundzüge einer Philosophie der Wissenschaften bei Ernst Cassirer*. Darmstadt: Wissenschaftliche Buchgesellschaft.

Ismael, J., and van Fraassen, B. (2003) 'Symmetry as a Guide to Superfluous Theoretical Structure', in K. Brading and E. Castellani (eds), *Symmetries in Physics: Philosophical Reflections*. Cambridge: Cambridge University Press, 2010: 371–92.

Itzkoff, S.W. (1997) *Ernst Cassirer: Scientific Knowledge and the Concept of Man*, 2nd edn. Notre Dame, IN: University of Notre Dame Press.

Jaffe, A., and Witten, E. (undated) 'Quantum Yang–Mills Theory', <http://www.claymath.org/millennium/Yang-Mills_Theory/yangmills.pdf>.

James, I.M. (ed.) (2006) *History of Topology*. Amsterdam: North-Holland.

Jannes, G. (2009) 'Some Comments on "The Mathematical Universe"', *Foundations of Physics* 39: 397–406.

Jantzen, B.C. (2011) 'No Two Entities Without Identity', *Synthese* 181: 433–50.

Jauernig, A. (2008) 'Kant's Critique of the Leibnizian Philosophy: contra the Leibnizians, but pro Leibniz', in D. Garber and B. Longuenesse (eds), *Kant and the Early Moderns*. Princeton: Princeton University Press: 41–63.

Jauernig, A. (2010) 'Disentangling Leibniz's Views on Relations and Extrinsic Denominations', *Journal of the History of Philosophy* 48 (2): 171–2.

Jenkins, C.S. (2011) 'Is Metaphysical Dependence Irreflexive?', *The Monist* 94 (2): 267–76.

Johansson, I. (1989) *Ontological Investigations: An Inquiry into the Categories of Nature, Man, and Society*. London: Routledge.

Johansson, I. (2000) 'Determinables as Universals', *The Monist* 83 (1): 101–21.

Johnson, W.E. (1921, 1922, 1924) *Logic*. Cambridge: Cambridge University Press. Vol. 1: 1921; vol. 2: 1922; vol. 3: 1924; for the hypertext version see <http://www.ditext.com/johnson/toc.html>.

Jones, J.H. (2008) 'Notes on the Price Equation', <http://www.stanford.edu/~jhj1/teachingdocs/Jones-PriceEquation.pdf>.

Jones, R. (1991) 'Realism About What?', *Philosophy of Science* 58: 185–202.

Jordan, F.Y., and Sudarshan, E.C. (1961) 'Lie Group Dynamical Formalism and the Relation between Quantum Mechanics and Classical Mechanics', *Reviews of Modern Physics* 33: 515.

Judd, B.R. (1993) 'Applied Group Theory 1926–1935', in A.S. Wightman (ed.), *The Collected Works of Eugene Paul Wigner, Part A, The Scientific Papers*, Vol. 1. Berlin: Springer Verlag: 17–33.

Kant, I. (1988) [1781; 2nd edn 1787] *Critique of Pure Reason*, trans. by Paul Guyer and Allen W. Wood. Cambridge: Cambridge University Press.

Kantorovich, A. (2003) 'The Priority of Internal Symmetries in Particle Physics', *Studies in History and Philosophy of Modern Physics* 34: 651–75.

Kantorovich, A. (2009) 'Ontic Structuralism and the Symmetries of Particle Physics', *Journal for General Philosophy of Science* 40 (1): 73–84.

Katzav, J. (2004) 'Dispositions and the Principle of Least Action', *Analysis* 64 (3): 206–14.

Katzav, J. (2005) 'Ellis on the Limitations of Dispositionalism', *Analysis* 65 (285): 92–4.

Keller, E.F. (2002) *Making Sense of Life: Explaining Biological Development with Models, Metaphors, and Machines*. Cambridge, MA: Harvard University Press.

Ketland, J. (2004) 'Empirical Adequacy and Ramsification', *The British Journal for the Philosophy of Science* 55: 409–24.

Kilmister, C.W. (1994/2005) *Eddington's Search for a Fundamental Theory: A Key to the Universe*. Cambridge: Cambridge University Press.

Kim, J. (1998) *Mind in a Physical World*. Cambridge, MA: MIT Press.

Kim, J. and Sosa, E. (eds) (1995) *Metaphysics. An Anthology*. Oxford: Blackwell.

Kistler, M., and Gnassounou, B. (eds) (2007) *Dispositions and Causal Powers*. Ashford: Ashgate.

Kitcher, P., Sterelny, K., and Waters, C.K. (1990) 'The Illusory Riches of Sober's Monism', *Journal of Philosophy* 87: 158–61.

Köhler, W. (1920) *Die physischen Gestalten in Ruhe und im stationären Zustand; Eine naturphilosophische Untersuchung*. Braunschweig: Friedr, Vieweg und Sohn.

Köhler, W. (1930) 'The New Psychology and Physics', in M. Henle (ed.), *The Selected Papers of Wolfgang Köhler*. New York: Liveright, 1971: 237–51.

Köhler, W. (1967) 'Gestalt Psychology', in M. Henle (ed.), *The Selected Papers of Wolfgang Köhler*. New York: Liveright, 1971: 108–24.

Korman, D. (2008) 'Review of *Austere Realism*', *Notre Dame Philosophical Reviews*, <http://ndpr.nd.edu/review.cfm?id=14508>.

Kraemer, E. (2008) 'Function, Gene and Behavior', available in the *PhilSci Archive*.

Krois, J.M. (2004) 'Ernst Cassirer's Philosophy of Biology', *Sign System Studies* 32: 1–19.

Kronz, F. (2004) 'Quantum Theory: von Neumann vs. Dirac', *The Stanford Encyclopedia of Philosophy*, Edward N. Zalta (ed.), <http://plato.stanford.edu/archives/fall2008/entries/qt-nvd/>.

Kroon, F., and Nola, R. (2001) 'Ramsification, Reference Fixing and Incommensurability', in P. Hoyhingen-Huene and H. Sankey (eds), *Incommensurability and Related Matters*. Dordrecht: Kluwer: 91–121.

Kuhlmann, M. (2006) 'Quantum Field Theory', *The Stanford Encyclopedia of Philosophy*, Edward N. Zalta (ed.), <http://plato.stanford.edu/archives/spr2009/entries/quantum-field-theory/>.

Kuhlmann, M. Lyre, H., and Wayne, A. (eds) (2002) *Ontological Aspects of Quantum Field Theory*. Singapore, London, and Hackensack, NJ: World Scientific Publishing Company.

Kuhn, T. (1978) *Black-Body Theory and the Quantum Discontinuity*. Oxford: Clarendon Press, 2nd edn. Chicago: University of Chicago Press.

Ladyman, J. (1998) 'What is Structural Realism?', *Studies in History and Philosophy of Science* 29: 409–24.

Ladyman, J. (2002) 'Science, Metaphysics and Structural Realism', *Philosophica* 67: 57–76.

Ladyman, J. (2007) 'Scientific Structuralism: On the Identity and Diversity of Objects in a Structure', *The Proceedings of the Aristotelian Society* 81 (1): 23–43.

Ladyman, J. (2007/2009) 'Structural Realism', *Stanford Encyclopedia of Philosophy*, Edward N. Zalta (ed.), <http://plato.stanford.edu/entries/structural-realism/>.

Ladyman, J. (2011) 'Structural Realism versus Standard Scientific Realism: The Case of Phlogiston and Dephlogisticated Air', *Synthese* 180: 87–101.

Ladyman, J., Ross, D., et al. (2007) *Every Thing Must Go: Metaphysics Naturalized*. Oxford: Oxford University Press.

Lam, V. (2007) 'The Singular Nature of Space-Time', *Philosophy of Science* 74: 712–23.

Lam, V. (2008) 'Structural Aspects of the Singular Feature of Space-Time', in D. Dieks (ed.), *Ontology of Spacetime*. Philosophy and Foundations of Physics Series. Vol. 2. Amsterdam: Elsevier: 111–31.

Landry, E. (2007) 'Shared Structure Need not be Shared Set-Structure', *Synthese* 158: 1–17.

Landry, E. (2011) 'How to Be a Structuralist All the Way Down', *Synthese* 179: 435–54.

Landry, E. (2012) 'Methodological Structural Realism', in E. Landry and D. Rickles (eds), *Structure, Object, and Causality*. The Western Ontario Series in Philosophy of Science. Dordrecht: Springer: 29–57.

Landry, E. and Rickles, D. (eds) (2012) *Structure, Object, and Causality*. The Western Ontario Series in Philosophy of Science. Dordrecht: Springer.

Landsman, N.P. (2007) 'Between Classical and Quantum', in J. Butterfield and J. Earman (eds), *Handbook of the Philosophy of Science: Philosophy of Physics Part A*. Amsterdam: North-Holland/Elsevier: 417–554.

Lange, M. (2000) *Natural Laws in Scientific Practice*. Oxford: Oxford University Press.

Lange, M. (2004) 'A Note on Scientific Essentialism, Laws of Nature and Counterfactual Conditionals', *Australasian Journal of Philosophy* 82: 227–41.

Lange, M. (2007) 'Laws and Meta-Laws of Nature', *Studies in History and Philosophy of Modern Physics* 38: 457–81

Lange, M. (2009a) *Laws and Lawmakers: Science, Metaphysics, and the Laws of Nature*. Oxford: Oxford University Press.

Lange, M. (2009b) 'Review of The Metaphysics Within Physics', *Mind* 118: 197–200.

Langton, R. (1998) *Kantian Humility. Our Ignorance of Things in Themselves*. Oxford: Oxford University Press.

Langton, R. (2004) 'Elusive Knowledge of Things in Themselves', *Australasian Journal of Philosophy* 82 (1): 129–36.

Langton, R. (2009) 'Ignorance and Intrinsicality', Pacific APA Conference, Vancouver.

LaPorte, J. (2004) *Natural Kinds and Conceptual Change*. Cambridge: Cambridge University Press.

Laudan, L. (1981) 'A Confutation of Convergent Realism', *Philosophy of Science* 48 (1): 19–49.

Laudan, L. (1984) 'Discussion: Realism Without the Real', *Philosophy of Science* 51: 156–62.

Le Poidevin, R. (2005) 'Missing Elements and Missing Premises: A Combinatorial Argument for the Ontological Reduction of Chemistry', *British Journal for the Philosophy of Science* 56: 117–34.

Leitgeb, H., and Ladyman, J. (2008) 'Criteria of Identity and Structuralist Ontology', *Philosophia Mathematica* 16 (3): 388–96.

Lewis, D. (1970) 'How to Define Theoretical Terms', *Journal of Philosophy* 67; reprinted in Lewis' *Philosophical Papers*, vol. 1. Oxford: Oxford University Press, 1983: 78–95.

Lewis, D. (1986) *On the Plurality of Worlds*. Oxford: Blackwell Publishers.

Lewis, D. (2009) 'Ramseyan Humility', in D. Braddon-Mitchell and R. Nola (eds), *Conceptual Analysis and Philosophical Naturalism*. Cambridge, MA: MIT Press.

Livanios, V. (2010) 'Symmetries, Dispositions and Essences', *Philosophical Studies* 148: 295–305.

Lloyd, E. (2005) 'Units and Levels of Selection', *The Stanford Encyclopedia of Philosophy*, Edward N. Zalta (ed.), <http://plato.stanford.edu/entries/selection-units/>.

Loettgers, A. (2007) 'Model Organisms and Mathematical and Synthetic Models to Explore Gene Regulation Mechanisms', *Biological Theory* 2: 134–42.

Loewer, B. (2011) 'Counterfactuals all the Way Down?', review symposium of Marc Lange: *Laws and Lawmakers: Science, Metaphysics, and the Laws of Nature*, *Metascience* 20: 34–9.

Logue, H. (2013) 'Experience of Natural Kind Properties: Is there Any Fact of the Matter?', *Philosophical Studies* 162: 1–12.

Lowe, E.J. (2005) 'Ontological Dependence', *The Stanford Encyclopedia of Philosophy*, Edward N. Zalta (ed.), <http://plato.stanford.edu/archives/spr2010/entries/dependence-ontological/>.

Lowe, E.J. (2012) 'Mumford and Anjum on causal necessitarianism and antecedent strengthening', *Analysis*. Doi: 10.1093/analys/ans108.

Lurgat, F. (ed.) (1967) *Cargese Lectures in Theoretical Physics*. New York: Gordon and Breach.

Lyre, H. (2001) 'The Principles of Gauging', *Philosophy of Science* 68 (3): 371–81.

Lyre, H. (2004) 'Holism and Structuralism in U(1) Gauge Theory', *Studies in History and Philosophy of Modern Physics* 35: 643–70.

Lyre, H. (2010) *Humean Perspectives on Structural Realism*, in F. Stadler (ed.), *The Present Situation in the Philosophy of Science*. Dordrecht: Springer: 381–97.

Lyre, H. (2011) 'Is Structural Underdetermination Possible?', *Synthese* 180 (2): 235–47.

Lyre, H. (2012) 'Structural Invariants, Structural Kinds, Structural Laws', in D. Dieks et al. (eds), *Probability, Laws and Structures*. Dordrecht: Springer: 169–82.

Machamer, P., Darden, L., and Craver, C. (2000) 'Thinking about Mechanisms', *Philosophy of Science* 67(1): 1–25.

Mackey, G.W. (1978) *Unitary Group Representations in Physics, Probability, and Number Theory*. Reading, MA: The Benjamin/Cummings Pub. Co.

Mackey, G.W. (1993) 'The Mathematical Papers', in A.S. Wightman (ed.), *The Collected Works of Eugene Paul Wigner*, Vol. 1. New York: Springer: 241–90.

Magnus, P.D. (2012) *From Planets to Mallards: Scientific Enquiry and Natural Kinds*. Basingstoke: Palgrave Macmillan.

Malament, D. (ed.) (2002) *Reading Natural Philosophy: Essays in the History and Philosophy of Science and Mathematics*. Chicago and LaSalle, IL: Open Court.

Manin, Yu I. (1976) 'Problems of Present Day Mathematics I: Foundations', in F.E. Browder (ed.), *Mathematical Problems Arising from Hilbert Problems*. Proceedings of Symposia in Pure Mathematics XXVIII. Providence: RI, AMS.

Marmodoro, A. (ed.) (2010) *The Metaphysics of Powers: Their Grounding and Their Manifestations*. London: Routledge.

Marquis, J.-P. (2007) 'Category Theory', *The Stanford Encyclopedia of Philosophy*, Edward N. Zalta (ed.), <http://plato.stanford.edu/archives/spr2011/entries/category-theory/>.

Marsonet, M. (ed.) (2002) *The Problem of Realism*. London and Aldershot: Ashgate.

Martin, C. (2003) 'On the Continuous Symmetries and the Foundations of Modern Physics', in K. Brading and E. Castellani (eds), *Symmetries in Physics*. Oxford: Oxford University Press: 29–60.

Martin, C.B. (1994) 'Dispositions and Conditionals', *The Philosophical Quarterly* 44: 1–8.

Massimi, M. (2009) 'Philosophy and the Sciences After Kant', *Royal Institute of Philosophy Supplement* 84 (65): 275–311.

Massimi, M. (2011a) 'From Data to Phenomena: A Kantian Stance', *Synthese* 182 (1): 101–16.

Massimi, M. (2011b) 'Kant's Dynamical Theory of Matter in 1755, and its Debt to Speculative Newtonian Experimentalism', *Studies in History and Philosophy of Science Part A* 42 (4): 525–43.

Maudlin, T. (1994) *Quantum Non-Locality and Relativity*. Oxford: Blackwell.

Maudlin, T. (2007/2009) *The Metaphysics Within Physics*. Oxford: Oxford University Press.

Maxwell, G. (1962) 'The Ontological Status of Theoretical Entities', in H. Feigl and G. Maxwell (eds), *Minnesota Studies in the Philosophy of Science* 3: 3–14.

Maxwell, G. (1970a) 'Structural realism and the meaning of theoretical terms', in S. Winokur and M. Radner (eds), *Analyses of Theories and Methods of Physics and Psychology*. Minnesota Studies in the Philosophy of Science: Volume 4. Minneapolis, MN: University of Minnesota Press: 181–92.

Maxwell, G. (1970b) 'Theories, Perception and Structural Realism', in R. Colodny (ed.), *The Nature and Function of Scientific Theories*. University of Pittsburgh Series in the Philosophy of Science: Volume 4. Pittsburgh: University of Pittsburgh Press: 3–34.

Maxwell, G. (1972) 'Scientific Methodology and the Causal Theory of Perception', in H. Feigl, R.W. Sellars, and K. Lehrer (eds), *New Readings in Philosophical Analysis*. New York: Appleton-Century Crofts: 148–77.

McCabe, G. (2006) 'Structural Realism and the Mind' (available in the *PhilSci Archive*).

McKay-Illari, P. (2011) 'Is Activity-Entity Dualism Ontic Structural Realism for the Biological Sciences?', Structuralism Conference, Bristol.

McKay-Illari, P. and Williamson, J. (2013) 'In Defence of Activities', *Journal of General Philosophy of Science* 44, 69–83.

McKenzie, K. (2011) 'Arguing Against Fundamentality', *Studies in History and Philosophy of Science Part B* 42 (4): 244–55.

McKenzie, K. (2012) 'Physics Without Fundamentality: A Study into Fundamentalism about Particles and Laws in High-Energy Physics'. PhD Thesis, University of Leeds.

McKenzie, K. (2014) 'Priority and Particle Physics: Ontic Structural Realism as a Fundamentality Thesis', *British Journal for the Philosophy of Science* 65: 353–80.

McKitrick, J. (2008) 'Review of Dispositions and Causal Powers', *Notre Dame Philosophical Reviews*, <http://ndpr.nd.edu/news/23655-dispositions-and-causal-powers/>.

McKitrick, J. (2010) 'Manifestations as Effects', in A. Marmodoro (ed.), *The Metaphysics of Powers: Their Grounding and Their Manifestations*. London: Routledge: 73–83.

Mehra, J. (1987) 'Niels Bohr's Discussions with Albert Einstein, Werner Heisenberg, and Erwin Schrödinger: The Origins of the Principles of Uncertainty and Complementarity', *Foundations of Physics* 17 (5): 461–506.

Melia, J., and Saatsi, J. (2006) 'Ramseyfication and Theoretical Content', *British Journal for the Philosophy of Science* 57: 561–85.

Mellor, D.H. (1974) 'In Defense of Dispositions', *The Philosophical Review* 83: 157–81.

Mertz, D.W. (1996) *Moderate Realism and its Logic*. New Haven: Yale University Press.

Mertz, D.W. (2002) 'Combinatorial Predication and the Ontology of Unit Attributes', *The Modern Schoolman* 79: 163–97.

Messiah, A.M.L., and Greenberg, O. (1964) 'Symmetrization Postulate and Its Experimental Foundation', *Physical Review* 136: B248.

Miller, A., (2010) 'Realism', *The Stanford Encyclopedia of Philosophy*, Edward N. Zalta (ed.), <http://plato.stanford.edu/archives/spr2012/entries/realism/>.

Mirman, R. (1969) 'The Physical Basis of Combined Symmetry Theories', *Progress of Theoretical Physics* 41: 1578–84.

Mirman, R. (1995) *Group Theory: An Intuitive Approach*. River Edge, NJ: World Scientific.

Mirman, R. (2005) *Group Theoretical Foundations of Quantum Mechanics*. Lincoln, Nebraska: iUniverse.

Mitchell, S. (2003) *Biological Complexity and Integrative Pluralism*. Cambridge: Cambridge University Press.

Mittelstaed, P., and Stachow, E.W. (eds) (1988) *Recent Developments in Quantum Logics*. Mannheim: Bibliographishes Institut.

Molnar, G. (1999) 'Are Dispositions Reducible?', *The Philosophical Quarterly* 49: 1–17.

Molnar, G. (2003) *Powers: A Study in Metaphysics*, ed. S. Mumford. Oxford: Oxford University Press.

Monton, B. (2011) 'Prolegomena to Any Future Physics-Based Metaphysics', in J.L. Kvanvig (ed.), *Oxford Studies in the Philosophy of Religion*. Oxford: Oxford University Press: 142–65.

Monton, B. (ed.) (2007) *Images of Empiricism: Essays on Science and Stances with a Reply from Bas C. van Fraassen*. Mind Association Occasional Series. Oxford: Oxford University Press.

Monton, B. and van Fraassen, B.C. (2003) 'Constructive Empiricism and Modal Nominalism', *British Journal for the Philosophy of Science* 54: 405–22.

Moore, W. (1989) *Schrödinger: Life and Thought*. Cambridge: Cambridge University Press.

Morgan, G.J. (2010) 'Laws of Biological Design: A Reply to John Beatty', *Biology and Philosophy* 25: 379–89.

Morgan, M. (2003) 'Experiments without Material Intervention', in H. Radder (ed.), *The Philosophy of Scientific Experimentation*. Pittsburgh: Pittsburgh University Press.

Morgan, M. and Morrison, M. (eds) (1999) *Models as Mediators*. Cambridge: Cambridge University Press.

Morganti, M. (2004) 'On the Preferability of Epistemic Structural Realism', *Synthese* 142 (1): 81–107.

Morganti, M. (2009) 'Tropes and Physics', *Grazer Philosophische Studien* 78: 185–205.

Morganti, M. (2011) 'Is There a Compelling Argument for Ontic Structural Realism?', *Philosophy of Science* 78: 1165–76.

Morganti, M. (2013) *Combining Science and Metaphysics.: Contemporary Physics, Conceptual Revision and Common Sense*. Basingstoke: Palgrave Macmillan.

Morrison, M. (1999) 'Models as Autonomous Agents', in M. Morrison (ed.), *Models as Mediators*. Cambridge: Cambridge University Press: 38–65.

Morrison, M. (2003) 'Spontaneous Symmetry Breaking: Theoretical Arguments and Philosophical Problems', in K. Brading and E. Castellani (eds), *Symmetries in Physics: Philosophical Reflections*. Cambridge: Cambridge University Press: 347–63.

Morrison, M. (2007) 'Spin: All is Not What it Seems', *Studies in History and Philosophy of Modern Physics* 38: 529–57.

Muller, F., and Saunders, S. (2008) 'Discerning Fermions', *British Journal for the Philosophy of Science* 59: 499–548.

Muller, F., and Seevinck, M. (2009) 'Discerning Elementary Particles', *Philosophy of Science* 76: 179–200.

Muller, F.A. (1997) 'The Equivalence Myth of Quantum Mechanics', published in two parts in *Studies in History and Philosophy of Modern Physics* 28 (1997): 35–61, 219–47, and an Addendum in 30 (1999): 543–5.

Muller, F.A. (2010) 'The Characterization of Structure: Definition versus Axiomatisation', in F. Stadler et al. (eds), *The Present Situation in the Philosophy of Science*. Berlin: Springer: 399–416.

Muller, F.A. (2011) 'Withering Away, Weakly', *Synthese* 180: 223–33.

Mumford, S. (1998) *Dispositions*. Oxford: Oxford University Press.

Mumford, S. (2004) *Laws in Nature*. London: Routledge.

Mumford, S. (2006) 'Looking for Laws: Authors' Reply', *Metascience* 15: 462–9.

Mumford, S. (2009a) 'Causal Powers and Capacities', in H. Beebee, P. Menzies, and C. Hitchcock (eds), *The Oxford Handbook of Causation*. Oxford: Oxford University Press: 265–78.

Mumford, S. (2009b) 'Passing Powers Around', *The Monist* 92: 94–111.

Mumford, S. (2011) 'Dispositional Modality', in C.F. Gethmann (ed.), *Lebenswelt und Wissenschaft, Deutsches Jahrbuch Philosophie 2*. Hamburg: Meiner Verlag: 380–94.

Murray, J.D. (2003) *Mathematical Biology*. Dordrecht: Springer.

Musgrave, A. (1992) 'Discussion: Realism About What?', *Philosophy of Science* 59: 691–7.

Myrvold, W. (2011) 'Nonseparability, Classical, and Quantum', *British Journal for the Philosophy of Science* 62 (2): 417–32.

Nambu, Y. (2008) 'Spontaneous Symmetry Breaking in Particle Physics: A Case of Cross Fertilization', <http://www.nobelprize.org/nobel_prizes/physics/laureates/2008/nambu_lecture.pdf>.

Ne'eman, Y. (1987) 'Hadron Symmetry, Classification and Competences', in M.G. Doncel, A. Hermann, L. Michel, and A. Pais (eds), *Symmetry in Physics (1600–1980)*. Barcelona: Bellaterra: 499–540.

Newman, M.H.A. (1928) 'Mr. Russell's Causal Theory of Perception', *Mind* 37: 137–48.

Newton, I. (1962) 'De gravitatione et aequipondio fluidorum', in A.R. Hall and Boas M. Hall (eds), *Unpublished Papers of Isaac Newton*. Cambridge: Cambridge University Press: 89–156.

Newton, T.D., and Wigner, E.P. (1949) 'Localized States for Elementary Systems', *Reviews of Modern Physics* 21 (3): 400–6.

Nola, R. (2012) 'Varieties of Structuralism', *Metascience* 21: 59–64.

North, J. (2009) 'The "Structure" of Physics: A Case Study', *Journal of Philosophy* 106: 57–88.

North, J. (2013) 'The Structure of a Quantum World', in D. Albert and A. Ney (eds), *The Wave Function*. Oxford: Oxford University Press: 184–202.

Norton, J.D. (2007) 'Causation as Folk Science', *Philosophers' Imprint* 3 (4), <http://www.philosophersimprint.org/003004/>; reprinted in H. Price and R. Corry (eds), *Causation, Physics and the Constitution of Reality*. Oxford: Oxford University Press, 2007: 11–44.

Nounou, A. (2003) 'A Fourth Way to the A-B Affect', in K. Brading and E. Castellani (eds), *Symmetries in Physics: Philosophical Reflections*. Cambridge: Cambridge University Press, 2010: 172–99.

Nounou, A. et al. (2010) 'A New Perspective on Objectivity and Conventionalism', *Metascience* 19: 3–27.

Nozick, R. (2003) *Invariances*. Cambridge, MA: Harvard University Press.

Odenbaugh, J. (2008) 'Models', in S. Sarkar and A. Plutynski (eds), *A Blackwell Companion to the Philosophy of Biology*. Oxford: Wiley-Blackwell: 506–24.

Odenbaugh, J. (2009) 'Models in Biology', in E. Craig (ed.), *Routledge Encyclopedia of Philosophy* (online).

Okasha, S. (2006) *Evolution and the Levels of Selection*. Oxford: Oxford University Press.

Okasha, S. (2009) 'Causation in Biology', in H. Beebee, P. Menzies, and C. Hitchcock (eds), *The Oxford Handbook of Causation*. Oxford: Oxford University Press, 707–25.

Okasha, S. (2011) 'Reply to Sobers and Waters', *Philosophy and Phenomenological Research* 82: 241–8.

Olby, R. (1966) *The Origins of Mendelism*. New York: Schocken Books.

O'Malley, M. (2014) 'Review of Thomas Pradeu's The Limits of the Self: Immunology and Biological Identity', *British Journal for the Philosophy of Science* 65: 179–83.

Oyama, S., Griffiths, P.E., and Gray, R.D. (eds) (2001) *Cycles of Contingency: Developmental Systems and Evolution*. Cambridge, MA: MIT Press.

Pagonis, C. (1996) 'Quantum Mechanics and Scientific Realism'. PhD Thesis, Cambridge University.

Papineau, D. (ed.) (1996) *Philosophy of Science*. Oxford: Oxford University Press.

Parsons, C. (1990) 'The Structuralist View of Mathematical Objects', *Synthese* 84: 303–46.

Pashby, T. (2012) 'Dirac's Prediction of the Positron: A Case Study for the Current Realism Debate', *Perspectives on Science* 20: 440–75.

Paul, L.A. (2009) 'Counterfactual Theories', in H. Beebee, P. Menzies, and C. Hitchcock (eds), *The Oxford Handbook of Causation*. Oxford: Oxford University Press: 158–84.

Paul, L.A. (2010) 'Mereological Bundle Theory', in H. Burkhardt, J. Seibt, and G. Imaguire, *The Handbook of Mereology*. Munich: Philosophia Verlag.

Paul, L.A. (preprint) 'A One-Category Ontology', <http://lapaul.org/papers/Paul-OneCategory.pdf>.

Perovic, S. (2008) 'Why Were Matrix Mechanics and Wave Mechanics Considered Equivalent?', *Studies in History and Philosophy of Science Part B* 39 (2): 444–61.

Pickering, A. (1984) *Constructing Quarks*. Edinburgh: Edinburgh University Press.

Piron, C. (1976) *Foundations of Quantum Physics*. Reading, MA: W.A. Benjamin.

Planck, M. (1909) 'Eight Lectures in Theoretical Physics', in P. Pesic (ed.), *Eight Lectures in Theoretical Physics*. Mineola, NY: Dover.

Plfaum, M.J. (2005) 'Deformation theory', <http://math.colorado.edu/~sbc21/seminars/deformations/PflDT.pdf>.

Poincaré, H. (1898) 'On the Foundations of Geometry', *The Monist* 9: 1–43.

Poincaré, H. (1905 [1952]) *Science and Hypothesis*. New York: Dover.

Poincaré, H. (1906) *The Value of Science*, translation (1913) from the original French edition by G.B. Halsted; reprint edition New York: Dover Publishing, 1958.

Poli, R. (2004) 'W. E. Johnson's Determinable-Determinate Opposition and His Theory of Abstraction', *Poznan Studies in the Philosophy of the Sciences and the Humanities* 82 (1): 163–96.

Pooley, O. (2006) 'Points, Particles and Structural Realism', in D. Rickles, S. French, and J. Saatsi (eds), *Structural Foundations of Quantum Gravity*. Oxford: Oxford University Press: 83–120.

Popper, K. (1959) *The Logic of Scientific Discovery*. London: Hutchinson.

Post, H.R. (1960) 'Simplicity in Scientific Theories', *British Journal for the Philosophy of Science* 11 (41): 32–41.

Post, H.R. (1971 [1993]) 'Correspondence, Invariance and Heuristics', *Studies in History and Philosophy of Science* 2: 213–55; reprinted in S. French and H. Kamminga (eds), *Correspondence, Invariance and Heuristics: Essays in Honour Of Heinz Post*. Boston Studies in the Philosophy of Science 148. Dordrecht: Kluwer Academic Press, 1993: 1–44.

Pradeu, T. (2012) *The Limits of the Self: Immunology and Biological Identity*, trans. by Elizabeth Vitanza. Oxford: Oxford University Press.

Price, H. (2009) 'Metaphysics After Carnap: The Ghost Who Walks', in D. Chalmers, R. Wasserman, and D. Manley (eds), Metametaphysics. Oxford: Oxford University Press: 320–46.

Price, H. and Corry, R. (eds) (2007) *Causation, Physics and the Constitution of Reality*. Oxford: Oxford University Press.

Price, H. and Weslake, B. (2009) 'The Time-Asymmetry of Causation', in H. Beebee, P. Menzies, and C. Hitchcock (eds), *The Oxford Handbook of Causation*. Oxford: Oxford University Press: 414–45.

Primas, H. (1983) *Chemistry, Quantum Mechanics and Reductionism*. Berlin: Springer.

Psillos, S. (1999) *Scientific Realism: How Science Tracks Truth*. London: Routledge.

Psillos, S. (2000) 'Carnap, the Ramsey-Sentence and Realistic Empiricism', *Erkenntnis* 52: 253–79.

Psillos, S. (2001) 'Is Structural Realism Possible?', *Philosophy of Science* 68: S13–S24.

Psillos, S. (2006) 'The Structure, the Whole Structure and Nothing but the Structure?', *Philosophy of Science* 73: 560–70.

Psillos, S. (2009) 'Regularity Theories', in H. Beebee, C. Hitchcock, and C. Menzies (eds), *The Oxford Handbook of Causation*. Oxford: Oxford University Press: 131–57.

Psillos, S. (2011) 'The Idea of Mechanism', in P. McKay, F. Russo, and J. Williamson (eds), *Causality in the Sciences*. Oxford: Oxford University Press: 771–88.

Psillos, S. (2012) 'Adding Modality to Ontic Structuralism: An Exploration and Critique', in E. Landry and D. Rickles (eds), *Structure, Object, and Causality*. The Western Ontario Series in Philosophy of Science. Dordrecht: Springer: 169–85.

Psillos, S. (2013) 'Semi-realism or NeoAristotelianism', *Erkenntnis* 78: 29–38.

Putnam, H. (1978) *Meaning and the Moral Sciences*. London: Routledge & Kegan Paul.

Quine, W.V.O. (1951) 'Two Dogmas of Empiricism', *The Philosophical Review* 60: 20–43.

Quine, W.V.O. (1976) 'Whither Physical Objects?', in R.S. Cohen et al. (eds), *Essays in the Memory of Imre Lakatos*. Dordrecht: Reidel: 497–504.

Radder, H. (ed.) (2003) *The Philosophy of Scientific Experimentation*. Pittsburgh: Pittsburgh University Press.

Radick, G. (forthcoming) 'The Professor and the Pea: Lives and Afterlives of William Bateson's Campaign for the Utility of Mendelism', in *Owning and Disowning Invention: Intellectual Property and Identity in the Technosciences in Britain, 1870–1930*, co-edited with Christine MacLeod. *Studies in History and Philosophy of Science, Special edition.*

Ramsey, F. (1978) *Foundations: Essays in Philosophy, Logic, Mathematics and Economics*, ed. by D.H. Mellor. London: Routledge and Kegan Paul: 103–4.

Ramsey, J.L. (2000) 'Realism, Essentialism, and Intrinsic Properties: The Case of Molecular Shape', in N. Bhushan and S. Rosenfeld, (eds), *Of Minds and Molecules: New Philosophical Perspectives on Chemistry*. Oxford: Oxford University Press: 104–24.

Reck, E.H., and Price, M.P. (2000) 'Structures and Structuralism in Contemporary Philosophy of Mathematics', *Synthese* 125 (3): 341–83.

Redhead, M. (1975) 'Symmetry in Intertheoretical Relations', *Synthese* 32: 77–112.

Redhead, M. (1995) *From Physics to Metaphysics*. Cambridge: Cambridge University Press.

Redhead, M. (2001) 'Quests of a Realist', *Metascience* 10 (3): 341–7.

Redhead, M. (2003) 'The Interpretation of Gauge Symmetry', in K. Brading and E. Castellani (eds), *Symmetries in Physics: Philosophical Reflections*. Cambridge: Cambridge University Press, 2010: 124–39.

Redhead, M. and Teller, P. (1991) 'Particles, Particle Labels, and Quanta: The Toll of Unacknowledged Metaphysics', *Foundations of Physics* 21: 43–62.

Redhead, M. and Teller, P. (1992) 'Particle Labels and the Theory of Indistinguishable Particles in Quantum Mechanics', *British Journal for the Philosophy of Science* 43: 201–18.

Reeder, N. (1995) 'Are Physical Properties Dispositions?', *Philosophy of Science* 62: 141–9.

Resnik, M.D. (1997) *Mathematics as a Science of Patterns*. Oxford: Clarendon Press.

Rickles, D. (2006) 'Time and Structure in Canonical Gravity', in D. Rickles, S. French, and J. Saatsi (eds), *Structural Foundations of Quantum Gravity*. Oxford: Oxford University Press: 152–95.

Rickles, D., French, S., and Saatsi, J. (eds) (2006) *Structural Foundations of Quantum Gravity*, Oxford: Oxford University Press.

Ritchie, A.D. (1948) *Reflections on the Philosophy of Sir Arthur Eddington*. Cambridge: Cambridge University Press.

Roberts, B.W. (2011) 'Group Structural Realism', *British Journal for the Philosophy of Science* 62 (1): 47–69.

Roberts, J.T. (2008) *The Law-Governed Universe*. Oxford: Oxford University Press.

Rosales, A. (preprint) 'The Metaphysics of Natural Selection: A Structural Approach', presented at the Annual Conference of the *BSPS* 2007.

Rosen, G. (2001) 'Abstract Objects', *The Stanford Encyclopedia of Philosophy*, Edward N. Zalta (ed.), <http://plato.stanford.edu/archives/spr2012/entries/abstract-objects/>.

Rosen, G. (2010) 'Metaphysical Dependence', in B. Hale and A. Hoffmann (eds), *Modality*. Oxford: Oxford University Press: 109–36.

Rosenberg, A. (2012) 'Why do Spatiotemporally Restricted Regularities Explain in the Social Sciences?', *British Journal for the Philosophy of Science* 63 (1): 1–26.

Ross, D. (2008) 'Ontic Structural Realism and Economics', *Philosophy of Science* 75: 731–41.

Rowbottom, D. (2009) 'Models in Biology and Physics: What's the Difference?', *Foundations of Science* 14: 281–94.

Rowbottom, D. (2010) 'Evolutionary Epistemology and the Aim of Science', *Australasian Journal of Philosophy* 88: 1–17.

Rudolph, E. (1994) 'Substance as Function: Ernst Cassirer's Interpretation of Leibniz as Criticism of Kant', in E. Rudolph and I.O. Stamatescu (eds), *Philosophy, Mathematics and Modern Physics. A Dialogue*. Heidelberg: Springer-Verlag, 1995: 235–42.

Rueger, A. (1988) 'Atomism from Cosmology: Erwin Schrödinger's Work on Wave Mechanics and Space-time Structure', *Historical Studies in the Physical and Biological Sciences* 18: 377–401.

Rueger, A. (1998) 'Local Theories of Causation and the A Posteriori Identification of the Causal Relation', *Erkenntnis* 48 (1): 27–40.

Rueger, A. (2006) 'Connection and Influence: A Process Theory of Causation', *Journal for General Philosophy of Science* 37: 77–97.

Ruetsche, L. (2002) 'Interpreting Quantum Field Theory', *Philosophy of Science* 69: 348–78.

Ruetsche, L. (2003) 'A Matter of Degree: Putting Unitary Inequivalence to Work', *Philosophy of Science* 70 (5): 1329–42.

Ruetsche, L. (2011) *Interpreting Quantum Theories*. Oxford: Oxford University Press.

Russell, B. (1912) *The Problems of Philosophy*. London: Williams and Norgate/New York: Henry Holt and Company; reprinted in New York and Oxford: Oxford University Press, 1997.

Russell, B. (1927) *The Analysis of Matter*. London: Kegan Paul, Trench, Trubner & Co.

Russell, B. (1948) *Human Knowledge*. New York: Simon and Schuster.

Russell, B. (1967, 1968, 1969) *The Autobiography of Bertrand Russell*, 3 volumes. London: George Allen and Unwin; Boston and Toronto: Little Brown and Company (vols 1 and 2); New York: Simon and Schuster.

Ryckman T.A. (1991) 'Conditio Sine Qua Non? Zuordnung in the Early Epistemologies of Cassirer and Schlick', *Synthese* 88 (1): 57–95.

Ryckman T.A. (1999) 'Einstein, Cassirer, and General Covariance—Then and Now', *Science in Context* 12: 585–619.

Ryckman T.A. (2003a) 'The Philosophical Roots of the Gauge Principle: Weyl and Transcendental Phenomenological Idealism', in K. Brading and E. Castellani (eds), *Symmetries in Physics: Philosophical Reflections*. Cambridge: Cambridge University Press, 2010: 61–88.

Ryckman T.A. (2003b) 'Surplus Structure from the Standpoint of Transcendental Idealism: The "World Geometries" of Weyl and Eddington', *Perspectives on Science* 11 (1): 76–106.

Ryckman T.A. (2005) *The Reign of Relativity, Philosophy in Physics 1915–1925*. Oxford: Oxford University Press.

Saatsi, J. (2005) 'Reconsidering the Fresnel-Maxwell Theory Shift: How the Realist Can Have Her Cake and EAT it too', *Studies in History and Philosophy of Science* 36: 509–38.

Saatsi, J. (2007) 'Living in Harmony: Nominalism and the Explanationist Argument for Realism', *International Studies in the Philosophy of Science* 21: 19–33.

Saatsi, J. (2008) 'Eclectic Realism—the Proof of the Pudding', *Studies in History and Philosophy of Science* 39: 273–6.

Saatsi, J. (2009) 'Form vs. Content-driven Arguments for Realism', in P.D. Magnus and J. Busch (eds), *New Waves in Philosophy of Science.* Basingstoke: Palgrave Macmillan, 2010: 8–28.

Saatsi, J. (2011) 'The Enhanced Indispensability Argument: Representational Versus Explanatory Role of Mathematics in Science', *British Journal for the Philosophy of Science* 62 (1): 143–54.

Saatsi, J. (2012) 'Scientific Realism and Historical Evidence: Shortcomings of the Current State of Debate', in W. de Regt Henk (ed.), *EPSA Philosophy of Science: Amsterdam 2009.* Dordrecht: Springer: 329–40.

Sanford, D.H. (2011) 'Determinates vs. Determinables', *The Stanford Encyclopedia of Philosophy*, Edward N. Zalta (ed.), <http://plato.stanford.edu/archives/spr2011/entries/determinate-determinables/>.

Saunders, S. (1993) 'To What Physics Corresponds', in S. French and H. Kaminga (eds), *Correspondence, Invariance, and Heuristics; Essays in Honour of Heinz Post.* Dordrecht: Kluwer, 2010: 295–326.

Saunders, S. (2003a) 'Physics and Leibniz's Principles', in K. Brading and E. Castellani (eds), *Symmetries in Physics: Philosophical Reflections.* Cambridge: Cambridge University Press, 2010: 289–307.

Saunders, S. (2003b) 'Structural Realism, Again', *Synthese* 136: 127–33.

Saunders, S. (2003c) 'Critical notice: Cao's "The Conceptual Development of 20th Century Field Theories", *Synthese* 136: 79–105.

Saunders, S. (2006a) 'On the Explanation for Quantum Statistics', *Studies in History and Philosophy of Modern Physics* 37: 192–211.

Saunders, S. (2006b) 'Are Quantum Particles Objects?', *Analysis* 66: 52–63.

Schaffer, J. (2003) 'The Problem of Free Mass: Must Properties Cluster?', *Philosophy and Phenomenological Research* 66: 125–38.

Schaffer, J. (2005) 'Quiddistic Knowledge', *Philosophical Studies* 123 (1–2): 1–32.

Schaffer, J. (2009) 'The Deflationary Metaontology of Thomasson's *Ordinary Objects*', *Philosophical Books* 50: 142–57.

Schaffer, J. (2012) 'Why the World has Parts: Reply to Horgan & Potrč' in P. Goff (ed.) *Spinoza on Monism.* Basingstoke: Palgrave: 77–91.

Schlick, M. (1932) 'Positivismus und Realismus', trans. by P. Heath, *Erkenntnis* 3: 1–31.

Schmidt, H.-J. (2008) 'Structuralism in Physics', *The Stanford Encyclopedia of Philosophy*, Edward N. Zalta (ed.), <http://plato.stanford.edu/entries/physics-structuralism/>.

Schrödinger, E. (1937) 'World Structure', *Nature* 140: 742–4.

Schrödinger, E. (1995) 'The Interpretation of Quantum Mechanics', *Dublin Seminars (1949–1955) and Other Unpublished Essays*, edited with introduction by Michel Bitbol. Woodbridge, CT: Ox Bow Press.

Schrödinger, E. (1996) *Nature and the Greeks and Science and Humanism*, with a foreword by Roger Penrose. Cambridge: Cambridge University Press.

Searle, J. (1959) 'Determinables and the Notion of Resemblance' (in symposium with Stephan Korner), *Proceedings of the Aristotelian Society*, Suppl. 33: 141–58.

Segal, I. (1967) 'Representation of Canonical Commutation Relations', in F. Lurgat (ed.), *Cargese Lectures in Theoretical Physics*. New York: Gordon and Breach: 107–70.

Shapere, D. (1977) 'Scientific Theories and Their Domains', in F. Suppe (ed.), *The Structure of Scientific Theories*. Urbana, IL: University of Illinois Press: 518–99.

Shapere, D. (1982) 'Reason, Reference, and the Quest for Knowledge', *Philosophy of Science* 49 (1): 1–23.

Shapiro, S. (1997) *Philosophy of Mathematics: Structure and Ontology*. Oxford: Oxford University Press.

Sider, T. (2001) *Four-Dimensionalism*. Oxford: Oxford University Press.

Simons, P. (1994) 'Particulars in Particular Clothing: Three Trope Theories of Substance', *Philosophy and Phenomenological Research* 54: 53–576.

Simons, P. (2000) 'Identity Through Time and Trope Bundles', *Topoi* 19: 147–55.

Skow, B. (2010) 'Deep Metaphysical Indeterminacy', *Philosophical* Quarterly 60 (241): 851–8.

Skyrms, B. (1984) 'EPR: Lessons for Metaphysics', *Midwest Studies in Philosophy IX*. Minneapolis, MN: University of Minnesota Press: 245–55.

Slater, M.J., and Haufe, C. (2009) 'Where No Mind Has Gone Before: Exploring Laws in Distant and Lonely Worlds', *International Studies in the Philosophy of Science* 23 (3): 265–76.

Slowik, E. (2012) 'On Structuralism's Multiple Paths Through Spacetime Theories', *European Journal for Philosophy of Science* 2 (1): 45–66.

Smeenck, C. (2009) 'Interpreting Gauge Theories', review of *Gauging What's Real* by Richard Healey, *Metascience* 18: 11–22.

Smith, A.D. (1977) 'Dispositional Properties', *Mind* 86: 439–45.

Smith, B. (ed.) (1988) *Foundations of Gestalt Theory*. Munich and Vienna: Philosophia.

Solomon, G. (1989) 'Discussion: An Addendum to Demopoulos and Friedman (1985)', *Philosophy of Science* 56: 497–501.

Stachel, J. (2002) '"The Relations Between Things" versus "The Things Between Relations": The Deeper Meaning of the Hole Argument', in D. Malament (ed.), *Reading Natural Philosophy: Essays in the History and Philosophy of Science and Mathematics*. Chicago and LaSalle, IL: Open Court: 231–66.

Stachel, J. (2005) 'Structural Realism and Contextual Individuality', in Y. Ben-Menahem (ed.), *Hilary Putnam*. Cambridge: Cambridge University Press: 203–19.

Stadler, F. (ed.) (2010) *The Present Situation in the Philosophy of Science*. Dordrecht: Springer.

Stanford, P. (2006) *Exceeding Our Grasp*. Oxford: Oxford University Press.

Stebbing, S. (1937) *Philosophy and the Physicists*. London: Methuen and Co.

Stein, H. (1977) 'On Space-Time and Ontology: Extract from a Letter to Adolf Grunbaum', in J. Earman, C. Glymour, and J. Stachel (eds), *Foundations of Space-Time Theories*. Minneapolis, MN: University of Minnesota Press, 1997: 374–402.

Stern, A. (2008) 'Anyons and the Quantum Hall Effect—A Pedagogical Review', *Annals of Physics* 323: 204–49.

Stewart, I., and Golubitsky, M. (1992) *Fearful Symmetry: Is God a Geometer?*, Oxford: Blackwell.

Straumann, N. (2008) 'Unitary Representations of the Inhomogeneous Lorentz Group and Their Significance in Quantum Physics', arXiv:0809.4942.

Strevens, M. (2008) *Depth: An Account of Scientific Explanation*. Cambridge, MA: Harvard University Press.

Suárez, M. (2004) 'An Inferential Conception of Scientific Representation', *Philosophy of Science* 71 (5): 767–79.

Suárez, M. and Cartwright, N. (2008) 'Theories: Tools versus Models', *Studies in History and Philosophy of Modern Physics* 39: 62–81.

Sudarshan, E.C.G., and Duck, I.M. (2003) 'What Price the Spin–Statistics Theorem?', *Pramana: Indian Academy of Sciences* 61: 645–53.

Suppe, F. (1989) *Scientific Realism and Semantic Conception of Theories*. Illinois: University of Illinois Press.

Suppes, P. (1957) *Introduction to Logic*. New York: Van Nostrand.

Suppes, P. (1960) 'A Comparison of the Meaning and Uses of Models in Mathematics and the Empirical Sciences', *Synthese* 12: 287–301.

Suppes, P. (2000) 'Invariance, Symmetry and Meaning', *Foundations of Physics* 30: 1569–85.

Suppes, P. (ed.) (1977) *The Structure of Scientific Theories*. Urbana, IL: University of Illinois Press.

Swoyer, C. (2000) 'Properties', *Stanford Encyclopedia of Philosophy*, Edward N. Zalta (ed.), <http://plato.stanford.edu/archives/win2011/entries/properties/>.

Tegmark, M. (2006) 'The Mathematical Universe', *Foundations of Physics* 38: 101–50.

Teller, P. (1983) 'Quantum Physics, the Identity of Indiscernibles and Some Unanswered Questions', *Philosophy of Science* 50: 309–19.

Teller, P. (1995) *An Interpretative Introduction to Quantum Field Theory*. Princeton: Princeton University Press.

Teller, P. (1998) 'Quantum Mechanics and Haecceities', in E. Castellani (ed.), *Interpreting Bodies: Classical and Quantum Objects in Modern Physics*. Princeton: Princeton University Press: 114–41.

Teller, P. (1999) 'The Ineliminable Classical Face of Quantum Field Theory', in T. Cao (ed.), *Conceptual Foundations of Quantum Field Theory*. Cambridge: Cambridge University Press: 314–23.

Teller, P. and Redhead, M. (2001) 'Is Indistinguishability in Quantum Mechanics Conventional?', *Foundations of Physics* 30: 951–7.

Thomasson, A. (2007) *Ordinary Objects*. Oxford: Oxford University Press.

Timpson, C.G. (2008) 'Philosophical Aspects of Quantum Information Theory', in D. Rickles (ed.), *The Ashgate Companion to the New Philosophy of Physics*. Ashford: Ashgate: 197–261.

Tobin, E. (2012) 'The Metaphysics of Determinable Kinds', in A. Bird, B. Ellis and H. Sankey (eds.), *Properties, Powers and Structures: Issues in the Metaphysics of Realism*. London: Routledge: 219–32.

Tonietti, T. (1988) 'Four Letters of E. Husserl to H. Weyl and Their Context', in W. Deppert et al. (eds), *Exact Sciences and Their Philosophical Foundations*. Frankfurt am Main: Peter Lang: 343–84.

Tooley, M. (2009) 'Causation, Laws and Ontology', in H. Beebee, C. Hitchcock, and C. Menzies (eds), *The Oxford Handbook of Causation*. Oxford: Oxford University Press: 368–86.

Torretti, R. (2010) 'Nineteenth Century Geometry', *The Stanford Encyclopedia of Philosophy*, Edward N. Zalta (ed.), <http://plato.stanford.edu/archives/sum2010/entries/geometry-19th/>.

USDA Forest Service (2003) 'Humongous Fungus a New Kind Of Individual', *ScienceDaily*, <http://www.sciencedaily.com/releases/2003/03/030327074535.htm>.

van Fraassen, B.C. (1980) *The Scientific Image*. Oxford: Oxford University Press.

van Fraassen, B.C. (1989) *Laws and Symmetry*. Oxford: Clarendon.

van Fraassen, B.C. (1991) *Quantum Mechanics: An Empiricist View*. Oxford: Oxford University Press.

van Fraassen, B.C. (1995) '"World" is not a Count Noun', *Nous* 29: 139–57.

van Fraassen, B.C. (1997) 'Structure and Perspective: Philosophical Perplexity and Paradox', in M.L. Dalla Chiara, K. Doets, D. Mundici, and J. van Benthem (eds), *Logic and Scientific Methods*. Dordrecht: Kluwer, 511–30.

van Fraassen, B.C. (2006) 'Structure: Its Shadow and Substance', *British Journal for the Philosophy of Science* 57 (2): 275–307.

van Fraassen, B.C. (2009) 'Objectivity, Invariance, and Convention: Symmetry in Physical Science', *Studies in History and Philosophy of Modern Physics* 40: 84–7.

Varadarajan, V.S. (1985) *The Geometry of Quantum Mechanics*. New York: Springer-Verlag.

Vetter, B. (2009) 'Review of Bird', *Logical Analysis and History of Philosophy* 8: 320–8.

Vickers, P. (2013) *Understanding Inconsistent Science*. Oxford: Oxford University Press.

Von Ehrenfels, C. (1890) 'Über 'Gestaltqualitäten', *Vierteljahrsschrift für wissenschaftliche Philosophie* 14: 242–92; English translation in B. Smith (ed.), *Foundations of Gestalt Theory*. Munich and Vienna: Philosophia, 1988: 82–107.

Von Meyenn, K. (1987) 'Pauli's Belief in Exact Symmetries', in M.A. Doncel et al. (eds), *Symmetries in Physics (1600–1980). Proceedings of the 1st International Meeting of the History of Scientific Ideas*. Barcelona: Bellaterra.

Votsis, I. (2003) 'Is Structure Not Enough?', *Philosophy of Science* 70 (5): 879–90.

Votsis, I. (2004) 'Tracing the Development of Structural Realism', in I. Votsis, 'The Epistemological Status of Scientific Theories: An Investigation of the Structural Realist Account'. PhD Thesis, London School of Economics, <http://www.votsis.org/PDF/Votsis_Dissertation.pdf>.

Votsis, I. (2005) 'The Upward Path to Structural Realism', *Philosophy of Science* 72: 1361–72.

Votsis, I. (2011) 'Structural Realism: Continuity and its Limits', in A. Bokulich and P. Bokulich (eds), *Scientific Structuralism*. Boston Studies in the Philosophy of Science. Dordrecht: Springer.

Votsis, I. (2012) 'How *Not* to Be a Realist', in E. Landry and D. Rickles (eds), *Structural Realism: Structure, Object and Causality*. Dordrecht: Springer: 59–76.

Walker, M. (2009) Ant Mega-Colony Takes Over World, BBC, July 2009.

Wallace, D. (2003) 'Everett and Structure', *Studies in History and Philosophy of Modern Physics* 34: 86–105.

Wallace, D. (2006) 'In Defence of Naïveté: On the Conceptual Status of Lagrangian Quantum Field Theory', *Synthese* 151: 33–80.

Wallace, D. (2011) 'Taking Particle Physics Seriously: A Critique of the Algebraic Approach to Quantum Field Theory', *Studies in History and Philosophy of Modern Physics* 42: 116–25.

Wasserman, R. (2009) 'Material Constitution', *Stanford Encyclopedia of Philosophy*, Edward N. Zalta (ed.), <http://plato.stanford.edu/entries/material-constitution/>.

Waters, C.K. (2006) 'A Pluralist Interpretation of Gene-centered Biology', in Stephen Kellert, Helen Longino, and C. Kenneth Waters (eds), *Scientific Pluralism*, vol. XIX. Minnesota Studies in the Philosophy of Science. Minneapolis, MN: University of Minnesota Press.

Waters, C.K. (2008) 'Beyond Theoretical Reduction and Layer-Cake Antireduction: How DNA Retooled Genetics and Transformed Biological Practice', in M. Ruse (ed.), *The Oxford Handbook of Philosophy of Biology*. Oxford: Oxford University Press: 238–62.

Waters, C.K. (2011) 'Okasha's Unintended Argument for Toolbox Theorizing', *Philosophy and Phenomenological Research* 82: 232–40.

Werkmeister, W.H. (1949) 'Cassirer's Advance Beyond Neo-Kantianism', in P. Schilpp (ed.), *The Philosophy of Ernst Cassirer*. Evanston, IL: Library of Living Philosophers: 759–98.

Weyl, H. (1931) *The Theory of Groups and Quantum Mechanics*, trans. from the second, revised German edition by H.P. Robertson. London: Methuen and Co. The first edition was published in 1928 in New York: Dover.

Weyl, H. (1952) *Symmetry*. Princeton: Princeton University Press; extract in E. Brading and E. Castellani (eds), *Symmetries in Physics: Philosophical Reflections*. Cambridge: Cambridge University Press, 2010: 21–2.

Weyl, H. (1963) *Philosophy of Mathematics and Natural Science*. New York: Atheneum.

Weyl, H. (1968) *Gesammelte Abhandlungen*, Vol. III. Berlin: Springer-Verlag.

Wightman, A.S. (ed.) (1993) *The Collected Works of Eugene Paul Wigner*: Part A, *The Scientific Papers*, Vol. 1. New York: Springer.

Wigner, E. (1927) 'Über nicht kombinierende Terme in der neueren Quantentheorie', *Zeitschrift für Physik* 40: 883–92.

Wigner, E. (1935) 'Symmetry Relations in Various Physical Problems', *Bulletin of the American Mathematical Society* 41: 306.

Wigner, E. (1937) 'On the Consequences of the Symmetry of the Nuclear Hamiltonian on the Spectroscopy of Nuclei', *Phys. Rev.* 51: 106–19.

Wigner, E. (1939) 'On Unitary Representations of the Inhomogeneous Lorentz Group', *Annals of Mathematics* 40 (1): 149–204.

Wigner, E. (1959) *Group Theory and Its Application to the Quantum Mechanics of Atomic Spectra*. New York: Academic Press.

Wigner, E. (1963) 'Oral History Transcript', interviews held at the Niels Bohr Library & Archives, Niels Bohr Library and Archive, Sessions II and III, <http://www.aip.org/history/ohilist/4963_1.html>.

Wigner, E. (2003a) 'Symmetry and Conservation Laws', in K. Brading and E. Castellani (eds), *Symmetries in Physics: Philosophical Reflections*. Cambridge: Cambridge University Press: 23–6.

Wigner, E. (2003b) 'The Role of Invariance Principles in Natural Philosophy', in K. Brading and E. Castellani (eds), *Symmetries in Physics: Philosophical Reflections*. Cambridge: Cambridge University Press: 369–70.

Williamson, J. (2009) 'Probabilistic Theories', in H. Beebee, P. Menzies, and C. Hitchcock (eds), *The Oxford Handbook of Causation*. Oxford: Oxford University Press, 185–212.

Wilson, A. (2009) 'Disposition-manifestations and Reference-frames', *Dialectica* 63 (4): 591–601.

Wilson, J. (2012) 'Fundamental Determinables', *Philosophers' Imprint* 12, <http://www.philosophersimprint.org/012004/>.

Wilson, M. (2006) *Wandering Significance: An Essay on Conceptual Behavior*. Oxford: Oxford University Press.

Wilson, R. (1996) 'Promiscuous Realism', *British Journal for the Philosophy of Science* 47: 303–16.

Wilson, R. (2007) 'The Biological Notion of Individual', *The Stanford Encyclopedia of Philosophy*, Edward N. Zalta (ed.), <http://plato.stanford.edu/archives/fall2008/entries/biology-individual/>.

Wilson, W. (1937) 'The Origin and Nature of Wave Mechanics', *Science Progress* 32: 209.

Winokur, S., and Radner, M. (eds) (1970) *Analyses of Theories and Methods of Physics and Psychology*, Minnesota Studies in the Philosophy of Science: Volume 4. Minneapolis, MN: University of Minnesota Press.

Wolff, J. (2012) 'Do Objects Depend on Structures?', *British Journal for the Philosophy of Science* 63: 607–25.

Wooley, R.G. (1998) 'Is There a Quantum Definition of a Molecule?', *Journal of Mathematical Chemistry* 23: 3–11.

Woodward, J. (2003) *Making Things Happen: A Theory of Causal Explanation*. Oxford: Oxford University Press.

Worrall, J. (1989) 'Structural Realism: The Best of Both worlds?', *Dialectica* 43: 99–124; reprinted in D. Papineau (ed.), *The Philosophy of Science*. Oxford: Oxford University Press, 1996: 139–65.

Worrall, J. (2007) 'Miracles and Models: Why Reports of the Death of Structural Realism May Be Exaggerated', in Anthony O'Hare (ed.), *Philosophy of Science* (Royal Institute of Philosophy 61). Cambridge: Cambridge University Press: 125–54.

Worrall, J. (2012) 'Miracles and Structural Realism', in E. Landry and D. Rickles (eds), *Structural Realism: Structure, Object and Causality*. Dordrecht: Springer: 77–98.

Wuthrich, C. and Lam, V. (2015) 'No Categorical Support for Radical Ontic Structural Realism', *The British Journal for the Philosophy of Science* 66: 605–34.

Yudell, Z. (2010) 'Melia and Saatsi on Structural Realism', *Synthese* 175: 241–53.

Zahar, E. (2001) *Poincaré's Philosophy: From Conventionalism to Phenomenology*. Chicago and La Salle, IL: Open Court.

Zahar, E. (2007) *Why Science Needs Metaphysics: A Plea for Structural Realism*. Chicago: Open Court Publishing Co.

Zimmerman, D. (1997) 'Immanent Causation', *Noûs* 31, Supplement: Philosophical Perspectives 11: Mind, Causation, and World: 433–71.

Index of Names

Index of Subjects